HANDBOOK
OF
PYROTECHNICS

by
Karl O. Brauer

1974
CHEMICAL PUBLISHING CO. INC.
NEW YORK, N.Y.

©1974

CHEMICAL PUBLISHING CO., INC.

ISBN: 0-8206-0349 X PBK

Contents

Introduction

During the last twenty years, the field of pyrotechnics, which was limited mainly to applications in military ordnance, fireworks, and rock blasting, has been developed to a highly advanced science and to a widely used technology. Today, pyrotechnics find extensive applications in spacecraft, aircraft and underwater vehicle systems, and also in production methods, for example for metal forming, cladding, and riveting.

Pyrotechnics are ideally utilized in applications where space is limited and where low weight is a major requirement, as for instance in aircraft crew ejection systems, or in external fuel tank release mechanisms. Optimum utilization of the small size, low weight and high reliability of pyrotechnics is made in spacecraft and missile applications. Many space missions would be impossible without pyrotechnic devices and pyro-technic systems. Escape tower release, stage separation, fairing release, recovery and landing systems, location aids, and flotation systems are some typical examples of ideal utilization of pyrotechnics in spacecraft. Pyrotechnic devices are widely used in missile systems for ignition, con-trol, booster separation, fairing release, and, in cases of malfunction, for destruction.

"Pyrotechnic" means "explosive-actuated" and refers especially to devices in which explosives are burned rather than detonated. In aerospace applications, all electrically fired explosive devices are referred to as "electro-explosive".

The main advantages of pyrotechnic devices are: High power-to-weight ratio, high reliability, small size, low operating current, simple circuit requirement, reasonably low cost, ability to deliver more energy in a shorter time than any other mechanical device, and precisely controllable force.

Another advantageous feature of some pyrotechnic devices is the possibility of providing time delays by placing elements with fixed burning times between sections of fuze, or by mounting time delay trains in the explosive devices themselves.

In recent years, it was difficult for engineers, technicians, designers and students to solve some problems connected with the design, development, testing and evaluation of pyrotechnic devices and systems because of lack of sufficient data, design rules, and experience. Sparse data about some explosive materials and only a few reports about the performance and reliability of a certain type of squibs and power cartridges were available. As a result of this lack of sufficient data, until a few years ago, a newly developed pyrotechnic device could only be qualified by a very extensive and costly series of tests to warrant reliable functioning. The author, who was instrumental in the development and design of several well-known revolutionary aircraft, missiles, spacecraft and underwater vehicles in Europe and in the United States, for example the first multi-engine turbojet aircraft, the first tactical transport aircraft Ar 232, the first turbojet bomber of the world, Ar 234, the Dolphin and Polaris missiles, drones, aircraft crew ejection systems, Paraglider systems, recovery and landing systems for the Gemini and Apollo spacecraft and instrument capsules, re-entry vehicles, and underwater vehicle systems, experienced this lack of design guides and data for the development and design of pyrotechnic devices and systems for many years.

It is the purpose of this handbook to provide useful data and information about theory and practical application of pyrotechnics for engineers, designers, technicians and students. The contents of this handbook are divided into six parts: Explosive Materials, Explosive-Actuated Devices, Pyrotechnic Systems, Reliability and Testing, Explosive Production Methods, and Appendix.

The handbook contains numerous charts, graphs and illustrations as useful aids. Theory, data, and practical applications are explained in detail. Valuable new information is presented in this handbook, as for example data about the effects of extreme environmental conditions on pyrotechnic materials and devices, hints and data for qualification testing, hints for the design and application of pyrotechnic systems, and data for the application of explosive methods in manufacturing processes.

It is recommended to use this handbook together with the book

"Military and Civilian Pyrotechnics" by Dr. Herbert Ellern, published by the Chemical Publishing Company, which contains more detailed information about the properties and production of pyrotechnic materials and an extensive manufacturing formulary.

In writing this handbook, the author made an attempt to cover the whole field of pyrotechnics technology and to present the newest and most complete data and information. The author would appreciate comments from users of this book, which may help to improve the contents of future editions of this handbook.

Thousand Oaks, California
      Karl O. Brauer

Part I

Explosive Materials

Explosive materials and compositions which are commonly used in pyrotechnic devices and systems are described in detail in the present Part I, Chapter 1 and 2, to provide the necessary information about the functioning, application, and physical and chemical properties of these materials. Only with the knowledge of the basic characteristics of these materials will it be possible for the user of this handbook to fully understand the functioning of pyrotechnics and to develop new pyrotechnic devices and systems.

1 Explosives and Compositions

Explosive materials and compositions are used in a great variety of pyrotechnic devices and systems mainly as producers of either gas, flame, heat, shock, smoke, light, or sound. Since explosives usually release energy in the form of heat and gas due to chemical disassociation in a way similar to gasoline, kerosene, butane, oil, and coal, they can generally be classified as fuels. The main difference of explosives as compared with conventional fuels is the characteristic that explosives contain oxygen and do not require atmospheric oxygen for combustion, and that the combustion rate of explosives is much faster than that of conventional fuels. The ability of explosives to perform work is based on the release of gas at high temperature and at high pressure. In most explosives, this pressure effect is aided by the quickness of the action which often may produce shock phenomena.

Explosive and pyrotechnic materials and compositions are classified, according to their characteristics, into two major categories:

A. Deflagrating or burning materials and compositions, which undergo combustion,

B. High explosives, which undergo detonation or high-order explosion, to differentiate it from low-order explosion, such as fast combustion of propellant.

All explosives, primers, propellants, and high explosives, are unstable materials by their very nature. While they are safe to handle and to process under controlled conditions, they must be treated and handled with great caution and the respect that their latent power deserves.

It is a not commonly known characteristic of some explosive materials that, the more powerful the explosive, the more difficult it is to initiate. This feature is the main reason for the arrangement of a minimum of two elements of explosive materials in an explosive train, in which the first material is a priming explosive to provide a reliable initiation of the second charge, which is the main charge, often also called base charge, and which has a high power output for performing the required work. The priming explosive is often also called "ignition bead" which, in many pyrotechnic devices, consists of a very small quantity of a highly sensitive explosive material and which is applied to the electric bridgewire in the device to be initiated by heat.

In some devices, a third or intermediate charge is used to aid in the transition to detonation. In these cases, initiating explosive materials are used for the ignition bead and for the intermediate charge, whereas a booster material is used for the main charge.

According to this commonly used arrangement of different explosive elements, high explosive materials can also be classified into the two major groups: Initiating or primary explosives and non-initiating or secondary explosives.

1. Initiating explosive materials which are initiated by an electric current, impact, or friction, are used to initiate the detonation of relatively insensitive explosive materials.
2. Non-initiating explosive materials are detonated by an initiating material which, during the detonation, generates the required shock to accomplish the necessary destruction for the actuation of the pyrotechnic device. Non-initiating explosives can be divided into three types:
    a. Boosters which can easily be initiated and which detonate at a high rate. For this reason, booster materials are not recommended for loading in large masses. A typical booster material is lead azide.
    b. Bursting charges which are initiated by a booster and which are suitable for loading in mass. A typical bursting charge material is TNT.
    c. Explosive materials which can only be used as ingredients in explosive mixtures, because they are either too sensitive or too insensitive to be used alone. A typical example for a too sensitive explosive material is nitroglycerin, and a typical example for a too insensitive explosive is ammonium nitrate.

A. *Priming Materials.*

Priming or "first fire" materials are the first explosive material initiated in an explosive train and are used to initiate the detonation of less sensitive high explosives. Only a few explosives which have a high degree of sensitivity to initiation through shock, friction, electric spark, or high temperature, but which are not too sensitive, are suitable as ideal priming explosives. They must be able to release very high temperature or shock, or both, upon disassociation. A high sensitivity of priming materials is necessary because they are usually initiated either by raising their temperature by means of an electrically heated wire, or by the spit of a powder fuse, or by shock caused by the striking of the explosive by a firing pin in a rifle or pistol cartridge or in a special mechanically-initiated device. It is the main function of primers, which are power producers, to initiate a second, more stable and better power generating explosive which requires the high temperature output from the priming explosive.

Typical commonly used priming explosive materials are:

Lead azide

Lead styphnate (lead trinitroresorcinate)

LMNR (lead mononitroresorcinate)

KDNBF (potassium dinitrobenzofurozan)

Barium styphnate

Zirconium potassium perchlorate

Other priming materials which are not frequently used are mercury azide, potassium chlorate, and mercury fulminate.

B. *Propellants*

Deflagrating material is divided into three groups, according to their physical shape:

1. Loose powders, fine grains, spherical shotgun grains
2. Pressed or extruded grains (size 1/4 inch or larger)
3. Solid propellant billets, as used in gas generators

Deflagrating materials are generally good gas or heat producers. Some of these materials burn in the same manner as solid rocket propellants, but more rapidly. Since deflagrating materials are more stable than primers, they are usually set off by an initiating material. The force exerted by the deflagrating material is produced by the vapors evolved during combustion, in the same manner as in a solid-

5

propellant auxiliary power unit (APU) gas generator. Similar to the conventional solid propellants, they burn evenly over their whole exposed surface area. Their burning rate depends mainly on the physical form of the deflagrating material. The highest burning rate is obtained in a charge consisting of thin flakes. The pressure rise rate decreases with an increase of the grain size of the material, and the burning rate is also affected by temperature and pressure.

Deflagrating powders, or propellant or pyrotechnic powders are used as the main charge in numerous impulse-type pressure cartridges which are utilized for the actuation of valves, thrusters, and guillotine devices. Important factors that influence the selection and sizing of the propellant powder for a certain application are: Maximum energy required, ignitibility, peak pressure, and burning time. The time from ignition of the propellant to its total combustion depends on the chemical composition of the propellant and the shape or form factor, and on the temperature and pressure. Very important for proper sizing of a propellant charge of a given shape for a certain required output and burning time is the ratio between surface area and total volume, since propellant powders burn only on the surface. When a very fast burning rate and a very high peak pressure is required, flake or flat powders are ideally used, whereas spherical or cylindrical powders are best used when a lower pressure for a longer period of time is required.

Rifle-type powders have such a fast burning rate and develop such a high pressure that the total combustion time may be only one millisecond. In applications where burning times of one second or longer are desirable, it is not advisable to use pressure cartridges with rifle-type smokeless powder which has a too fast reaction time, but to use a gas generator propellant instead. The propellant normally used in gas generators, which is similar to the propellant used in rocket engines, has a much slower burning rate than smokeless powder. The propellant used in a typical gas generator may have a burning rate of 0.1 inch per second at a pressure of 2000 psi, whereas the burning rates of smokeless powders are measured in inches per second. For this reason, the gas generator propellant can ideally be used in applications where burning times of many seconds are required. Gas generator propellants are more difficult to ignite than smokeless powders, and some of these propellants require relatively high pressures and temperatures to be maintained in order to sustain combustion.

Other typical burning explosives are $BKNO_3$, double-base powders, Hi-Temp (RDX), and zirconium potassium perchlorate.

C. *High Explosives*

High explosives are chemical compositions, which when initiated by a suitable stimulus, disassociate almost instantaneously into other more stable components. This reaction is known as high-order or low-order detonation. As a result of the explosive decomposition, detonating explosives produce some gas and high temperatures. The detonation is so fast that a shock wave is generated that acts on its surroundings with great brisance, or shattering effect, before the pressure of the exerted gas can take effect. This type of explosive material is used to pulverize rock and to sever steel beams. The most sensitive detonating explosives are so unstable that they can be set off by the slightest vibration, friction, or heating, and for this reason, such type of detonating materials is useless.

A different type of detonating explosive, which is relatively stable, is used in explosive devices and systems. Some of these high explosives are so stable that rifle bullets can be fired through them or they can be set on fire without detonating. They are set off only by the severe shock of an initial detonation, which is usually supplied by the explosion of a more sensitive explosive material. The more stable explosives which detonate at very high velocities of up to 9000 meters per second exert a much greater force during their detonation that the explosive materials used to initiate them.

Typical high explosives which are widely used in explosive devices are Tetryl, TNT, RDX, and PETN. For the initiation of the most commonly used materials, RDX and PETN, a shock impact of 3000 to 5000 meters per second is required which can be obtained only from another explosive material. For this reason, explosive devices in which RDX or PETN is used as a main charge, contain an explosive train consisting of three units: a primer, a booster, and the high explosive main charge. An exception are EBW (explosive bridgewire) devices, where RDX and PETN is directly initiated by the bridgewire.

The following example may explain the tremendous shock generated by a typical high explosive: If an explosive charge of lead azide is set off against one end of a three foot long and one foot diameter cylinder consisting of RDX, the shock from the disassociation of the lead azide is sufficient to cause decomposition of the face of the RDX cylinder which occurs with release of gas, shock, and increase in temperature. This zone of decomposition travels at an enormous speed down the length of the cylinder to the opposite end. This speed is called the detonation rate, which is different for each type of explosive material and depends also to a high degree on the density and confinement. In RDX, for instance, pressed to a density of 1.6 grams per cubic-centimeter, the detonation rate is 8200 meters per second, or over 18,000 miles per hour. This means that the time required for the decomposition to travel from end to end of the three foot long cylinder is about 0.0001 second. At this high decomposition rate, the byproducts of gas and particles cannot escape from the decomposition area, which results in the generation of extremely high shock pressures. The gases and particles follow the detonation wave from one end to the other as it proceeds through the cylinder. As the decomposition zone reaches the far cylinder end, it is not only traveling at a speed of 18,000 miles per hour, but it has also developed a peak shock pressure of about 4,000,000 pounds per square inch. This combination of velocity and shock results in the tremendous shattering effect of high explosives, which is utilized to break bolts, shatter steel parts, and deform sturdy metal structures.

A recently developed new type of high-performance detonating explosives are the Astrolite materials, which were developed from high-performance rocket propellants. The Astrolite explosives are not considered for use in conventional explosive devices, but mainly for special applications. They are explosives in liquid or putty form which combine the advantages of high performance and safety characteristics of solid explosives and the versatility and handling flexibility of liquid explosives. The sensitivity of Astrolite explosives to impact shock or adiabatic compression is considerably less than the sensitivity of RDX. It can be dropped from low-flying aircraft, and it can withstand 30-caliber rifle fire without detonating or burning. A standard military detonator is only required for initiation of Astrolite explosive material.

Astrolite-A which is intended mainly for demolition work and special applications requiring very high performance, is an extremely powerful explosive in liquid or putty form. This explosive will perform up to five times the work of an equal amount of TNT, and it has three to five times the power of commercial dynamite and Composite C-4, as proven in field tests.

A different Astrolite composition, Astrolite-G, is one of the highest velocity liquid explosives, with a detonation velocity of 8600 meters per second. This explosive is more powerful than Composition C-4 in shaped charges, and it has excellent propagation properties. These characteristics make Astrolite-G ideally suited for advanced weapons concepts.

A third type of Astrolite compositions, Astrolite-P, is a high-velocity explosive putty which sticks to any surface and which can be molded in any desired shape. It can be used for general purpose demolition, metal cutting, and similar applications.

D. *Explosive Cord*

A long column of explosive material encased in a continuous sheath is known as explosive cord. In recent years, various types of explosive cords have been developed to meet different requirements. Explosive cords are ideally used for the transfer of a detonation wave, for rapid ignition of a propellant, and also for cutting and separating metal parts.

The most commonly used explosive cords are:

Mild-detonating fuse (MDF),

Confined detonating fuse (CDF),

Flexible linear shaped charge (FLSC), and

Time-delay line.

Properties of commonly used explosive cord are listed in Table I.

1. *Mild-detonating fuse* (MDF) consists of a small-diameter metal tubing which contains a core of detonating material. It can be used for severance and fracturing of thin-walled metal structures, or as a transfer line for a detonation wave, when confinement of the explosive forces is not required. A mild-detonating fuse equipped with a sheath of a suitable metal and a plastic covering, which is capable of confining the detonation, can also be utilized for transmission of a detonation wave past other explosive material, personnel, or fragile equipment. A detonator is normally used to initiate MDF.

The average tubing diameter of mild-detonating fuse is 0.10 inches. MDF is available with a sheath of lead, aluminum, copper, and silver. For special applications, a polyethylene jacket can be drawn over the metal sheath.

2. *Confined detonating fuse* (CDF) is mainly used to transmit a detona-

Table 1.1

Properties of Explosive Cord

Cord Type	MDF Mild-Detonating Fuse	CDF Confined Detonating Fuse	FLSC Flexible Linear Shaped Charge	Small Column Insulated Delay (SCID)	Deflagrat. Small Col. Insul. Delay (HIVELITE)
Cross-sect. Shape	Tube	Tube	Chevron-Shaped	Tube	Tube
Size	0.040" dia. & up	Core 0.040" to 0.072" dia.		0.040" to 0.080" dia.	0.040" to 0.080" dia.
Material	Lead Aluminum Copper Silver + Polyethyl.	Lead Aluminum Silver + Fibergls. & Nylon	Lead Aluminum Silver	Lead with or without Fiberglass or Plastic	Woven Fiberglass
Explosive (Grain/ft)	1 thru 20 (1 thru 1000)	1 thru 5	5 thru 1000	2 thru 16 detonat.	1 thru 10 deflagrat.
Detonation Velocity (meters/sec)	5000 - 8000	5000 - 8000	5000 - 8000	0.008-1.27	152 - 254
Temperature Range	$-267°C$ to $+371°C$	$-267°C$ to $+371°C$	$-267°C$ to $+371°C$	$-73°C$ to $+535°C$	$-267°C$ to $+535°C$
Reliability	.9998	.9998	.9998	.9998	.9998
Application	Transfer Lines, Destruct, Severance, Fracturing	Transfer Lines	Metal Cutting, Destruct	Delay Fuse	Transfer Lines Rocket Ignition, Separation
Notes	0.242" O.D. f. 1 grain/ft. type; 0.320" O.D. f. 2 gr/ft.			Product of McCormick-Selph	Product of McCormick-Selph

tion wave from one point to another. CDF contains a very small amount of explosive material extruded in a metal sheath which is covered with external layers of nylon and fiberglass braid or similar material which has the purpose of retaining all products of the detonating high explosive core.

A confined detonating fuse is available with a flexible metal tubing of 0.040 inch to 0.072 inch diameter. The CDF is manufactured in continuous lengths of up to 1000 feet. Tests have proven that a typical CDF line has a tensile strength in excess of 250 pounds.

The confined detonating fuse has the advantageous feature that it is capable of propagating a detonation through tight knots, loops, twists or kinks that may be introduced into the assembly. Quick-disconnect bayonet-type end closures are sometimes mounted on both ends of a CDF line to facilitate assembly. Standard detonator cartridges with completely contained mild primers can be used to initiate CDF transfer lines. The explosives PETN and RDX are commonly used in CDF lines. A confined detonating fuse (CDF) containing an explosive charge of one grain per foot has an outside diameter of 0.242 inches, and a CDF containing a charge of two grains per foot has a diameter of 0.320 inches.

3. *A flexible linear shaped charge* (FLSC) is a flexible cord containing a long column of explosive material in an inverted Vee- or chevron-

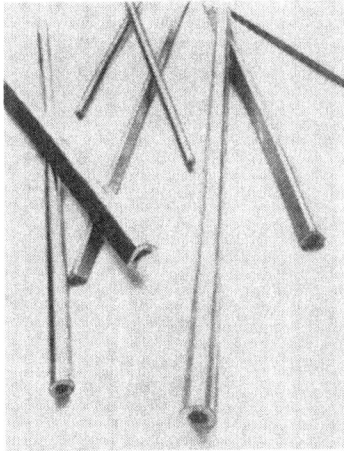

Fig. 1:1 A Mild-Detonating Fuse (MDF) and a Flexible Linear Shaped Charge (FLSC) (Courtesy of General Precision Systems, Inc., Link Group).

11

shaped seamless metal sheath. Soft aluminum or lead are normally used for the sheath. The main applications of FLSC are explosive cutting through thick metal and emergency cutting of metal straps, webbing, and similar material for a quick separation. The charge weight, standoff distance, and metal sheath design must be carefully controlled to achieve reliable cutting of materials. The FLSC can be bent into any desired shape without affecting the reliable functioning of the explosive. Blasting caps or initiators can be used to initiate FLSC. During the detonation, the shape of the explosive charge directs the shock wave through the open face of the Vee of the shaped charge, providing the desired cutting action.

A typical flexible linear shaped charge (FLSC) in four different sizes, together with four sizes of mild-detonating fuse (MDF) is shown in Figure 1:1.

4. *A small-column insulated delay* (SCID) is, with regard to construction and size, similar to mild-detonating fuse with the difference that a special evenly deflagrating time-delay charge contained in a lead sheath is used. Small-column insulated delay (SCID) has the advantageous feature that it can be tightly coiled, bent or flattened to oval or ribbon sections without affecting its propogating ability. Small-column insulated delay (SCID) cord is available in sizes from 0.040 inch diameter and in lengths up to 250 feet.

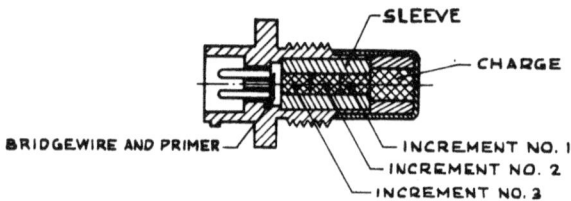

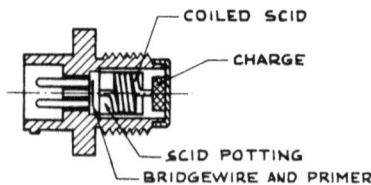

Fig. 1:2 Comparison of Delay Cartridges (Courtesy of McCormick-Selph).

A specific time delay in a device is achieved simply by cutting the SCID cord length according to the predetermined burning time of the charge. Time delays from 0.005 seconds to more than 15 minutes can be obtained with SCID.

Figure 1:2 shows a comparison of a time-delay cartridge, in which small-column insulated delay cord SCID is utilized, with a conventional delay cartridge. The advantage of space and weight reduction achieved in the SCID cartridge are obvious.

5. *A deflagrating small-column insulated delay cord* is a flexible explosive cord which resembles SCID in construction and size. The main difference is that a different pyrotechnic core material is used in the deflagrating small column insulated delay which propagates at a greater rate and which produces a deflagrating output. A typical delay cord of this type is known as "HIVELITE", by McCormick-Selph.

Other flexible explosive cords which are used in numerous applications are "Primacore" (tradename by Ensign-Bickford), "Pyrocord" and "Tacot" cord (trademark by Du Pont), which are similar to mild-detonating fuse (MDF) and to confined detonating cord (CDF). A flexible explosive cord which has some functional similarity with flexible linear shaped charge (FLSC) is the extruded Du Pont EL-506 cord. This explosive cord consisting of PETN and an elastomeric binder has no metal cutting and destruct systems applications, and it is available in a variety of cross-sectional shapes and sizes.

E. *Explosive Sheet*

A high explosive in flexible sheet form, "Detasheet", developed by Du Pont, finds applications mainly in destruct devices, demolition, and in metal cutting and hardening techniques. "Detasheet" consists of a mixture of PETN and an elastomeric binder. It is available with two different explosive contents and in a thickness range from 0.042 inches to 0.333 inches. This flexible explosive material can easily be cut and can be applied to any complex-shaped surface by using an adhesive. Since "Detasheet" is waterproof, it is also ideally suited for underwater applications. It has a detonation rate of 7000 to 7200 meters per second. "Detasheet" C, which contains 63 percent PETN, can be initiated with a detonator No. 8 (6.9 grains PETN) or, to provide greater safety to personnel, the detonator Du Pont E-94 with a base charge of 2 grains PETN, can be used. For protection in radio frequency fields, Du Pont detonator X-570 is recommended for initiation of "Detasheet" C.

F. *Time Delay Compositions*

In many cases, a certain delay between initiation and the time of the detonation or deflagration of the main charge is required in pyrotechnic devices and systems. Time delay elements are used in these applications, mainly in systems in which certain functions must be completed before others begin. This can be accomplished in such a way that each device triggers the next one.

In systems where a function time interval is not required, the various devices can be equipped with charges of different column lengths of slow-burning pyrotechnic materials which are initiated simultaneously. Such a type of an arrangement can result in a lightweight system.

A time delay element usually consists of a casing containing several different charges: an initiator charge, a delay column, and an output charge. The initiator charge is ignited by an electrically-heated bridge-wire. The forward end of the delay column, which is arranged adjacent to the initiator charge, is made from an easily ignitable and fast-burning metal oxidant mixture designed to ignite the main part of the delay column, which is made of a slower burning delay composition. The output charge normally consists of a pressure generating composition designed to provide the required power to perform some work.

KDNBF and LMNR are used as initiator charge materials. Suitable materials used for the igniter end of the delay column consist of a mixture of deflagrating fine metal powders and oxidants which are to provide oxygen for combustion. Frequently used metal powders are molybdenum, zirconium, nickel-zirconium, titanium, boron, and tungsten.

Suitable oxidants for the metal powder-oxidant mixture are nitrate, chromate, chlorate, or perchlorate. These materials supply oxygen during combustion, without generating an undesirable amount of gaseous products, which is an important characteristic for application in time delay columns, because they occupy the same volume before and after burning. By avoiding the generation of dangerous pressure, the possibility of premature firing of the explosive charge at

Table 1.2

**Influence of Storage Life on Delay Periods
of Pyrotechnic Time Delay Elements**

Nominal Delay (sec)	0.3	0.7	1.0	2.0	4.0	10	15	20
Test Temperature (°C)	−59 71	−59 71	−59 71	−59 71	−59 71	−59 71	−59 71	−59 71
Original Delay Period (Average sec)	0.386 0.338	0.802 0.643	1.156 0.871	2.421 1.585	4.30 3.18	10.50 9.13	16.38 13.52	22.69 17.39
Number of Months Stored at ambient Temperature	29 29	40 40	21 21	34 34	38 38	15 15	18 18	31 31
Delay Period after Storage (Average sec)	0.426 0.323	0.834 0.626	1.191 0.826	2.672 1.682	4.976 3.151	11.37 9.12	16.19 13.02	23.70 18.12
Percent Change	−10.4 +4.4	−4.0 +2.6	−3.0 +5.2	−10.4 −6.2	−15.8 +1.0	−8.3 0.0	+1.1 +3.7	−4.7 −4.2

15

the output end of the delay column, and by keeping the shrink of the column to a minimum, the contact between column and output charge is assured, as required for reliable functioning of the delay element.

A very reliable metal powder-oxidant composition suitable for igniting time delay columns, is zirconium barium chromate, which has a high zirconium content. Other suitable compositions, which have

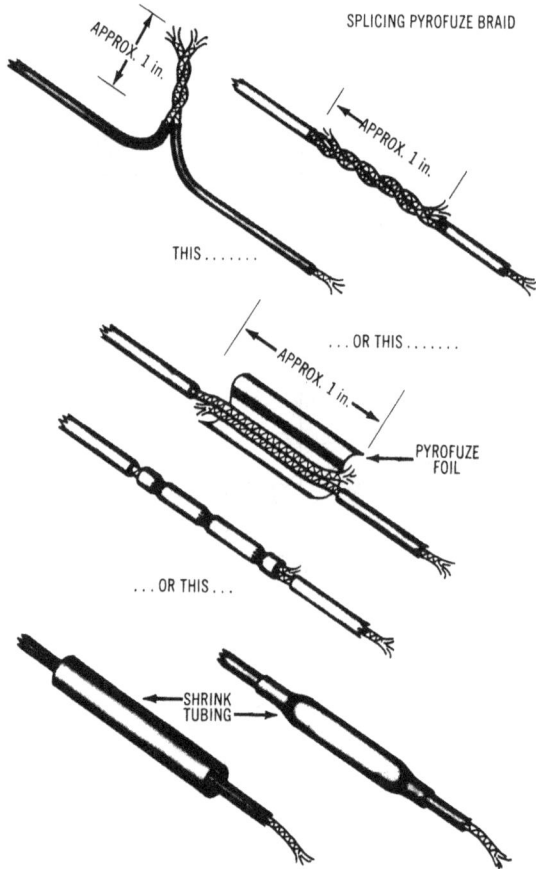

Fig. 1:3 A Pyrofuze time delay braid (Courtesy of Pyrofuze Corp.).

uniform burning rates, are boron-red lead, boron-barium chromate, and a mixture of manganese, iron oxide and diatomasceous earth.

A metal-oxidant composition is also used for the main part of the delay column, which must have a uniform slow burning rate. Nickel-zirconium is often used as the metal constituent, and potassium perchlorate or barium chromate as oxidizer in the composition. In addition to releasing oxygen, these oxidizers also catalyze the metal oxidizing reaction.

The burning rate of these compositions can easily be varied by changing the proportion of the metal constituents. A typical blend of these metals consists of 70% nickel and 30% zirconium plus the oxidant. This composition has a burning rate of 12 seconds per inch. Another blend, consisting of 30% nickel and 70% zirconium plus the oxidant has a burning rate of 25 seconds per inch.

Metal-oxidant compositions consisting of nickel-zirconium with potassium perchlorate or barium chromate are less affected by high ambient temperature than most other compositions.

Important factors that control the delay period of a pyrotechnic device are the type of the composition, particle size, packing density, and ambient temperature during storage and operation. The time for the composition to burn depends on these factors. Another influential factor is the heat sink effect which will reduce the burning time of the time delay column when the pyrotechnic device is in contact with metal.

To obtain time delay functioning with a high degree of accuracy and to minimize one variable in the burning rate of the column, the charge composition is placed in small increments into a grooved ferule and pressed to a high uniform density. The grooves of the ferule eliminate the possibility of shifting of the charge composition which might be caused by temperature changes. Since the grooved surface also provides a good heat sink, the burning time for a given amount of charge material is increased. This type of design has the additional advantage of the threads containing the slag after burning.

It was found that long storage affects the delay periods of time delay elements. Results of storage life tests are listed in Table 1.2.

An unusual type of time delay material is a braid made from a bi-metallic exothermically alloying composition, known under the trade name "Pyrofuze", which is described in Chapter 1, Section J in this book. These palladium-aluminum compositions are non-explosive, and they are initiated by normal squib voltages and amperages.

The braid is fabricated from eight single strands of Hi-R Pyrofuze wire woven into an interlocking braid with a coaxial configuration in cross-section. Braids can also be fabricated from less than eight strands and in larger multiples, such as 32 and 64 strands. The larger sizes are mainly used for retention and holding devices.

Pyrofuze braid can be initiated by the spit from a percussion or stab primer or by passage of electric current. Pyrofuze braids have been used successfully as time delay in fire transferral systems, particularly in systems requiring a high degree of precision and reliability. This braid material has the advantages of design simplicity, low weight, safety, and ease of handling.

Standard diameters of Pyrofuze braid range from 0.002 to 0.010 inch. The burning rate for 0.004 inch diameter 8-strand bare Pyrofuze braid is one foot per second. A fiberglass covering reduces this reaction about 10 percent. A 0.005-inch 8-strand Pyrofuze braid burns at an approximately 30 percent slower rate than the 0.004-inch material, and a 0.003-inch 8-strand braid burns at an approximately 30 percent faster rate than the 0.004-inch braid. For millisecond delay and as starter sections, the 0.002 and 0.003-inch diameter braids are advantageously used, whereas for long-term unprotected delays, the 0.004, 0.005 and 0.006-inch bare Pyrofuze braids find proper applications. For long-term protected delays, 0.005 and 0.006-inch fiberglass-covered Pyrofuze braid is recommended. The higher gage braids of 0.007, 0.008, 0.009 and 0.010 inch diameter are mainly used for mechanical restraint. Typical Pyrofuze delay braids and splice methods for braids are shown in Fig. 1:3.

G. *Heat-Producing Materials*

Heat-producing materials are used in a variety of applications, as for example, as "first fires" in pyrotechnic devices to ignite other materials, as primers, as heat generators in pyrotechnic heaters, as propellants in gas generators and rocket motors, and as incendiary materials in destruct systems.

Heat-producing materials used as primers are often formulated to generate heat and shock. These materials are usually very sensitive and easily ignited. Because of their high sensitivity, these materials are preferably applied to the main charge or booster material in the form of a soft paste, eliminating the danger involved in applying it by pressing it on in a dry condition.

Compositions used in heater devices should have a uniform burning rate, and they should not generate a large amount of gaseous products or dangerous pressure, which is achieved by suitable oxidants in the mixture. Thin-walled and light-weight heater designs result from avoiding internal pressure build-up. Another important consideration in the selection and formulation of compositions especially for application in heaters is the required burning time.

Depending on the application and on the work to be performed, materials with either high or low sensitivity are used. The sensitivity of heat-producing mixtures, which should be high for primers, but considerably lower for heater applications, is greatly affected by the oxidizer in the mixture. A high sensitivity is obtained by a mixture of zirconium powder with lead dioxide or lead tetraoxide as oxidizer, for example.

In more stable heat-producing metal-powder mixtures for applications in heaters and similar devices, the desirable lower sensitivity is obtained by nitrates, chromates or chlorates as oxidizers. In cases where a very low sensitivity is required, a mixture of aluminum powder with ferrous oxide or ferric oxide can be used.

More detailed descriptions of priming and propellant materials are given in Part I, Chapter 1 of this book.

H. *Smoke Generating Materials*

Smoke generating devices are used in civilian applications mainly as a ground wind direction indicator on auxiliary airfields, as location aids for spacecraft recovery, and for special effects in movie scenes, simulating explosion, fire, or battle effects. In military applications, smoke generating devices are used for marking and signalling from shells, pots, and rockets, and for camouflaging and screening of troop and ship movements.

Generally, smoke generating devices are manufactured in the form of pots and squibs. The standard sizes of pots range from 0.5 inch diameter, 4 inches long to 1 in. diameter, 12 in. long. Ignition of these smoke generating devices is accomplished either by fuze or by electric current. A variety of color smokes can be produced: White, black, gray, yellow, red, green, blue, purple, orange, and pink.

1. White Smoke Producers

    a. Dense white smoke can be produced by burning a mixture of zinc dust and hexachlorethane. Zinc dust is available in the form of a bluish-white powder. The specific gravity of zinc dust is 7.14 at 20°C. Its melting point is 419.4°C, and its boiling point is at 907°C.

    b. White phosphorus can also be used for producing dense white smoke, which consists of phosphorous pentoxide and phosphorous acid and which has an extremely high obscuring power. White phosphorus is spontaneously inflammable in air at normal temperature.

      White phosphorus, which has the symbol P_4 is a pale yellowish, translucent, crystallizable solid of waxy consistency. It has a specific gravity of 1.82 at 20°C. Its melting point is 44.1°C., and its boiling point is 280°C.

    c. White smoke can also be generated by exposing sulfur trioxide to air. On contact with air, sulfur trioxide fumes and generates white clouds consisting of very small droplets of sulfuric acid.

      Sulfur trioxide (composition SO_3) is available in the form of colorless prisms. It is soluble in concentrated sulfuric acid and in chlorsulfuric acid. It has a specific gravity of 1.92. Its melting point is 16.8°C., and its boiling point is at 46°C.

    d. Another chemical suitable for producing white and gray smoke is naphthalene. When it is heated, it generates inflammable vapors.

      Naphthalene, composition $C_{10}H_8$, is available in the form of white scales. It is insoluble in water. Its melting point is 80.2°C., and its boiling point is at 218°C.

2. Black Smoke Producers

    a. A mixture of sulfur, potassium nitrate and pitch or rosin, which has to be ignited, can be used in torches for producing black smoke.

    b. Pitch without any other constituents can also be used to produce black smoke. Pitch consists of a mixture of bituminous or resinous substances, and it is usually a dark colored, tenacious, fusible, thick, more or less solid material.

3. Grey Smoke Producers

    a. Dark grey smoke can be generated by burning a mixture of zinc dust, hexachlorethane, and naphthalene or anthracene.

    b. For producing light-grey screening smoke, silicon-tetrachloride, combination $SiCl_4$, can be used. In combination with ammonia vapor, it generates smoke which is similar to natural fog and which can be ideally used for camouflaging troop or ship movements.

        Silicon tetrachloride is a colorless fuming liquid which has a suffocating odor. It decomposes with the moisture of the air, forming a dispersion of silicic acid and hydrogen chloride.

        The smoke is generated from a funnel which is connected with two cylinders, one containing liquid ammonia, and the other charged with silicon tetrachloride containing about ten percent carbon dioxide under a maximum pressure of 550 psi at $55°C$.

        Silicon tetrachloride has a specific gravity of 1.5. Its melting point is $-70°C$., and its boiling point is $57.6°C$.

    c. Gray smoke can also be produced by the simple method of high boiling of coal-tar hydrocarbon, composition $C_{14}H_{10}$. This method was commonly used in trench warfare.

4. Yellow Smoke Producers

    a. For producing yellow-colored smoke, chrysoidine, which is an aniline azo-m-phenylenediamine chlorhydrate, can be used. It is available in the form of crystal powder or large black crystals.

    b. Another composition for producing yellow-colored smoke consists of about 35% auramine, about 25% lactose, about 32% potassium chlorate, and 8-9% chrysoidine. The yellow color of the smoke is caused by the auramine, which is a bright yellow dye.

    c. A Military Standard composition used for producing yellow smoke consists of about 40% yellow dye, about 24% potassium chlorate, 27% baking soda, and 9% sulfur.

5. Red Smoke Producers

    a. For producing red smoke, paranitraniline red, also known as Paratoner or Para Red, composition $O_2NC_6H_4N_2C_{10}H_6OH$, can be used. It is available in the form of red powder.

b. Another mixture used for producing red smoke consists of 50% rhodamine red, 25% potassium perchlorate, 20% antimony sulfide, and 5% gum arabic, or a similar suitable binder.

c. A military Standard composition used for producing red smoke consists of about 40% red dye, about 27% potassium chlorate, 22% baking soda, and about 11% sulfur.

6. Green Smoke Producers

a. A mixture of about 15% auramine, 25% indigo, 35% potassium chlorate, and 25% lactose can be used for producing green smoke.

b. Another composition for generating green smoke consists of 45% malachite green, 27% potassium chlorate, 23% antimony sulfide, and 5% gum arabic, or a similar binder.

c. A Military Standard composition used for producing green smoke consists of about 43% green dye, 22% potassium chlorate, 25% baking soda, and 10% sulfur.

7. Blue Smoke Producers

a. A mixture of about 40% indigo, 35% potassium chlorate, and 25% lactose can be used for generating blue smoke. Indigo, composition $C_{16}H_{10}O_2N_2$, is available in the form of dark blue lustrous powder. It can be sublimed at 300°C, and it decomposes at 392°C.

b. Another composition used for producing blue smoke consists of 40% methylene blue, 25% potassium perchlorate, 20% antimony sulfide, and 5% gum arabic, or a similar binder.

8. Various Colored Smoke Producers

Smoke of purple, orange, pink or similar other colors can be produced using similar mixtures as described for green or blue smoke, with the main difference being that a dye of the desired smoke color is used in compositions with the described constituents.

The burn time of smoke pots, as commonly used for producing smoke, depends mainly on the length of the device. For example, a nine inch long smoke pot has a burn time of approximately one minute, whereas the burn time of a six inch long smoke pot is approximately one half minute, and the burn time of a twelve inch long smoke pot is about one and one half minutes.

I. *Sound Producing Materials*

Sound producing materials are mainly used in pyrotechnic devices for signalling, for simulating booby traps or military ammunition bursts in training devices, and for entertainment. Two basically different types of sound producing material compositions are generally used in these devices: First, a composition that produces a loud, dark noise of short duration, similar to the detonation of a bomb or a cannon shot; second, a composition that produces a shrill whistle sound of longer duration. Both types of sound are caused by the sudden release of gases from the exploding or burning composition.

Devices used for producing a loud, dark noise of short duration are high nitrogen type gas producers. A typical composition used in these devices consists of 79.5% ammonium nitrate, 9% potassium nitrate, 6.9% ammonium oxalate, 5.6% ammonium dichromate, and an addition of about 1.6 parts of china clay.

Another typical composition commonly used for producing a loud, dark noise consists of a mixture of 67% potassium chlorate, 27% red phosphorus, 3% sulfur, 3% calcium carbonate, and a binder.

A typical composition used in devices for producing a whistling noise of longer duration consists of about 75% potassium chlorate, about 22% gallic acid, and about 3% red gum.

A rhythmic burning of the mixture in a rapid sequence seems to cause the whistling sound of a certain duration. The size and length of the container tube greatly influence the pitch of the whistling noise.

J. *Bi-Metallic Exothermically Alloying Composition*

Bi-metallic compositions, the elements of which will alloy exothermically when heated to operating temperatures and subsequently deflagrate without the support of oxygen, are used in special applications as initiators, rocket igniters, time delay elements and destruct systems. Palladium and aluminum are generally used in these compositions. Initiation by heating causes exothermic reaction in which temperatures in excess of the boiling point of the constituents are reached. The reaction continues until the alloying of the material is complete, unless it is cooled below the operating temperature. These bi-metallic compositions are insensitive to shock, impact and vibration. A typical bi-metallic composition, known under the trade name "Pyrofuze", is fabricated into various shapes, as for example:

wire, braid, foil, granules, and tubing. The wire, ribbon, and foil-shaped compositions, of which cross-section photo-micrographs are shown in Fig. 1:4, consist of a Palladium shell and an inner core of

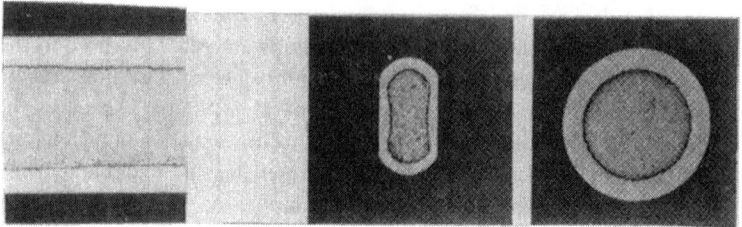

Fig. 1:4 Bimetallic composite material shapes (Courtesy of Pyrofuze Corp.).

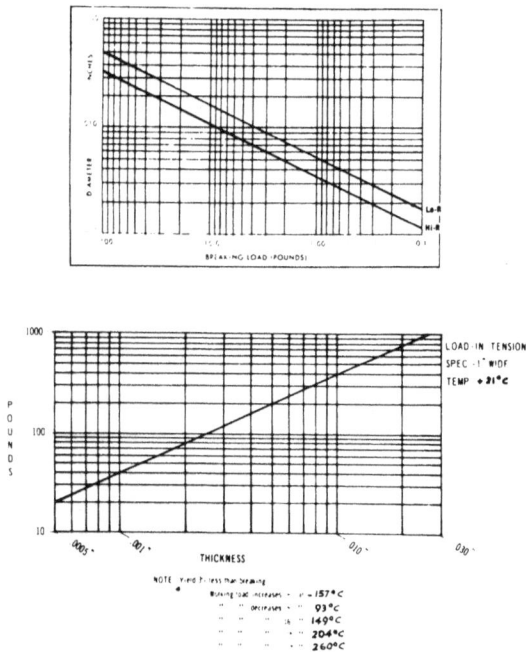

Fig. 1:5 Properties of Bimetallic "Pyrofuze" material (Courtesy of Pyrofuze Corp.).

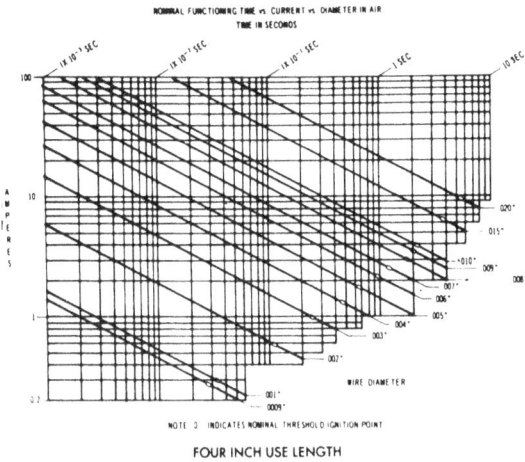

NOMINAL FUNCTIONING TIME vs. CURRENT vs. DIAMETER IN AIR
TIME IN SECONDS

NOTE: ⊙ INDICATES NOMINAL THRESHOLD IGNITION POINT

FOUR INCH USE LENGTH

Fig. 1:6 Electrical data for "Pyrofuze" wire (Courtesy of Pyrofuze Corp.).

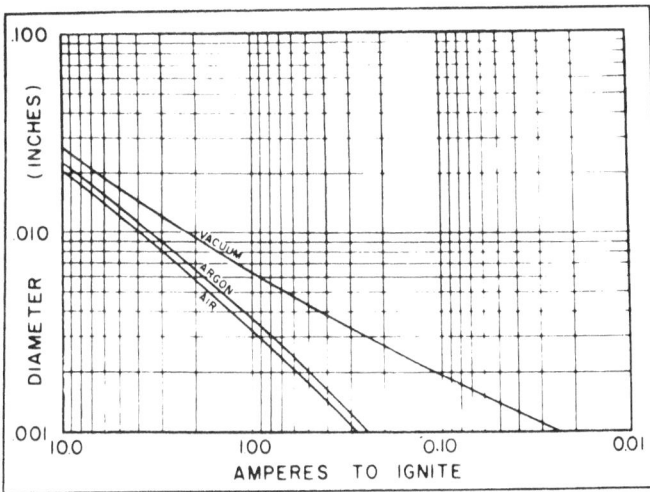

Fig. 1:7 Threshold ignition currents for "Pyrofuze" wire in selected media (Courtesy of Pyrofuze Corp.).

25

aluminum. It is an important requirement for these bi-metallic compositions that the ratio of both metals stays constant over the entire size range.

The reaction of these compositions is not of an explosive nature, but only thermal energy, 325 calories per gram and 2890 calories per cubic centimeter at a minimum temperature of 2800°C, is released, and there is no shock or detonation.

Strength properties of bi-metallic "Pyrofuze" material are shown in the graphs in Fig. 1:5. Electrical data for Pyrofuze wire are presented in Fig. 1:6, and threshold ignition currents for Pyrofuze wire in selected media are shown in Fig. 1:7.

K. *Explosive Materials for Future Space Applications*

It can be expected that for some future space missions, a sterilization of all spacecraft components will be required, especially for spacecraft that are to be used for extra-terrestrial landing operations. Such a sterilization requirement will include explosive materials which are to be used in pyrotechnic devices in these spacecraft systems. The explosive materials, as listed below, can be expected to be suitable for these future aplications while passing sterilization conditions:

Tetranitrobenzene, ammonia picrate, DAT (N) B, DIPAM, EL-511, Hexite, Beta-HMX, HNO, HNS, LDNR, lead azide, nitroguanidine, NONA, TACOT, tetranitro carbazol, TNO, trinitro naphtalene, magnesium and teflon, metal powder/oxidant mixture, such as B/KNO_3.

Explosive materials listed as questionable for sterilization, but subsequently proven as acceptable by actual testing are:

Black powder, LMNR, and lead styphnate.

2   Properties of Explosives

The present chapter is intended to provide detailed information about physical and chemical properties of the explosive materials that are commonly used in explosive devices and systems. Mainly the properties of priming materials, deflagrating materials, and high explosives are presented. It was deemed useful to describe also some materials which are used as components in explosive mixtures.

The properties of heat-producing materials, smoke generating, and sound producing materials are presented in the related sections in Chapter 1.

A. *Priming Materials*

1. Lead Azide

Lead Azide, $Pb(N_3)_2$, a commonly used initiating material, has very good stability in high temperature storage. It is a salt of hydrazoic acid, HN_3, and it is usually dextrinated to prevent the formation of long, unstable crystals and to improve its stability. Lead azide is one of the few commercially produced explosives which do not contain oxygen.

The explosive temperature of lead azide is $390°C$, which is higher than the explosive temperature of most other initiating explosive materials. Because of its high temperature of ignition, it is used in some compositions in combination with lead styphnate which ignites at a lower temperature. Lead azide is less sensitive to impact than mercury fulminate, and too insensitive to be used alone in cases where it is to be initiated by the impact of a firing pin.

Lead azide is also used in explosive rivets, with addition of tetracene or silver acetylide, to lower the ignition temperature.

Properties: Lead azide is available in the form of white to buff-colored crystalline powder. Its specific gravity is 4.80. In contact with copper, lead azide forms a supersensitive explosive.

2. Lead Styphnate

Lead styphnate, composition lead trinitroresorcinate, $C_6H(NO_2)_3$ (O_2Pb), is a lead salt of styphnic acid. It has good temperature stability. It is used as a primer and in priming composition and blasting caps, in combination with lead azide to facilitate its ignition.

Properties: The explosion temperature of lead styphnate is $310°C$. It has the disadvantageous feature that it is more sensitive to impact than lead azide.

3. LMNR

LMNR, known as lead mononitroresorcinate, is also a lead salt which is often used as a priming material. It is obtained by treating resorcinal in sulphuric acid and nitric acid followed by hydrolysis.

It is widely used in pyrotechnic devices because of its excellent stability, reliability, and uniform performance.

Note: All these lead salts, lead azide, lead styphnate, and LMNR, which have some ideal initiator characteristics, have the disadvantageous feature that, after firing, they leave a residue of metallic lead deposit in the pyrotechnic device. These conductive residues can provide a path for current leakage or battery drainage in electro-explosive devices.

New improved explosive materials which do not leave a conductive deposit in the pyrotechnic device have been developed. One of these improved materials is KDNBF.

4. KDNBF

KDNBF, known as dinitrobenzofuroxane, is less sensitive to impact than lead styphnate. KDNBF is obtained by treating o-nitroaniline with alcohol, acids, and potassium carbonate.

5. Barium Styphnate

Barium styphante, composition $(C_6HN_3O_8Ba·H_2O)$, has excellent stability at high temperatures and under vacuum, and it has an explosion temperature higher than lead styphnate. Barium styphnate is obtained by treating the acid with magnesium carbonate and barium chloride.

6. Barium Nitrate

Barium nitrate, composition $Ba(NO_3)_2$, is used in blasting explosives, percussion primers, and in pyrotechnic compositions, and also in some propellants. It is used in percussion primers instead of potassium chlorate with the object of producing noncorrosive primers.

Barium nitrate Grade A with a purity of 99.8% is used in priming compositions, and Grade B with a purity of 99.0% is used in pyrotechnic compositions.

Properties: Barium nitrate is available in the form of colorless crystals or white powder. Its specific gravity is 3.244 at a temperature of $23°C$ (room temperature). Its melting point is $575°C$. Barium nitrate burns with a green flame.

7. Mercury Fulminate

Mercury fulminate, composition $Hg(ONC)_2$, has been used universally as a primer for initiating the detonation of high explosives and for firing propellant powders. It is commonly used in the production of caps and detonators for initiating explosives for industrial, military and sporting purposes.

Mercury fulminate is frequently used in combination with potassium chlorate and with similar materials which cause a more prolonged blow and a bigger flame than mercury fulminate alone. In reinforced booster-type detonators, the effectivity of mercury fulminate is greatly improved by a main charge of a sensitive and powerful high explosive, for example tetryl.

8. Mercury Azide

Mercury azide, also known as mercurous azide, composition HgN_3, has been widely used in initiating mixtures. It has been stated that mercury azide does not give rise to a supersensitive explosive when it is in contact with copper, as is the case with lead azide.

9. Potassium Sulfate

Potassium sulfate, composition K_2SO_4, is mainly used as a component material in priming compositions. Properties: Potassium

29

Table 2.1

Sensitivity Characteristics of Priming Explosives

Test	Lead Styphnate	Explosive Material		Barium Styphnate
		L M N R	K D N B F	
80 gram falling Ball Test (Drop Height at which no Detonation occured in 10 seconds)	8 inches		14 inches	12 inches
Sand Friction Test (Minimum Force at which action occurred)	80 pounds		80 pounds	400 pounds
Friction Pendulum Test (Drop Height at which no Detonation occurred in 10 successive tests)	50 inches		30 inches	50 inches
2 kg Drop Weight Test (Min. Drop Height at which detonation occurred)	1.97 inches	17.7 inches	0.78 inches	47.24 inches
Explosion Temperature	270°C at 0 sec; 250°C at 90 sec	255-285°C at 5 sec	250°C at 5 sec	341°C at 5 sec
Static Electricity Test (Min. Electricity at which action occurred)	0.00020 joule		0.00020 joule	2.0 joule

sulfate is available in the form of colorless or white crystals, granules, or powder. Its specific gravity is 2.66, and its melting point is at 1072°C.

The sensitivity characteristics of the most commonly used priming materials are listed in Table 2.1.

B. *High Explosives*

1. PETN

PETN, known as pentaerythritol tetranitrate, also called Penthrite, or tetranitropentaerythritol, or pentaerythrite tetranitrate, has the composition $C(CH_2ONO_2)_4$. PETN is produced by treating pentacrytritol first in nitric acid and then in acetone.

PETN is mainly used in detonating compositions. It is a very effective high explosive for demolition purposes and is used in blasting caps in combination with diazodinitrophenol, with lead azide. It is often used as a charge material in explosive cords.

Properties: PETN is available in the form of a fine granular white or light-buff colored powder. Its melting point is at $141°C$. It has an explosion temperature of $215°C$, and it is easily initiated. Its detonation rate at a density of d = 1.63 is 8000 to 8300 meters per second.

PETN is very sensitive to blows and is more readily detonated than tetryl, and it is less sensitive to impact than nitroglycerin.

2. RDX

RDX, known as cyclotrinethylenetrinitramine, also known under the name Cyclonite, is widely used as a detonating material in explosive devices. RDX is made from a product of formaldehyde and ammonia, treated in nitric acid.

RDX is relatively insensitive to electric sparks. It has a higher storage-life temperature than PETN and is less sensitive to impact than PETN. RDX has an explosion temperature of $257°C$.

3. TNT

TNT, known as trinitrotoluene, also called trilite, tritol, triton, alpha-trinitrotoluene, Sym-trinitrotoluene, and known in Great Britain as Trotyl, in France as Trolite, and in Germany as Sprengmunition –02, has the composition $C_6H_2(CH_3)(NO_2)_3$.

TNT is used as high explosive charge material, and also in detonation fuse and as a component of smokeless powder. It is relatively insensitive to shock. In crystalline form it can easily be detonated with mercury fulminate, but in compressed form, it requires a more powerful initiator. When it is to be used in cast form, a booster material, such as tetryl, is required for its initiation.

Properties: TNT is available in the form of yellow crystals. Its specific gravity is 1.654, and its melting point is 80.8°C. It explodes at 240-270°C. The detonation rate, which increases with the density, is 6700 meters per second at a density of d = 1.6. In the presence of alkalies, TNT forms unstable and dangerous compounds.

4. Tetryl

Tetryl, known as trinitrophenylmethylnitramine, also called Tetralite or Pyronite, has the composition $C_6H_2(NO_2)_3(NCH_3NO_2)$. It is a sensitive high explosive. It is stable, but more sensitive to shock and friction than TNT.

Tetryl is loaded by pressing. It can be compressed into pellets for use as a booster charge. It is used in reinforced detonators, as standard booster explosive material for high explosive shells, and as the base for service tetryl caps which are necessary for positive detonation of TNT. Usually, a mixture of mercury fulminate and potassium chlorate is included in the cap to insure detonation of the tetryl.

Properties: Tetryl is available in the form of yellow crystals or granules. Its melting point is at 129 to 130°C. Tetryl contains 24.4% nitrogen. Its detonation rate is 7200-7300 meters per second at a density of d = 1.7.

5. Hexanitrodiphenylaminoethylnitrate

This has the composition $[C_6H_2(NO_2)_3]_2 N(CH_2)_2 ONO_2$, and is used as a component of reinforced detonators or as a booster. It is slightly less sensitive to impact than tetryl or pentyl, but it has about the same sensitivity to detonation. Its explosive strength is about the same as that of tetryl, and 20% greater than that of TNT. The explosive strength of this composition can be enhanced by the addition of an oxygen-bearing selt, such as potassium chlorate.

Properties: This composition ignites at 390-400°C when it is rapidly heated. Its specific gravity is 1.69.

6. Hexanitrodiphenyloxide

This has the composition $(O_2N)_3C_6H_2OC_6H_2(NO_2)_3$, and is used as a stable mixture of a very low sensitivity in detonating compositions. It is more powerful than picric acid.

Properties: This material is available in the form of white plates. Its melting point is 269°C.

7. Tetranitromethane

This has the composition $C(NO_2)_4$, has a large excess of oxygen and gives very powerful explosives when mixed with other nitro high explosives which are deficient in oxygen. It is used in detonating compositions and in blasting explosives.

Properties: Tetranitromethane is available in the form of a colorless liquid. Its boiling point is $126°C$.

8. Manganese Dioxide

MnO_2 is a good oxygen carrier and is used in detonating compositions and for pyrotechnic purposes. It decomposes on heating.

Properties: Manganese dioxide, known in its mineral form as pyrolusite, is available in the form of a brown or black powder. Its specific gravity is 5.03.

9. "Tacot"

This explosive is a special composition which has the explosive power equivalent to TNT and which has the advantageous feature that it can withstand temperatures of up to $316°C$ for a time period of up to nine hours. "Tacot" can be used in high density charges and it can be used in plastic-bonded form in flexible cords and sheets.

Note: "Tacot" is a registered Du Pont trademark.

Other newly developed explosive compositions which can withstand temperatures in excess of $260°C$ are DIPAM, HNAB, diamino-hexanitrobiphenol, and HNS, hexanitrostilbene, which are used as charge materials in explosive cord, mainly in the mild-detonating fuse (MDF) and in the flexible linear shaped charge (FLSC).

The properties of some of the most commonly used high explosives are listed in Table 2.2.

C. *Deflagrating Materials*

1. Nitrate Ester Compositions

One group of the most widely used chemicals for the formulation of solid propellants are the nitrate esters. Nitroglycerin is the most commonly known composition of the nitrate esters. More recently developed nitrate ester compositions are TMETN, trimethylol ethane trinitrate, and TEGDN, triethylene glycol dinitrate.

Proper‌

Explosive	Formula	Condition	Loadi‌ Densi‌
Composition A-3	RDX/Wax; 91/9	pressed	1.6‌
Composition B	RDX/TNT/Wax; 60/39/1	cast	1.6‌
EDNA	$C_2H_4(NHNO_2)_2$	pressed	1.5‌
EDNATOL (55-45)	EDNA/TNT; 55/45	cast	1.6‌
Explosive D	$C_6H_2(NO_2)ONH_4$	pressed	1.54
H B X	Comp.B/TNT/AL/D-2; 70/12/18/5	cast	1.7‌
Pentolite (50-50)	PETN/TNT; 50/50	cast	1.6‌
Pentolite (50-50)	PETN/TNT; 50/50	pressed	1.6‌
P E T N	$C(CH_2ONO_2)4$	pressed	1.6‌
Picratol (52-48)	Explos.D/TNT; 52/48	cast	1.6‌
Picric Acid	$C_6H_2(NO_2)_3OH$	cast	1.6‌
P T X - 2	RDX/PETN/TNT; 43.2/28/28.8	cast	1.70
P T X - 2	" " " " " "	pressed	1.61
R D X	$(CH_2)_3(NNO_2)_3$	pressed	1.60‌
Tetryl	$C_6H_2(NO_2)_3N(CH_3)NO_2$	pressed	1.55
T N T *)	$C_6H_2(NO_2)_3CH_3$	cast	1.60‌
T N T	" " "	pressed	1.55
Torpex-2	Comp.B/TNT/AL; 70/12/18	cast	1.80
Tritonal	TNT/AL; 80/20	cast	1.75

* Arbitrarily assigned standard values to which other values in column are referenced

Ref.: "The Properties of High Explosives", Holex Inc.

Explosives

sance	Detonation Rate (m/sec)	Relative Power	Impact Sensitivity	Booster Sensitivity	Chem. Name of Explos.
21	8200	135	43	1.70	
32	7700	134	41	1.39	
13	7550	137	25	2.09	Ethylene-Dinitramine
12	7330	119	47	1.28	
87	7050	99	150	1.27	Ammonium Picrate
06	7400	138	66	1.25	
21	7450	126	19	1.93	
				2.36	
26	7920	145	(10)		Pentaerythritol-Tetranitrate
00	6915	100	120	1.00	
41	7900	138	20	1.87	
				2.32	
31	8235	150	18		Cyclotrimethylene Trinitramine
12	7375	128	24	2.01	Trinitrophenylmethyl-Nitramine
06	7080	112	52		Trinitrophenol
00)	6850	(100)	(100)	0.82	Trinitrotoluene
20	7530	134	38	1.67	
93	6770	124	71	0.58	

2. Nitrocellulose Compositions

Another group of chemicals used in many solid propellants are the nitrocellulose compositions. A typical frequently used nitrocellulose composition is PNC, plastisol nitrocellulose, which is available in bead form. The median diameters of the spherical particles range from 10 to 200 microns. The size and distribution of the particles to specific requirements is controlled by a unique production process.

PNC plastisol nitrocellulose can be used alone as a propellant, and for special applications in solid propellants it can also be combined with high explosives in percentages from 70 to 90%. In other propellant compositions, PNC can also be combined with nitrate ester or with a nitrate ester and a high explosive component.

3. Fast Burning Propellant

In applications where very fast rise times and very high pressure outputs are required, a fast burning propellant powder can be used. These powders have the capability of reaching pressures of 15,000 to 50,000 psi in 1 to 2 milliseconds, without the hazard of detonations that could be experienced with conventional propellant powders. These powders can be compressed into a grain of extremely high strength and can withstand pressures of 20,000 psi without cracking. Because of these advantageous characteristics, the burning process of the fast burning propellant continues uniformly, even at extremely high pressures.

D. *Component Materials for Explosive Compositions*

1. Sodium Chlorate

Sodium chlorate, which has the composition $NaClO_3$, is used as oxidizer in pyrotechnic compositions. A disadvantageous characteristic is its hygroscopicity which can be overcome by a coating.

Properties: Sodium chlorate is available in the form of colorless crystals. Its specific gravity is 2.5, and its melting point is at 250°C. Sodium chlorate begins to release oxygen at about 350°C.

2. Sodium Nitrate

Sodium nitrate, also known as Chile salpeter, which has the composition $NaNO_3$, is used as a component of permissible explosives, pyrotechnic compositions, explosive powders, for example in pellet powder, blasting powder B, and in ammonia dynamites. Sodium nitrate has the disadvantageous feature that it is deliquescent in moist air.

Properties: Sodium nitrate is available in the form of colorless crystals, white granules or powder. Its specific gravity is 2.26, and its melting point is at 308°C. It decomposes at a temperature of about 380°C.

3. Potassium Sulfate

Potassium sulfate, K_2SO_4, is a common constituent of potassium salt minerals. It is frequently used as a component in priming compositions.

Properties: Potassium sulfate is available in the form of white crystals, granules, or powder. Its specific gravity is 2.66, and its melting point is at 1072°C.

4. Sulfur

Sulfur, S, also known as brimstone, is used as a component of black powder and other ignitable mixtures to facilitate ignition. It is also frequently used in aluminum and magnesium flares, and in torches in combination with potassium nitrate and rosin and pitch.

Grade A sulfur is used in the production of black powder, Grade B sulfur in the manufacture of pyrotechnic compositions, and Grade C sulfur as a component of some dynamite compositions and ignition mixtures for blasting caps. Sulfur has been used to increase the sensitivity of tetryl to shock, and also to raise the temperature of detonation of Hexogen. It decreases the temperature of detonation of TNT and TNB, picric acid, tetryl, and pentrite.

Properties: Sulfur is available in the form of a yellow powder. Its specific gravity is 2.05, its melting point is at about 120°C., and its boiling point is at 444.6°C. The ignition temperature of sulfur is about 250°C.

E. *Development Trends*

Conventional explosive materials as currently used in explosive-actuated devices are generally rated for reliable performance at ambient temperature from −54°C. to +72°C., and they are tested to withstand exposure to a temperature of +150°C. for 30 minutes.

The application of pyrotechnic systems in extreme environmental conditions required considerable improvements in basic materials and in explosive compositions. In recent years, new improved explosives have been developed which meet extremely high performance requirements and perform reliably at ambient temperatures of −195°C. to +205°C. Some of these new materials are capable of reliable functioning after an exposure to a temperature of +260°C. for two hours.

The thermal stability and insensitivity to impact of high detonation pressure explosives which yield detonation pressures of 200 to 300 kilobars are constantly being improved. Since these high-detonation pressure explosives must have a high density, improvements in mixing and loading processes will have to be made.

These parallel improvements and developments of high-performance explosive compositions and highly refined, safe and economical manufacturing processes will finally result in optimum products.

Part II

Explosive-Actuated Devices

3 Initiators

Pyrotechnic devices can be initiated by electrical, mechanical, or laser initiator units. Pneumatic initiators have also been used in some systems, but they are not commonly considered for pyrotechnic systems. Since electrical explosive initiators are most commonly used as primary stimulus components for the actuation of pyrotechnic devices, this type of initiator is described first in this chapter.

A. *Electrical Explosive Initiators*

In electrical explosive initiators, an electrical input, which causes an exothermic reaction in the charges within the initiator cartridge, is converted into chemical or kinetic energy in the form of heat, pressure and shock waves.

The use of electrical initiators offers the highest reliability and the greatest power-to-weight ratio. Other important advantages of these initiators are their excellent environmental characteristics and their long shelf life.

Electrical explosive initiators are divided into three major categories:
1. Squibs and power cartridges
2. Igniters and electric matches
3. Detonators

When classified according to low- or high-energy systems, two different types of electrical explosive initiators are considered:
* Hot wire initiators, which are low-energy devices,
* Exploding bridgewire (EBW) initiators, which are high-energy devices.

41

The following explosive materials are frequently used in initiator cartridges:

LMNR (lead mononitroresorcinate)

KDNBF (potassium dinitrobenzofuroxan)

Barium styphnate.

In most initiator cartridges, two different charges are used: A primary explosive and a base charge. A small amount of highly sensitive explosive material, which is applied to the resistance wire or bridgewire and which is initiated by heat, is used as a primer. The base charge is normally made from another explosive material from the initiator category as listed above.

Generally, a detonator cartridge (detonator) consists of a body containing a resistance wire and an explosive charge which is detonated by electric current or impact and which subsequently detonates larger, less sensitive charges.

In a hot wire detonator, a primary explosive material, ignited by an electrically heated resistance wire, subsequently initiates a high-energy shock-producing intermediate charge which, in turn, is used to initiate a base charge. Such arrangements are commonly used in detonation devices.

An ideal explosive material for the intermediate charge is lead azide which, during the detonation, generates high shock waves as required for initiating a base charge, consisting of PETN, RDX, HMX, or tetryl.

A cross-section through a typical hot-wire initiator cartridge is shown in Fig. 3:1.

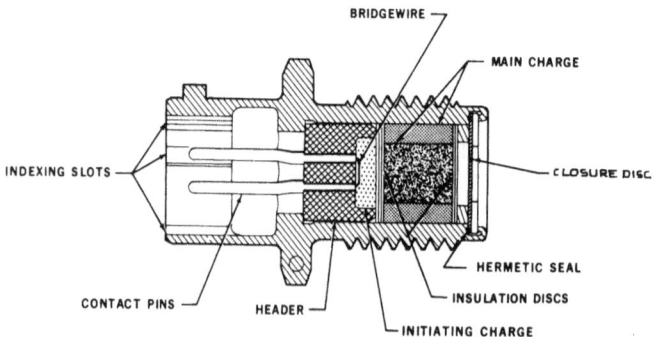

Fig. 3:1 An initiator cartridge (Courtesy of Hi-Shear Corp., Ordnance Div.).

The stainless steel body of this initiator cartridge has one threaded end for mounting in a pyrotechnic device, and one end fitted with indexing slots to mate with a standard electrical pigmy connector. The gold-plated contact pins are mounted in a header which is designed to withstand internal cartridge pressures of up to 50,000 psi. The header is made from alumina which has high shear strength, high thermal and low electrical conductivity. This results in high electrical insulation values, stable and reproducible dimensional properties and a low thermal expansion compatibility with the stainless steel body. The bridgewire is made from 0.002 inch diameter Tophet A material and has a resistance of 1.0 ohm at 21°C.

The stainless steel closure disc at the output end of the cartridge is circumferentially welded to the cartridge body to obtain a hermetically sealed joint. The closure disc is structurally reinforced by a square hole washer from stainless steel which is retained by a curl crimp at the end of the cartridge body. The curl crimp provides a cushioning action and eliminates direct shear areas. The hermetic seal has a leakage rate of less than 1×10^{-6} cc He (STP)/sec at 1 atmosphere differential.

The most important features of these initiators are reproducible bridgewire resistance within ±0.1 ohm, reproducible outputs, and the ability to function under extreme environmental conditions. These initiator cartridges have an all-fire and no-fire reliability in excess of 99.99% with a 95% confidence level.

In a typical hot wire initiator cartridge, a firing current of 5.0 ampere to one or to each bridgewire results in an everage functioning time of 1.2 milliseconds. The minimum all-fire current is 3.5 amperes. The no-fire reliability is based on 1-watt, 1-ampere, 5-minute pulse no-fire tests. These initiator cartridges have a minimum resistance of 100 megohms when 500 volt dc is applied for one minute between the contact pins shorted together and the cartridge body. No ignition occurs when a 500 micro-microfarad capacitor charged with 9000 volts is discharged between the contact pins shorted together with the cartridge body .

Hot wire initiators, as described, have been developed for peak pressures of 250 to 11,500 psi. These initiator cartridges function reliably in a temperature range from −184°C. to +150°C., and in a wide range of altitude and ocean depth. Their storage life exceeds five years.

In an exploding bridgewire (EBW) cartridge, an electric exploding wire initiates a secondary explosive charge directly, which results in extremely short function times. As compared to hot wire initiators, explosive bridgewire initiator cartridges have the additional advantage that their simultaneity is high, they are relatively insensitive to shock, and they are less sensitive explosive items than hot wire initiators.

According to requirements in a great variety of applications, various types of hot wire initiators and exploding bridgewire initiators have been developed. The first category, squibs and power cartridges, is described in the following section.

1. *Explosive Squibs and Power Cartridges*

    a.   Squibs

    A squib is usually a simple and small initiator which consists of a cylindrical body containing an explosive charge and bridgewires, and, when energized, generates flame and gas, but no sharp, shattering high explosive effect. Consequently, squibs are used for either their igniting or pressurizing effect to initiate other pyrotechnics, such as igniters for solid propellants in rocket motors, or to initiate smokeless powder or a gas cartridge whose charge, immediately after ignition, pressurizes a vessel or a flotation device.

    A great variety of squib types is available, for example:
    1. Open match squibs
    2. Thin-bottom squibs

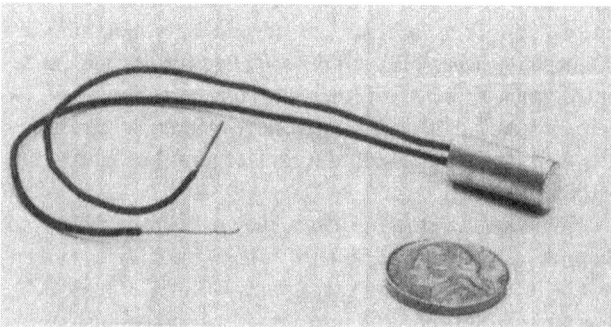

Fig. 3:2 A Squib (Courtesy of Hi-Shear Corp., Ordnance Div.).

3. Side-burning squibs

All these types of squibs are available with charge materials for specific effects, such as end flash, jet flame, hot slag, brisk flame burst, coruscating slag, and hot gas.

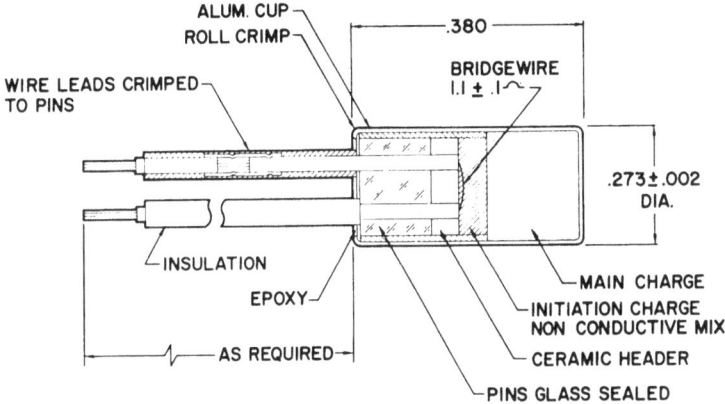

Fig. 3:3 Cross-section through a squib (1 amp/1 watt) (Courtesy of Hi-Shear Corp., Ordnance Div.).

In a typical squib, as shown in Fig. 3:2 and in Fig. 3:3, a bridgewire embedded in a priming composition mounted inside a metal housing ignites an explosive main charge which fires out of the housing to ignite or pressurize the pyrotechnic device. A single bridgewire, or when greater reliability is required, a dual bridgewire may be used in squibs. A typical bridgewire resistance is 1.0 ohm, a standard all-fire current is 4.5 amperes, and a standard no-fire current is 0.1 ampere. Squibs are generally used with a bridgewire resistance ranging from 0.1 ohm to 4.0 ohms: Simple, light-weight squibs are equipped with pigtail leads, as shown in Fig. 3:2, whereas other types have wire leads which are crimped to connector pins, as shown in Fig. 3:3. The lead pins of the squib are hermetically sealed by a glass-to-metal seal which can withstand blow-back pressure of up to 30,000 psi. The output end of the squib is solder-sealed and crimped for protection of the explosive charge from deterioration during storage.

Squibs with redundant bridgewire circuits have three wire connections, of which one is being shared, or four connections.

The two circuits can be energized from one electrical source or from two separate sources. Standard squibs operate reliably at a temperature range from $-54°C$. to $+150°C$. Squibs for higher and lower temperature ranges can be manufactured according to special requirements. Squibs with built-in time delays from milliseconds to 30 seconds have been developed for single or multi-step delayed action and for programmed sequencing of operations. Common sizes of squibs are from 0.194 inch diameter and 0.170 inch long to 0.750 inch diameter and various lengths to accommodate delay trains of various lengths for different delay times. A great advantage of squibs is their light weight which ranges from 0.02 lbs. to 0.14 lbs.

The mounting of a squib having a simple cylindrical housing, as shown in Fig. 3:2 and Fig. 3:3, can be accomplished by confining this small device between the surfaces of an assembly in contact with the material to be ignited or initiated.

b.   Power Cartridges

Power cartriges are initiators in which a propellant composition is used for initiation and for generating heat and pressurized gas for a short duration. The function principle of power cartridges is similar to that of squibs. The main difference of power cartridges as compared to squibs is in the design of the body, headers, contacts and seals.

Power cartriges are ideally used for the actuation of separation nuts and bolts, valves, ejection systems, pressure-actuated pumps, switches and piston devices, such as pyrotechnic cutters for cables, straps, reefing lines, hoses and tubes.

A cross-section through a typical power cartridge with a glass header is shown in Fig. 3:4. In many power cartridges, a header from alumina is used, as shown in Fig. 3:1.

Power cartridges contain an ignition primer and a main charge of a pressure producing propellant, mounted in one casing. The ignition primer has the purpose of initiating the main propellant charge which consequently generates gas in the required quantity and of the desired temperature. The gas thus produced can be used for numerous work functions, for example, for pushing a piston, inflating a flotation device, or fracturing a diaphragm.

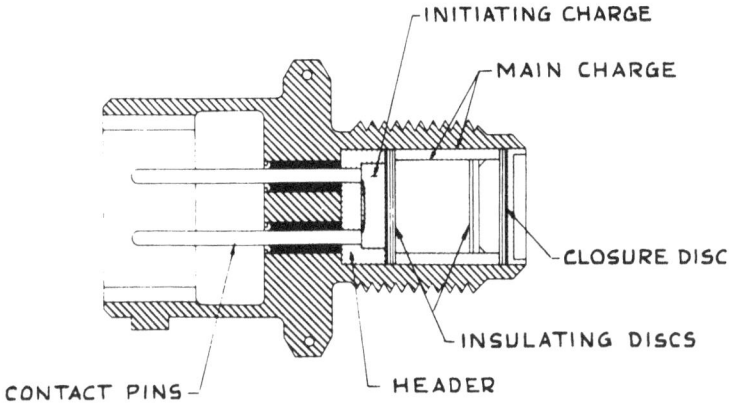

Fig. 3:4 A power cartridge, glass header type (Courtesy of Hi-Shear Corp., Ordnance Div.).

The amount of the generated gas is relatively small. The major energy results from the gas temperature. It is therefore important that pyrotechnic devices, which are initiated by power cartridges, perform their work within a very short time, because the efficiency of the cartridge can be considerably reduced by heat transfer from the hot gases to the surrounding material. A typical power cartridge produces a pressure of 4400 psi in a

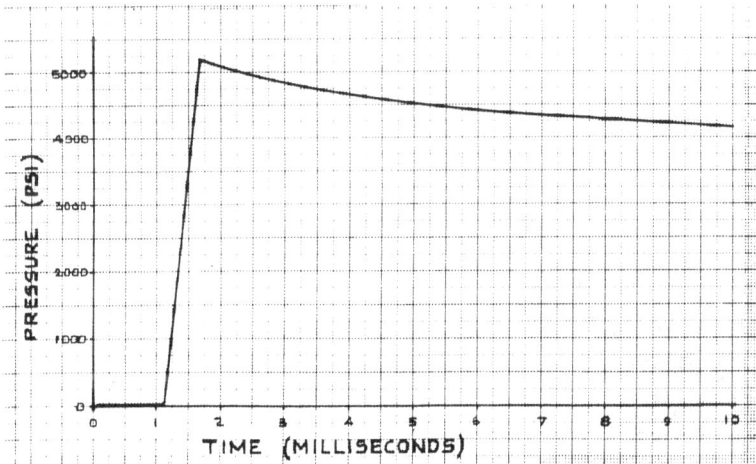

Fig. 3:5 Pressure vs. Time-typical for a cartridge in a 10cc. closed test chamber with a 50 ampere firing pulse (Courtesy of Hi-Shear Corp., Ordnance Div.).

closed test chamber of 10 cc volume in five to ten milliseconds after initiation of the cartridge, as shown in the Pressure-Time diagram in Fig. 3:5.

Pressure ranges in these cartridges are generally from 100 psi to 10,000 psi. Power cartridges with a pressure capacity of 70,000 psi have been developed for special applications.

For conventional applications, power cartridges in a great variety of capacities and sizes are available as standard shelf items. But sometimes a power cartridge has to be designed according to specific requirements, for which no standard item is readily available. For an optimum power cartridge design, the following data are to be determined:

- The required total energy
- The rate at which the energy must be delivered, for example, it is possible to deliver the energy at a constant rate, or at a low initial but increasing rate, or at a high initial rate which decreases gradually
- The type of work to be performed, such as inflating a flotation device, puncturing a diaphragm, or driving a piston in a guillotine cutter, valve, thruster, or pin puller.
- The duration of the required work performance
- The electrical firing characteristics, such as the required no-fire current, all-fire current, and special requirements with regard to the actuation delay time

The required amount of propellant for the actuation of a certain pyrotechnic device can best be determined by tests. For a preliminary calculation of the required amount of propellant for a power cartridge, the following formula can be used:

$$W_p = (1.15 \times 10^{-4}) P V$$

where W_p = weight of propellant in grams
P = pressure in psi
V = volume of cavity in cu.in.

The precise amount of propellant required for a certain power capability can then be determined by several cartridge test firings. The above listed formula should only be applied for approximate sizing, because heat transfer, time and friction are not taken into account.

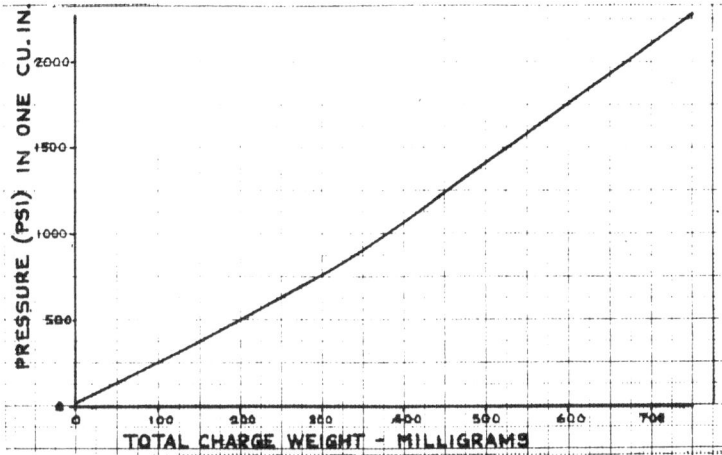

Fig. 3:6 Power Cartridges-average peak pressure vs. total charge weight (Courtesy of Hi-Shear Corp., Ordnance Div.).

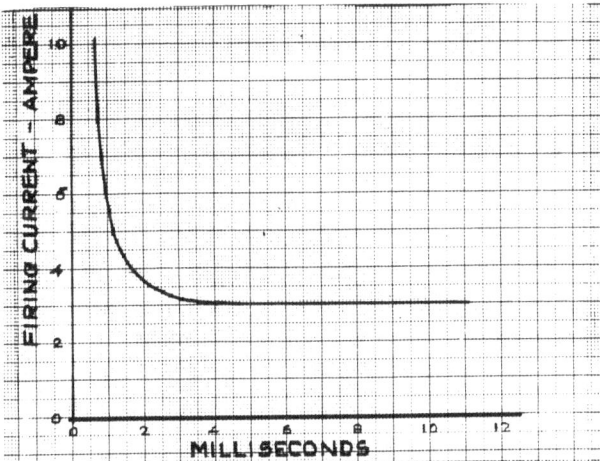

Fig. 3:7 Time from application of current to first indication of pressure or bridge-wire break. Function time vs. current for power cartridges. (Courtesy of Hi-Shear Corp., Ordnance Div.).

Average peak pressures, as measured in a test chamber of 1.0 cubic inch volume, in relationship to total charge weights of a typical propellant used in power cartridges are shown in the graph in Fig. 3:6.

49

It is recommended to determine the pressure that is insufficient to actuate the pyrotechnic device, and to establish the precise no-function pressure, the highest pressure, and a specific safety factor to obtain reliable function of the power cartridges.

Function times in relationship to firing currents for typical power cartridges are shown in the graph in Fig. 3:7, and firing currents in relationship to rupture times for single and dual bridgewire firing of power cartridges are shown in the graph in Fig. 3:8.

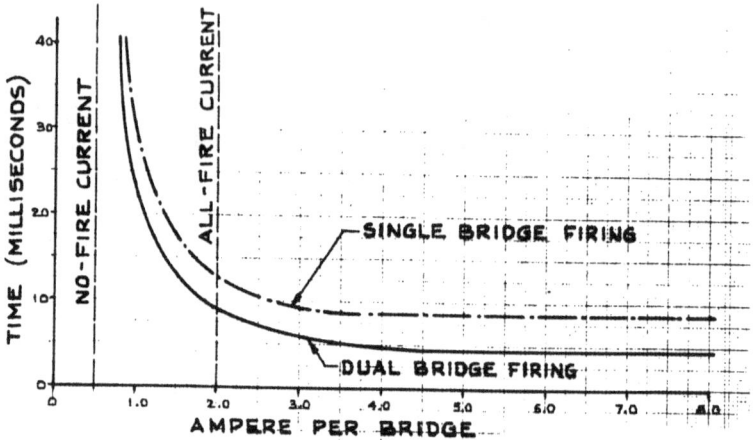

Fig. 3:8 Current vs. rupture time (Ref. Gould Laboratories).

A great variety of power cartridges has been developed for spacecraft and missile systems. For these applications for high altitude and extreme environmental conditions, the bodies of the cartridges are made from stainless steel stabilized against intercrystalline corrosion.

Power cartridges are ideally used in spacecraft and missile systems for umbilical release, airborne emergency separation systems, thrusters, pin pullers, drogue chute mortars, line cutters, thrust reversal systems, termination devices, and separation systems for fairings and nose cones. These cartridges may be considered as the "power pack" in the pyrotechnic systems field.

A typical power cartridge for applications as listed above uses a firing current of 5.0 amperes applied to one bridgewire or to each bridgewire in a dual bridge system, which results in an average function time of 1.2 milliseconds. The maximum all-fire current is 3.5 amperes to one bridgewire or to each bridgewire. The cartridges are tested for no-fire reliability by subjecting them to a 1-watt, 1-ampere, 5-minute pulse without a degradation of their performance. A no-fire reliability of 99.99% at 95% confidence level is normally required for power cartridge applications in spacecraft and missile systems.

The minimum insulation resistance of these power cartridges is 100 megohms when 500 volts dc is applied for one minute between the contact pins shorted together and the cartridge body. During dielectric strength tests, the cartridges have to withstand 500 volts ac applied between the contact pins shorted together and the cartridge body for one minute without failure. Electrostatic discharge tests have to prove that no ignition occurs when a 500-micromicrofarad capacitor charged with 9000 volts is discharged between the contact pins shorted together and the cartridge body. Recently developed power cartridges are capable of sustaining discharges of up to 25,000 volts. These cartridges can also withstand radio frequency hazards. Test cartridges were exposed to a range of 8 megacycles to 10,000 megacycles with a power input of 0.10 watt to each bridgewire. All cartridges tested passed without any degradation as evidenced by subsequent firing tests.

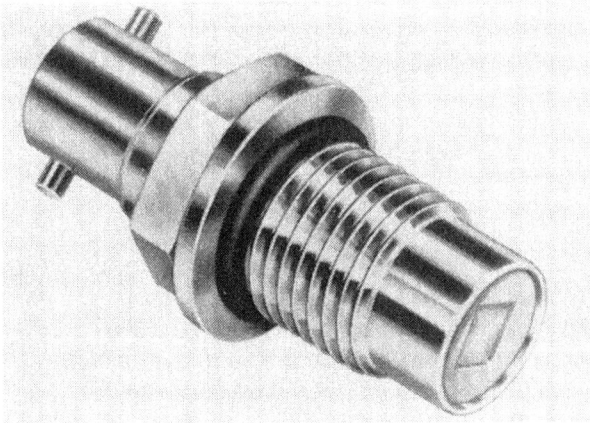

Fig. 3:9 An initiator-booster assembly (Courtesy of Hi-Shear Corp., Ordnance Div.).

During extensive radiation tests, power cartridges were irradiated in a gamma-radiation source to a total dosage of 10^9 erg/g (C) and performed within allowable limits during firing tests.

Numerous applications and tests proved that these power cartridges function reliably in a temperature range of $-184°C$ to $+150°C$, in extreme environmental conditions and at an altitude of 550,000 feet.

A special type of power cartridge is an initiator-booster assembly, which is designed on a buildingblock concept, utilizing initiator and booster components, only for gas pressure applications. A typical initiator-booster assembly is shown in Fig. 3:9. This typical power cartridge, as shown, produces an output of 12,000 psi in a 0.2 cubic inch test pressure chamber. The cartridge is designed to function reliably in a temperature range of $-62°C$ to $+85°C$ and to withstand shocks of 100 G in each of the three orthogonal axes.

Extensive tests proved that these power cartridges without a voltage blocking device cannot be initiated by direct application to the connector pins of 50 volts dc from a low impedance source. With a voltage blocking device, they remain operable after subjecting them to 220 volts dc applied to the connector pins. Tests also proved that these initiator-booster cartridges will not fire due to radio frequency at present and proposed levels.

Reliable functioning of these cartridges at altitudes in excess of 420,000 feet and in environments closely approaching complete vacuum can be expected, as numerous simulation tests proved.

To develop standard universal power cartridges for multiple applications appears to be a worthwhile goal.

2. *Igniters*

Pyrotechnic igniters are flame-producing devices which are often electrically initiated. They are used for primary ignition, for example, to ignite solid-propellant boosters, gas generators, and liquid-powered motors, to heat thermal batteries, or to activate other similar devices. Two different types of igniters are commonly used: Electric matches, and igniter or initiator cartridges.

a. Electric Matches

Matches are the simplest type of conventional igniters. The main difference of electrical matches as compared to igniter-initiators is in the burning time and in the design. Matches are not housed in a metal body, but they consist of a small roll of resin-impregnated paper, laminated on each side with brass-foil, and filled with a small amount of explosive. The charge used in matches usually has a higher sensitivity than that used in igniter cartridges. A bridgewire and initiation leads, which are either soldered or welded, are mounted in the paper roll. The soldered leads are often preferred for large quantities, because they can be made more economically by high-production manufacturing processes than welded leads. Welded leads in matches are mainly used in special designs, which have a limited production number.

Electric matches are used with bridgewire resistance from 0.25 to 4.5 ohms. Their operating current ranges from 0.58 ampere all-fire and 0.10 ampere no-fire to 3.5 ampere all-fire and 1.0 ampere no-fire. Matches are, similar to igniter cartridges, relatively insensitive to shock, vibration, and acceleration. The mounting of matches presents no problem, because they can simply be embedded in the material to be ignited.

b. Igniter and Initiator Cartridges

Igniters and initiators constitute a great improvement over low-voltage squibs. They contain pyrotechnic materials which are selected for their flame-producing properties. Typical igniters generate high temperatures, but only small amounts of gas or no gas. Their time-temperature performance curves usually have a sustained plateau. Conventional igniters and initiators are safe with up to 1-ampere 1-watt current. For special applications, igniters with a safe current of 5 amperes and 5 watts have been developed.

The firing data of a typical igniter of initiator cartridge are: No-fire at 71°C, at 0.05 amperes for 5 seconds maximum; all-fire at −54°C, at 0.60 amperes for 100 milliseconds maximum. The insulation resistance between the terminals and the case is 100 megohms minimum at 500 volts dc. The size of this typical small igniter is 1/8 inch diameter and 0.4 inch long, and its weight is only 0.5 grams.

Low-voltage type igniters are used to initiate solid-propellant gas generators and similar devices. Their all-fire current is 1.5 amperes, and their no-fire current is 0.25 amperes. A typical igniter or ignitiator cartridge develops a peak pressure of 2000 psi in a 30-cc volume test chamber in 45 milliseconds.

A typical safe-arm igniter used to initiate boron-potassium pellets is designed to withstand temperatures of 1480°C and pressures of 3500 psi. It functions reliably in a temperature range of −59°C to +77°C, and its auto-ignition temperature is rated at 177°C.

A sustained generation of hot particles, flame, and gas pressure is a major requirement for reliable low-temperature ignition of solid propellants. The pyrotechnic material combinations to be used in these igniter-initiator cartridges are selected and tested according to this requirement.

The igniters used for solid-propellant rocket units are almost exclusively of the pyrotechnic type. In some cases, the igniter is put into the forward end of the chamber so that the ignition gases will sweep past the propellant charge before reaching the nozzle. In some other cases, the igniters are mounted within the propel-

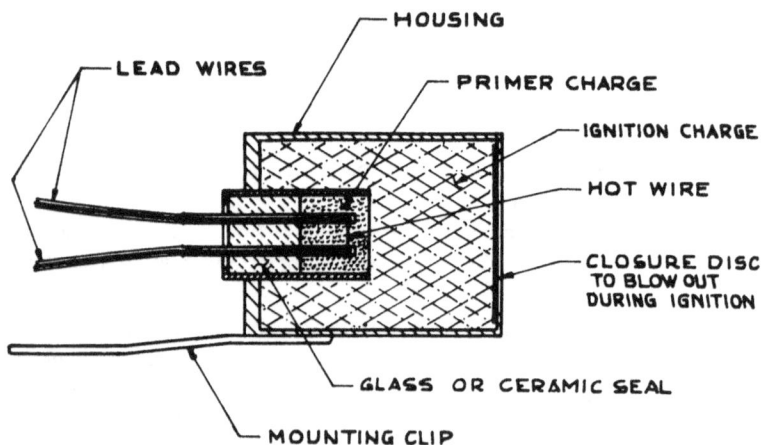

Fig. 3:10 An igniter.

lant charge, and the electric wires are connected to it through the nozzle. Where wire and igniter fragments, which are ejected in starting, may cause damage, the igniter is built into the chamber wall, and the wires are introduced through a pressure-tight seal. Some designs of igniters built into the nozzle have been satisfactory.

An igniter, as shown in Fig. 3:10, contains an electrically heated wire, which is embedded in a small amount of primer. Normally, an amount of less than 1 gram of primer is sufficient. The main charge, arranged adjacent to the primer, produces a hot flame which ignites the rocket propellant. In some rocket motor applications, igniters were used whose bodies were made from plastic so that they would burn and would not obstruct the gas flow. It is a major requirement that the igniters, like the propellant material, are resistant to moisture and capable of storage and operation over a wide range of environmental conditions.

Main considerations in the design of igniters are the properties of the igniter propellant, the design type and characteristics of the main charge, and the nozzle closure rupture pressure. Other important parameters are the functioning time, the characteristics of the igniter gases, the electrical energy limitations, the temperature, pressure and functioning characteristics of the rocket motor, and the igniter functioning variation with temperature. In case the rocket chamber pressure is raised by the igniter gases to a too high level or at a too high rate, the main propellant grain may be fractured, or the rocket nozzle closure may blow out, if it is equipped with a closure.

An empirical approach has generally been used in the design of rocket ignition systems, because it has not been possible to firm up a valid ignition theory taking into account these many variables. Other methods used for the ignition of rocket propellants are electric heater and hypergolic ignition.

The design of a special igniter for a specific solid-propellant motor is greatly influenced by the nozzle size and shape, burning rate, chamber pressure, diaphragm burst pressure, chamber volume, temperature sensitivity, temperature limits, and pressure rise-time delay. It is obvious that these many influencing factors can result in a great variety of igniter types and designs.

In an igniter design for a liquid propellant system, the type of propellant, hypergolic properties, mass flow, decomposition temperature, chamber volume, nozzle size, pressure control system, heat transfer conditions, and burning time are major influencing factors.

A special type of igniter is the rocket engine starter cartridge as used in the Titan missile. To provide redundancy, the cartridge is equipped with two squibs with independent firing circuits. The initiation of either or both squibs can actuate the igniter cartridge. Lead styphnate is used as ignition material, and the main charge consists of 95 ± 5 milligrams aluminum-potassium perchlorate/boron potassium nitrate.

The no-fire current for this igniter is 1.0 ampere, and the all-fire current is 3.0 amperes. The functioning time is less than 20 milliseconds at a temperature of $2^{\circ}C$. This cartridge is equipped with a glass-to-metal seal to provide maximum insulation. Tests proved that this igniter will not fire when exposed to a radio-frequency power density of 100 watts/m^2 over the 40,000 mc/sec frequency range, or when subjected to stray voltage input of 25 mv for four hours continuously, or when subjected to a static discharge of 0.01 joule by means of a 0.04 microfarad capacitor.

Before concluding this section on igniter and initiator cartridges, it may be noted that failures of pyrotechnic devices that have been encountered were not malfunctions of the main explosive charge, but of the ignition system, mainly of the igniter or initiator.

In the following two sections, some special initiating devices and components for igniter systems will be described:

- Through-bulkhead initiators
- Confined detonating fuse for igniter systems.

c. Through-Bulkhead Initiators

Through-bulkhead initiators are used to transmit a detonation shock wave through a solid steel bulkhead and initiate a deflagration on the opposite side without perforating the bulkhead or otherwise destroying the integrity of the hermetic seal formed by the bulkhead.

Some through-bulkhead initiators are designed to accommodate a standard transfer line bayonet connector. Detonation of the donor transmits a shock wave through the integral steel bulkhead and initiates a receptor charge within the initiator. The output of the receptor charge is attenuated and used to ignite an ignition charge, which can consist of boron potassium nitrate. The hot gas and burning particles generated by the deflagrating through-bulkhead initiator easily ignite any standard rocket ignition pellets or gas generating charges, such as black powder, smokeless powder, granular composite propellants, and metal oxidant ignition mixtures. The unit consisting of the through-bulkhead initiator and the transfer line is non-fragmenting and features a quick-disconnect to facilitate assembly.

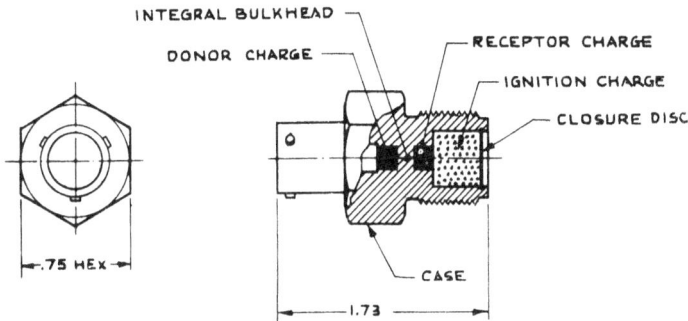

Fig. 3:11 A through-bulkhead initiator (Courtesy of McCormick-Selph).

A typical through-bulkhead initiator, as shown in Fig. 3:11, consists of a stainless steel case which contains a donor charge at the connector side and a receptor charge at the far end, separated by the integral bulkhead. An ignition charge, which is covered with a welded closure, is arranged adjacent to the receptor charge. The steel bulkhead in a fired unit is capable of withstanding a differential pressure in excess of 50,000 psi. The function time is 0.01 millisecond at 21°C from input of detonation stimulus to start of pressure rise. The peak pressure output is 4200 psig in a 10.0 cc chamber is 0.4 milliseconds.

Through-bulkhead initiators have the following advantageous features:

- Their need for glass-to-metal seals is eliminated.
- The need for functional ground checkout, except for

simple visual inspection, is eliminated.

- Their insensitivity permits installation during production assembly.

- They are completely insensitive to radio frequency, stray currents, or any induced electrical currents, because they are non-electric.

- Since they incorporate all activation energy in the form of chemical reaction, the need for supplementary electric power is eliminated.

- The rapid response results in excellent ignition reproducibility and simultaneity between multiple through-bulkhead igniters, which would be very difficult to accomplish by using conventional electro-explosive igniters.

d. Confined Detonating Fuse for Initiation Systems

A special type of igniter, as used for initiation of solid propellant in rocket motors, is the confined detonating fuse (CDF), a conductive transfer line, which detonates at a velocity of approximately 25,000 feet per second. It is an advantageous feature of the CDF line that it completely contains all potentially hazardous shrapnel and gases produced during the detonation. A standard detonating cartridge can be connected at one end of the CDF line, and the ignition or initiation output is obtained at the far end of the transfer line.

The confined detonating fuse system can be used ideally for simultaneous initiation of multiple pyrotechnic devices, such as explosive bolts, destruct charges, shaped charges, or for multiple ignition of solid propellant. Time delays can be built into a CDF line system without great difficulty.

A dual CDF igniter which functions with a time delay of 1.8 seconds after a current of 1 ampere, 1 watt has been applied, is shown in Fig. 3:12.

More detailed information about confined detonating fuse (CDF) is given in Part I, Chapter 1, Section D.

e. Pyrocore Igniter

For initiation of solid propellant in rocket motors, a special type of igniter, the "Pyrocore" igniter cord, can be used. It contains a

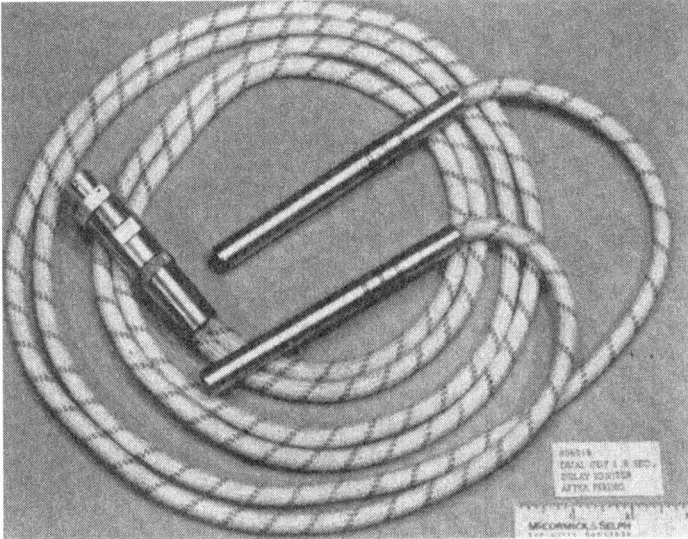

Fig. 3:12 A dual CDF igniter (Courtesy of McCormick-Selph).

high-explosive core designed to promote ignition at the speed of detonation. When initiated, this core produces hot particles which penetrate and ignite a propellant charge at high speed. "Pyro-core" is often used together with other igniter materials when the speed of the core is essential, but the duration of the flame provided by a larger igniter charge is to be used.

"Pyrocore" is a Du Pont trademark.

3. *Detonating Cartridges*

Detonating cartridges are high-explosive devices, producing maximum brisance. They are designed to provide shock waves and high pressure energies which are often utilized to initiate other high explosives. Detonator cartridges are used for linear or rotary actuation of explosive valves, thrusters, pin pullers, guillotine cutters, explosive bolts, ejection devices, for the separation or release of structures or components, such as stages of multi-stage vehicles, fairings, para-chutes, ejection seats, launching pads, sleds, and similar jettisonable equipment, and also for demolition devices. They are also frequently used for the actuation of mild detonating fuses (MDF) and linear shaped charges (FLSC). The cartridges can withstand the destructive effects of detonations which are capable of end and side initiations of linear shaped charges and mild detonating fuses. This feature permits the use of these detonating cartridges in contained applications where the explosive effects of a device must be limited directionally.

Three different explosive charges are normally used in detonating cartridges; ignition drop (or primer), intermediate charge, and base charge. The intermediate charge has the purpose of aiding in the transition to detonation. Both the ignition drop and the intermediate charge are initiator compositions, whereas the base charge is a booster composition. Lead azide and RDX or PETN are often used for the main charge.

The explosive materials used in these cartridges are chosen and mixed for a specific maximum shattering effect, and they are loaded to optimum density. Column heights and density of the explosive charges are carefully controlled to insure reproducible detonation and reliable propagation.

The output end of a typical detonator cartridge is sealed with a 0.003 inch thick closure disc of stainless steel which is welded to the cartridge body. The header assembly pins are brazed into the ceramic material. The bridgewire is made from 0.002 inch diameter Tophet material and has a nominal resistance of 1.0 ± 0.1 ohm at $21°C$. The insulation disc in these cartridges is made from Isomica material, which has ideal properties.

The function time of a typical detonating cartridge is 3 milliseconds, applying a firing current of 4.5 amperes to each bridgewire at $21°C$ or a current of 5.0 amperes at $-73°C$. Extensive no-fire current tests have been conducted, using the Bruceton method. A current of 1.8957 amperes was applied to each of two bridgewires at $21°C$ to establish a 99.9% no-fire probability with a 95% confidence level. In tests at elevated temperatures, detonator cartridges were subjected to a temperature of $177°C$ for 21 minutes, and during the last five minutes, a 1-ampere current was applied to each bridgewire. The cartridges were then returned to ambient temperature and fired without evidence of a reduction of performance. In addition to these tests, numerous extended exposure tests were conducted, during which a 1-ampere current was applied to each of two bridgewires for one hour without firing of the detonator cartridge and without any damaging effects.

These detonating cartridges function reliably in a temperature range of $-196°C$ to $+160°C$ and at altitudes in excess of 420,000 feet, and they can withstand shocks of 100 G in each of the three orthogonal axes. Several detonator cartridges have also been tested to shock pulses of 1200 G repeatedly without evidence of degradation. A storage life of ten years can be expected.

A cross-section through a typical squib-type detonator is shown in Fig. 3:13.

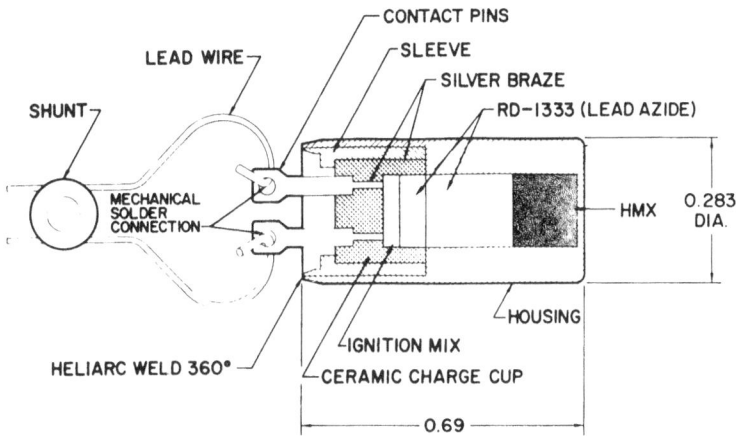

Fig. 3:13 A detonator cartridge (Courtesy of Hi-Shear Corp., Ordnance Div.).

4. *Selection of Firing Current and Sizing of Bridgewire*

When an initiator or another explosive actuated device is to be specified, the best suitable firing current can be selected. A high firing current can be used for a device which has to function reliably in an area where radio interference could cause problems. For such an application, the initiator or pyrotechnic device should be equipped with a bridgewire sized to prevent a low radio-frequency field from inducing a current flow through the bridgewire which might be sufficient to actuate the initiator or pyrotechnic device.

A high-resistance bridgewire can be considered, if only a small amount of current is available to operate the pyrotechnic device. This makes it possible for a small amount of electric current to actuate the initiator or device. In general, it is the tendency to consider the lowest permissible current rating to avoid radio-frequency interference, to conserve electricity, and to minimize the space and weight requirements for the electrical battery or generator systems.

Of great importance for optimum design and reliable functioning of an initiator or pyrotechnic device is also the decision whether to use low voltage or high voltage. In a conventional low-voltage pyrotechnic system, a heat-sensitive explosive material is pressed against a bridgewire of high resistance, for example 0.5 to 10 ohms. The bridgewire is sufficiently heated by low-energy pulses, about 0.03 joules or less, to initiate the heat-sensitive primary explosive charge. It occurred that persons, who generated sufficient static potential and handled these low-voltage explosive devices, accidentally ignited them. Other cases are known where stray current or voltage of low amplitude (a few volts or a few milliamperes) developed sufficient energy to trigger low-voltage pyrotechnic systems accidentally. Frequently, the primary explosives used in low-voltage systems are sensitive to impact and can be triggered by shock. Complex and heavy safety devices are often required in such systems, especially in missile systems.

In an exploding bridgewire high-voltage system, secondary explosives, which are insensitive, are pressed against a bridgewire of low resistance (less than 1 ohm). The bridgewire is exploded by a high-energy pulse of about 2 joules provided by a discharge of 2000 volts from a capacitor of one microfarad. This provides a rapid surge of energy, which is sufficient to initiate the insensitive secondary explosive. The use of high voltage for the explosive bridgewire system makes an accidental firing by static electricity or by low-amplitude stray currents impossible. In such a high voltage system, out-of-line blocking devices are usually not required, which results in a relatively simple system of reasonable size and weight.

In connection with these explanations and considerations about bridgewire systems and firing current selection, the question might be asked:
"How does an exploding bridgewire function?"

The exploding bridgewire received its name from the phenomenon that occurs when a large jolt of electric energy is applied to a fine wire in a very short time period. As the result of the application of the electric energy, the fine wire vaporizes with explosive violence. The magnetic field induced by the current as it passes through the fine wire, holds the vapor together momentarily. Then the vapor dis-integrates violently and creates a shock wave as well as extremely high heat which is characteristic for the metal vapor. The shock wave and high temperature set off the relatively stable secondary explosive charge.

Initiators and similar pyrotechnic devices are available with bridgewires made from nichrome, platinum, platinum-iridium, platinum rhodium, Tophet A, tungsten, and similar materials, in a variety of wire diameters and lengths. For a given bridgewire, the higher the applied electric current, the shorter the time period for the current flow to cause explosive vaporization. The material, diameter and length of a bridgewire for a specific application are determined according to the desired resistance and Watt density. A sensitive bridgewire can have a diameter of 0.00013 inch and requires only a current of 50 to 100 milliamperes to cause initiation, which would be equivalent to 500 ergs of energy. In contrast to this example, a relatively insensitive bridgewire could have a diameter of 0.002 inch, and it would require 3 to 5 amperes for 10 milliseconds to cause initiation, which would be equivalent to several thousands of ergs of energy.

Common basic ratings of bridgewires are firing current levels for actuation assuming a standard firing time of 10 milliseconds. Normally, initiators and other pyrotechnic devices are rated with a minimum all-fire current and a minimum no-fire current.

The minimum current that will always fire the initiator or pyrotechnic device is defined by the term "all-fire", whereas "no-fire" is the term for the maximum current that may be applied for a specified time period without firing the initiator or device. The sensitivity of the explosive charge, the resistance of the bridgewire and the heat-absorption of the device or system are the main factors that affect the length of time that "no-fire" current may be safely applied. A "no-fire" current below the specific maximum can be used for a proportionately longer time period.

All-fire currents of an unlimited magnitude can be applied. The functioning time is considerably reduced by higher firing currents, as the graphs of the characteristics of a typical initiator in Fig. 3:14 shows. The all-fire current of this initiator is 1.0 ampere. Initiators tested at this current functioned in 1.3 milliseconds. When a current of 0.5 ampere was applied, the average function time was 4.5

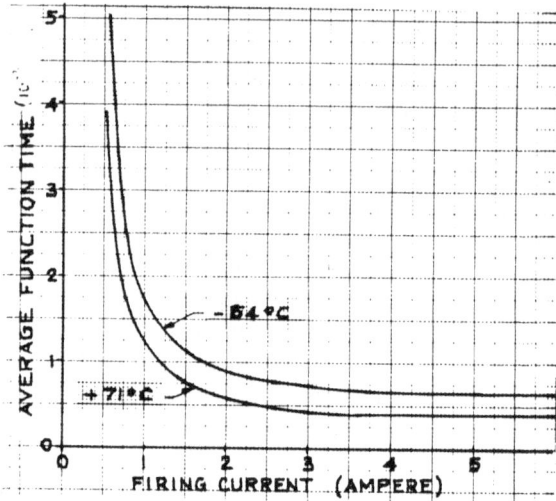

Fig. 3:14 Firing current vs. function time for microminiature Pyro switches at low and high temperatures (Courtesy of Atlas Chemical Industries, Inc.).

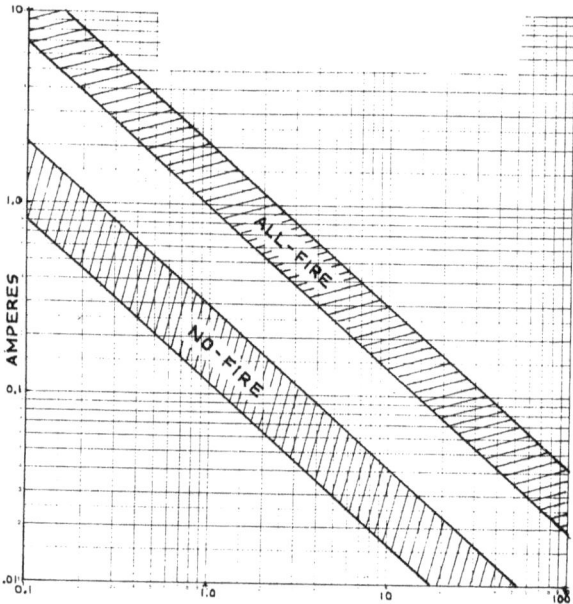

Fig. 3:15 Bridge resistance (ohms) All-fire and no-fire current vs. bridgewire resistance (Courtesy of Atlas Chemical Industries, Inc.).

milliseconds, and at 2.0 amperes, the function times averaged about 0.75 milliseconds. This example shows that no considerable performance increase can be expected by increasing the firing current more than two or three times.

Test data in the form of all-fire and no-fire current as a function of bridgewire resistance for initiators are shown in the graph in Fig. 3:15. The cross-hatched bands in these graphs are the results of plotting the all-fire and no-fire characteristics of more than thirty production initiators and other pyrotechnic devices. However, some electro-explosive devices, mainly 1-ampere, 1-watt no-fire 1-ohm bridgewire devices have slightly different all-fire and no-fire characteristics than shown in this graph.

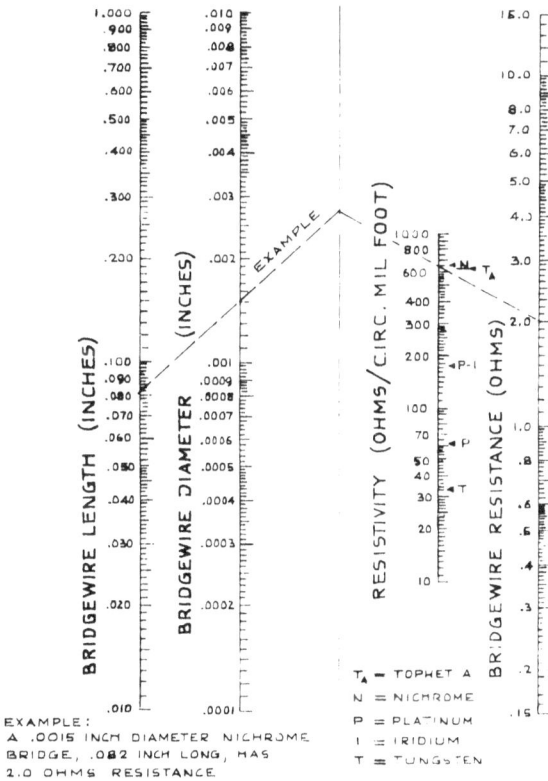

Fig. 3:16 Bridgewire Nomogram (Courtesy of Hi-Shear Corp., Ordnance Div.).

The all-fire and no-fire characteristics of a given bridgewire resistance depend on the bridgewire material and the explosive charge material used in the initiator or device.

The bridgewire nomogram presented in Fig. 3:16 provides a useful aid for determining the resistance and for sizing the diameter and length of a bridgewire from a certain material for a specific application. The example in the nomogram shows that a bridgewire of nichrome with a diameter of 0.0015 inch and a length of 0.082 inch has a resistance of 2.0 ohms.

The accuracy of firing characteristics, based on bridgewire resistance, is normally within 10%, which is generally acceptable and is suffucient to provide a wide spread between all-fire and no-fire levels.

Current Sources

An important factor for selecting a certain bridgewire material and for determining the bridgewire size is the firing current source. A typical source voltage is 28 volts dc from a battery or generator. A current of 110 volts would not be recommended, because it could in some cases damage the pyrotechnic device or destroy it.

If a capacitor discharge is applied for firing, a voltage of 1.5 to 300 volts or higher may be used. The capacitance and the voltage can

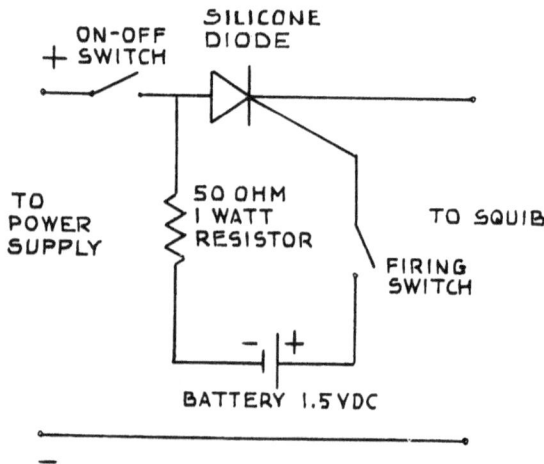

Fig. 3:17 A switching circuit 15 amp. max. (Courtesy of Hi-Shear Corp., Ordnance Div.).

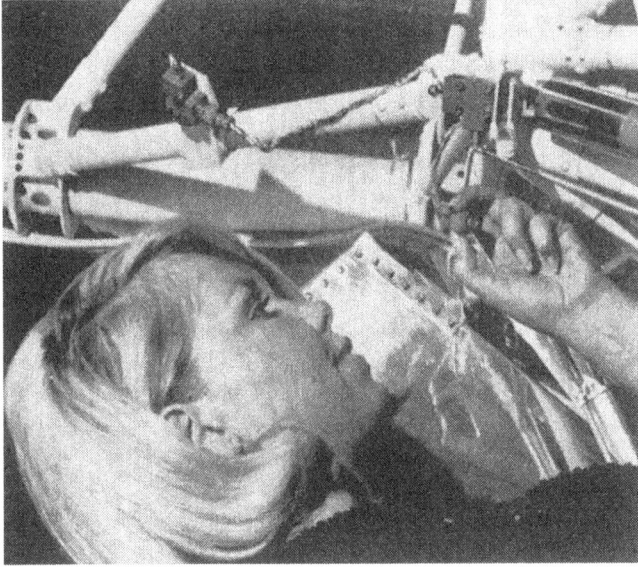

Fig. 3:19 Mounting of power cartridges in "Surveyor" spacecraft (Courtesy of Hi-Shear Corp., Ordnance Div.).

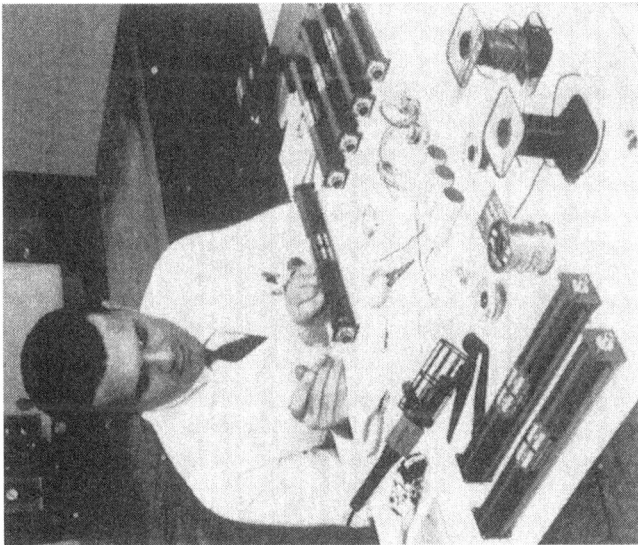

Fig. 3:18 Assembly of EBW initiator command modules (Courtesy of Hi-Shear Corp., Ordnance Div.).

be tailored according to the sensitivity of the pyrotechnic device. A rough conversion from generator or battery source requirement to that of a capacitor would be: A 15-ampere charge delivered to a 10-microfarad capacitor for 30 milliseconds results in a discharge capability of about 30 volts.

For the firing of a sensitive pyrotechnic device, a relatively low electrical input is required. It is also typical for sensitive devices that they fire faster than less sensitive ones. Less sensitive devices have the advantage that they are safe to handle and that the possibility of accidental firing by stray current is very small.

An example of a switching circuit for an initiator is shown in Fig. 3:17.

The assembly of exploding bridgewire initiator command modules is shown in Fig. 3:18.

The mounting of power cartridges in the "Surveyor" moon soft landing vehicle is presented in Fig. 3:19.

Power cartridges mounted in the "Surveyor" spacecraft are depicted in Fig. 3:20.

Fig. 3:20 Power cartridges in "Surveyor" spacecraft (Courtesy of Hi-Shear Corp.).

5. *Non-Electric Stimulus Transfer Systems*

Flexible explosive cords, mainly confined detonating fuse (CDF), mild detonating fuse (MDF), and small-column insulated delay (SCID), find ideal applications in non-electric stimulus transfer systems. These NESTS systems consist of several explosive cord transfer lines connected by standard couplings and manifolds, which require only one initiator to activate multiple pyrotechnic devices.

Typical applications for standard non-electric stimulus transfer systems and components are:

- Simultaneous multiple solid-propellant ignition with microsecond simultaneity.
- Simultaneous initiation of multiple explosive bolts, linear shaped charges, destruct charges, or any other explosive charges with microsecond simultaneity.
- Detonation or ignition through a solid steel bulkhead without jeopardizing the hermetic qualities of the seal formed by the bulkhead.
- A reduction in the number of electro-explosive components in any given explosive ordnance system to one, or for redundancy two, with the advantage of adding or subtracting any number of functions without modifying the systems electrical system, i.e. wiring, batteries, generators, heat batteries, etc.
- A reduction in the number of exploding bridgewire firing modules in any given exploding bridgewire system to one, or for redundancy two, while realizing increased reliability and safety against inadvertent initiation due to radio frequency or induced electrical energies.
- Elimination of the need for primary high explosives in any explosive ordnance system by combining non-electric stimulus transfer systems (NESTS) with any exploding bridgewire initiator or a reduction in the number of components containing primary explosive to one, the system initiator.

Standard manifolds for non-electric stimulus transfer systems which can accommodate two to six transfer lines, can be handled, shipped and stored as inert metal parts, since they contain no explosives. For applications requiring simultaneity on the order of

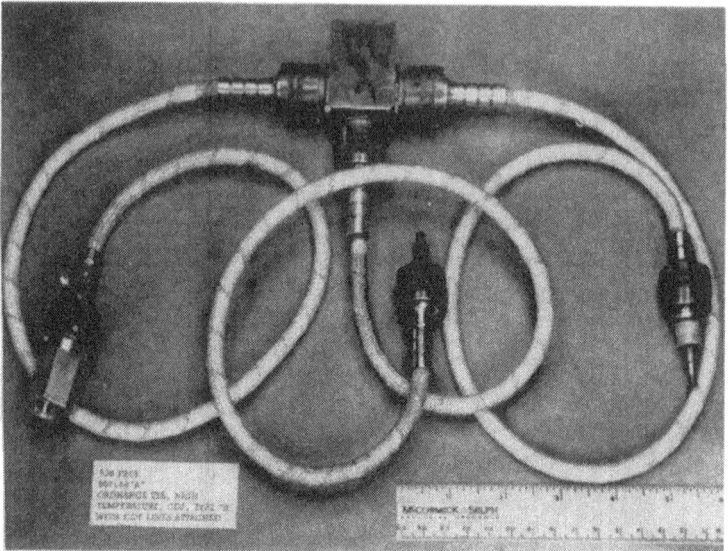

Fig. 3:21 NESTS system with tee joint and CDF lines (Courtesy of McCormick-Selph).

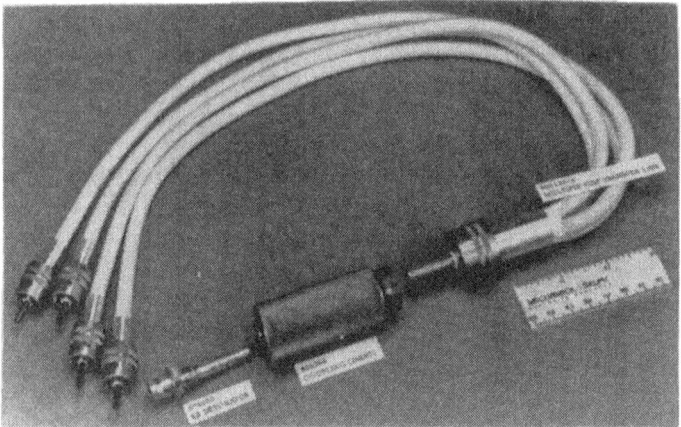

Fig. 3:22 Multiple CDF transfer line with quick disconnect and EBW detonator (Courtesy of McCormick-Selph).

Fig. 3:23 EBW NESTS ignition system with two initiators and six manifolds (Courtesy of McCormick-Selph).

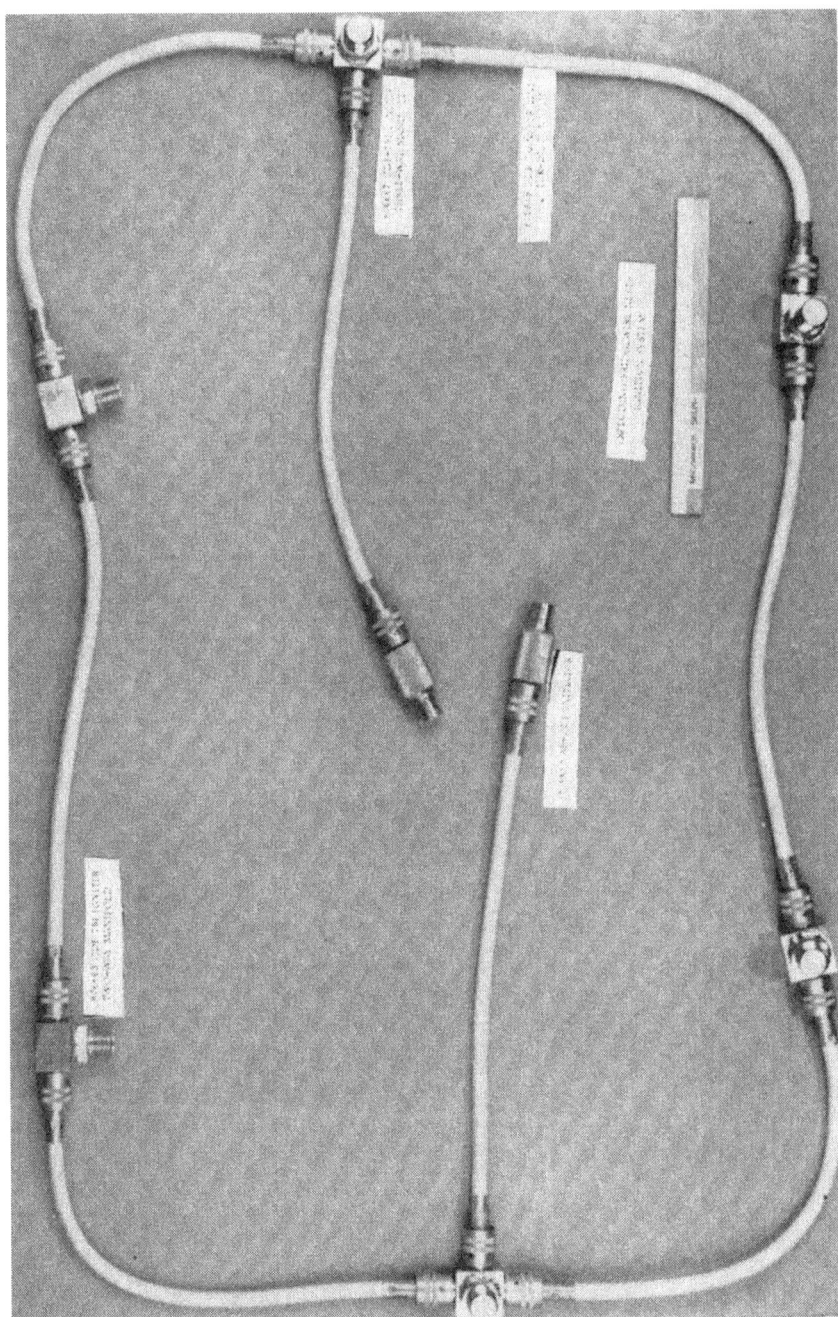

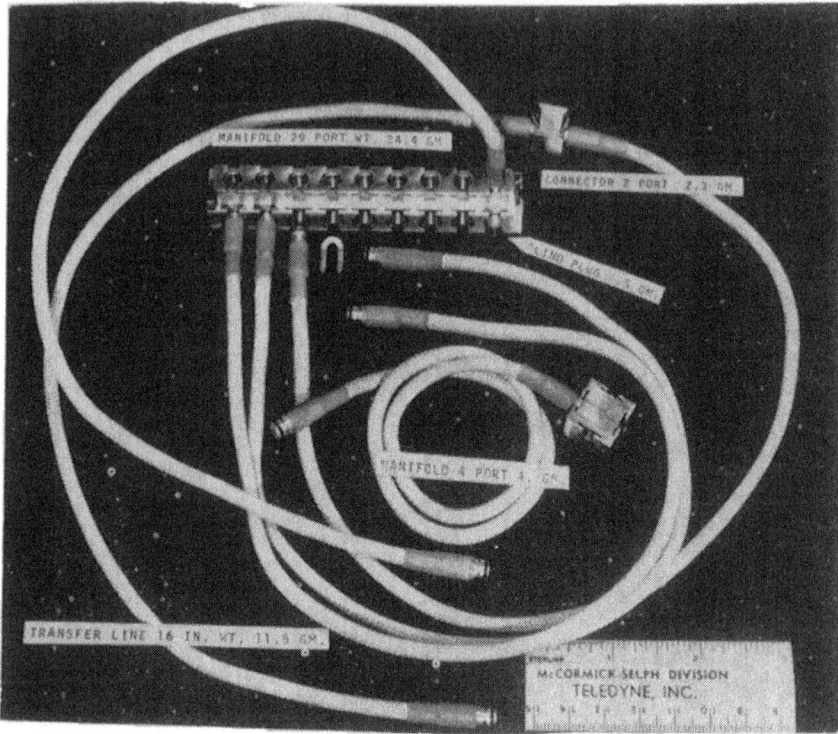

Fig. 3:24 Hivelite deflagrating transfer systems (Courtesy of McCormick-Selph).

± 0.5 microseconds, or for initiating a single function using redundant CDF transfer lines, a multiple transfer line manifold is recommended.

Non-electric stimulus transfer systems components are standardized and are completely interchangeable. For example, RDX-core CDF transfer lines, or RDX-loaded NESTS explosive bolts, through-bulkhead initiator (TBI), etc., can be interchangeably initiated by standard PETN-loaded CDF transfer lines. This building block design approach provides obvious advantageous flexibility in pyrotechnic systems design.

Typical non-electric stimulus transfer systems known as "NESTS" systems utilizing flexible explosive cords, as transfer lines and standard NESTS manifolds and couplings are shown in Fig. 3:21, 3:22 and 3:23.

HIVELITE deflagrating transfer systems with a four-port manifold and with a 29-port manifold are shown in Fig. 3:24.

B. *Mechanical Explosive Initiators*

A different category of explosive initiators, as mentioned in the introduction to this chapter, but not described yet, are the mechanical initiators. Generally, mechanical initiators or primers are used to initiate another explosive charge, and for various actuation purposes. The power output of a mechanical initiator is usually greater than that of a squib, but lower than that of a detonator.

The most commonly used mechanical initiators are:
1. Percussion Primers
2. Stab Primers

1. Percussion Primers

Percussion primers are flame-producing devices which are actuated by impact or friction. They are mainly used to ignite the smokeless powder charge in gun cartridges, for initiation of black powder, pressed delay train, or the primary explosive in a detonator. Since most types of percussion primers are standardized, mass-produced, commercially available, and not commonly used in pyrotechnic devices, they are only briefly described in this chapter.

Generally, a typical percussion primer, as shown in Fig. 3:25, consists of a tubular metal housing which contains a small impact-sensitive explosive charge. At one end, the housing is closed with a disc from thin sheet metal or paper, and at the other end with a metal anvil, which is a concave-shaped thick disc, backing the explosive charge. The anvil has the purpose of causing the generation of a crushing force between primer housing and anvil. When the round-end firing pin hits the percussion primer and dents the housing, it initiates without blowing back or rupturing at the indentation point. An important feature of percussion primers is the containment of the gaseous products of reaction which is desirable in obturated devices, and which results in increased efficiency of fire transfer. This capability is a requirement for time delay columns that must function accurately and reliably independent of ambient conditions, which necessitate leak-proof sealing of the percussion primer. The explosive charge compositions in percussion primers are selected according to

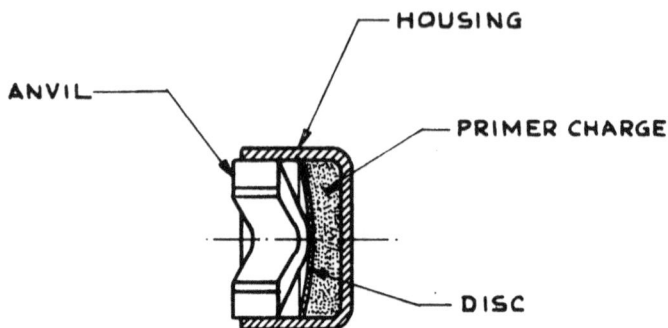

Fig. 3:25 Percussion primer.

the intended application. Low brisance and small gas generation are desirable characteristics of charge compositions in percussion primers to be used for pyrotechnic ignition. Some charge compositions are suitable for both ignition and detonation.

Lead styphnate, inorganic fuels, and oxidizer salts are normally used in charge compositions for percussion primers. A typical composition consists of potassium perchlorate (53%), lead sulfur cyanate (25%), antimony sulfide (17%), and a small amount of TNT (5%), or of mercury fulminate (40%), barium nitrate (25%), antimony sulfide (25%), barium carbonate (6%), and ground glass (4%). In other percussion primer compositions, lead peroxide is used in lieu of lead sulfur cyanate.

Percussion primers, as described, are capable of withstanding a temperature of up to 200°C for about 2 months. A new charge composition of greatly improved stability is the patented high explosive Tacot, a Du Pont product, which can withstand a temperature of up to 320°C.

2. Stab Primers

Stab primers are highly sensitive mechanical initiators of simple design which are preferably used in devices and systems where the available mechanical energy is very small. Stab primers, also known as stab detonators, are frequently used for initiating detonations. Ignition and mechanical work can also be accomplished with stab primers, using a suitable charge composition.

The main difference of stab primers as compared to percussion primers is in their basic design. In a stab primer, as shown in Fig. 3:26, the explosive charge is contained in a cylindrical housing which, in most cases, does not include an anvil, but is closed only with a circular flat cover at each end. During firing, the end of the firing pin which has the shape of a sharply pointed cone or of a pyramid, pierces a hole in the thin metal end cover and in the sensitive end of the charge.

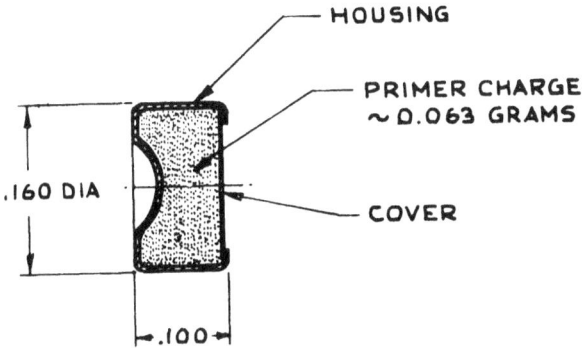

Fig. 3:26 Stab primer (MK 102).

A typical explosive composition used in stab primers consists of lead azide (29%), potassium chlorate (33%), antimony sulfide (33%), and a small amount of carborundum (5%). In some compositions, mercury fulminate or lead sulfocyanate is used in lieu of lead azide. However, since recently, mercury fulminate is not commonly used in civilian applications. A special charge mixture, NOL-130, contains a major percentage of barium nitrate instead of chlorate. The sensitivity of a stab primer can be improved by adding tetracene to the mixture.

It is interesting to note, that the Army M-26 stab primer is suitable for use as an igniter. A different type of stab primer, the M-41, can be used as a detonating device, since it contains a lead azide intermediate charge and a tetryl base charge.

A different type of mechanical initiators are the friction primers, which are described in detail in Chapter 9, Section A "Flares".

C. *Unconventional Initiators*

Recently, a new type of power cartridge, which is actuated by a laser beam, has been developed. A typical laser actuated cartridge, as shown in Fig. 3:27, consists of a metal housing which contains a propellant in about one half of its volume. The aft end of the charge cavity is closed with a metal disc, and the forward end is hermetically sealed with an optical glass window which admits laser light to the propellant. The absence of connector pins, pin headers, ceramic cups and bridgewires makes this power cartridge immune to static discharge, radio frequency fields and eliminates the need of electric power.

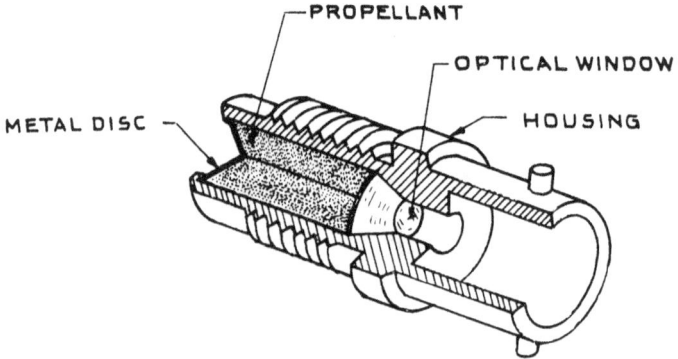

Fig. 3:27 Laser-actuated power cartridge (Courtesy of Hi-Shear Corp., Ordnance Div.).

The laser beam used for initiation is transmitted through fiber optics consisting of a bundle of approximately 250 glass fibers of 2.5 mils diameter each, which are enclosed in a flexible steel covering. The ends of the bundle are epoxy-potted into steel tubes of 0.125 inch diameter and ground and polished for optimum transmission characteristics. Standard bayonet connectors are fitted to the bundles to mate with the laser head and the cartridges.

An advantageous feature of this system is the ability to split the fiber optic bundle to provide multiple outputs, which can be used to initiate several cartridges simultaneously.

The laser-actuated cartridges are extremely reliable and safe. They are custom-built to specific requirements, and they are available as standard initiators, igniters, pressure cartridges, detonators, and gas generators.

The laser recommended for actuation of these cartridges has a wave length of 1.06 microns and a pulse length of 30-40 micro-

PYROFUZE BRAID

Single Bridge
Series (In Line)
Main Braid Common

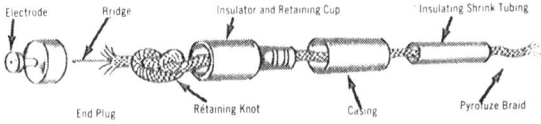

Electrode — Bridge — Insulator and Retaining Cup — Insulating Shrink Tubing
End Plug — Retaining Knot — Casing — Pyrofuze Braid

Double Bridge (Series)
Main Braid Common
or Single Bridge
(Detail Above)

Treble Bridge Series
(In Line) Main Braid Common
or Double With One Common Leg
(Detail Above)

Quadruple Bridge Series (In Line)
Main Braid Common or Double
(Detail Above)

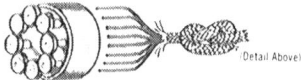

Eight Bridge (Series) Main Braid
Common or Quadruple Bridge
(Detail Above)

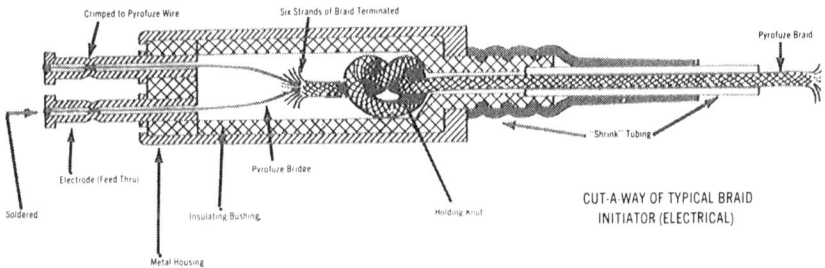

Crimped to Pyrofuze Wire — Six Strands of Braid Terminated — Pyrofuze Braid
Electrode (Feed Thru) — Pyrofuze Bridge — "Shrink" Tubing
Soldered — Insulating Bushing — Holding Knot
Metal Housing

CUT-A-WAY OF TYPICAL BRAID
INITIATOR (ELECTRICAL)

Fig. 3:28 "Pyrofuze" braid initiators (Courtesy of Pyrofuze Corp.).

77

seconds. Its energy input is 80 joules, and its energy output is 0 to 1.0 joule. The rod of the laser head consists of Neodymium glass. A 3.5-inch helical lamp is used in this unit. The power is supplied by a 12-volt 0.5 ampere silver-zinc battery, which has a recycle time of 7 seconds and a capacitance of 10 mfd.

Other unconventional initiator types are the bi-metallic exothermically alloying initiators known under the tradename "Pyrofuze". The functional characteristics of the bi-metallic composition, as used in these initiators, consisting of palladium and aluminum, are described in Chapter 1, Section J of this book.

Three different types of Pyrofuze initiators are used:
- Pyrofuze braid initiators
- Pyrofuze bridgewire initiators
- Pyrofuze granule initiators

Braid initiators with a single bridgewire and with multiple bridgewires are shown in Fig. 3:28. A typical Pyrofuze bridgewire initiator in which the bare Pyrofuze bridge is buried into the relatively insensitive pyrotechnic main charge is shown in Fig. 3:29. This design has the advantage that the Pyrofuze material itself is utilized as first fire and that "beading" compounds or other first fire compounds are

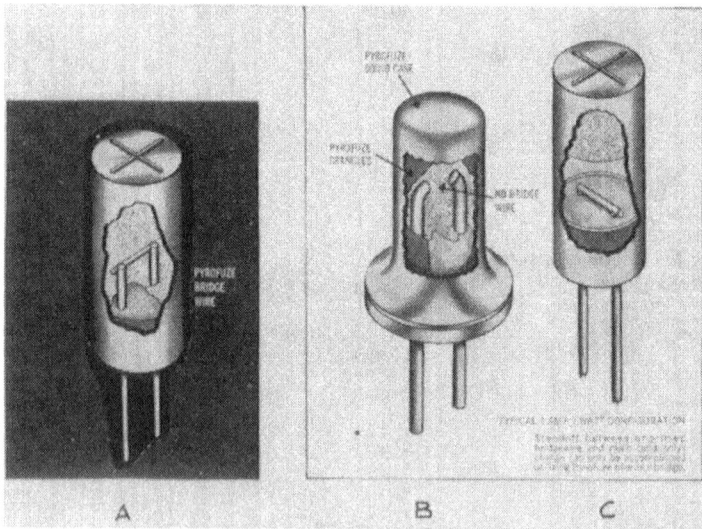

Fig. 3:29 "Pyrofuze" initiators (Courtesy of Pyrofuze Corp.).

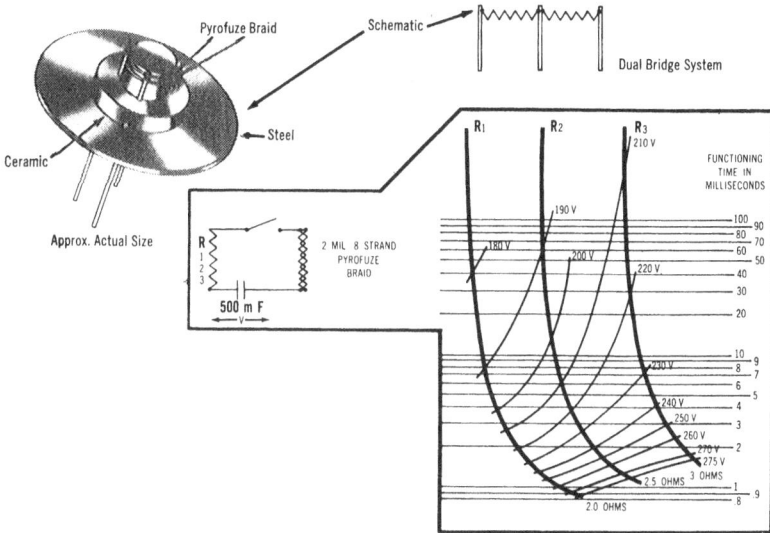

Fig. 3:30 "Pyrofuze" ignition scheme (Courtesy of Pyrofuze Corp.).

Fig. 3:31 "Pyrofuze" bridge during firing (Courtesy of Pyrofuze Corp.).

79

not required. In another bridgewire initiator, as shown in Fig. 3:29C, the standoff between the unprimed bridgewire and the main charge is accomplished by using Pyrofuze wire as a bridge. This initiator meets the 1 amp 1 watt requirement. In a Pyrofuze granule initiator, as shown in Fig. 3:29B, Pyrofuze granules are utilized for the bridge, prime and main charge, eliminating a bridgewire.

A unique ignition scheme for Pyrofuze braid initiators is shown in Fig. 3:30, and a Pyrofuze bridge during firing is shown in Fig. 3:31.

4   Piston and Bellows Devices

Actuation by explosives is utilized in numerous different types of devices, applying energy with a linear or rotary motion. Such devices are used to push or pull loads, for releasing a mechanism to permit motion, for locking a mechanism to prevent linear or rotary motion, to operate a latch, switch or relay, to puncture a membrane, or to rotate a shaft, and for many other applications. The function of these devices is based on the principle of internal combustion engines, with the only difference that these devices are one-shot engines with only a single charge of fuel-oxidizer.

In the category of linear motion devices, the piston and the bellows principles are applied most frequently. The piston principle is employed mainly in actuators, thrusters, pin pullers, cutters for cable, hose, wire bundles, reefing lines and webbing, release mechanisms, valves, switches, and in numerous special devices.

Usually, an explosive-actuated piston device consists of a cylindrical body or housing containing a propellant or pyrotechnic charge, a bridge-wire and lead wires in one end, and a piston in the opposite end. The lead wires pass through a hermetic seal to the bridgewire, similar to the arrangement in initiators. Some special types of piston devices are initiated by a percussion primer rather than by an electrically fired igniter. When fired, the propellant or pyrotechnic charge pressurizes the volume in the housing behind the piston end, and the piston driven by the output force of the high-pressure gases delivers the required thrust and travels a limited length at high velocity. According to the required function, the outer end of the piston part can be equipped with a clevis, a lug, a threaded end, or with a smooth cylindrical end, as for example in thrusters and pin pullers, or with a cutter blade in propellant-actuated cutters.

81

An important and critical design and manufacturing problem is to provide the O-ring seals in the dynamic interface between the piston and the cylinder body tight enough to prevent leakage of the explosive gases which could result in a loss of thrust in the device and in a contaminated atmosphere; yet, the dynamic interface has to be loose enough that it does not prevent full output and full stroke from being delivered. Extreme precision in design and manufacture is mandatory to attain this important balance and reliable functioning of the piston devices.

Two basically different types of piston devices are commonly used:

1. Devices in which the piston is normally retracted and extends upon firing,

2. Devices in which the piston is normally extended and retracts upon firing.

A special type of piston device has the feature of locking in the fired position, which is desirable, for example, in pin pullers and similar devices, because shock or vibration cannot cause the retracted pin to jump back into the released component or to interfere with it.

A more complex piston device is the telescoping type. It consists of several cylinders, each one sliding inside the adjacent one, which requires a set of dynamic seals in each interface and which complicates design and manufacturing. Only in cases where a very long stroke is required and where the available mounting space is too small for a conventional non-telescoping device, a telescoping device should be considered.

Piston devices with thrust outputs ranging from a few pounds to several thousand pounds have been developed. The stroke length is limited only by the size of the device and the application conditions. The thrust and the velocity of the operation of a piston device can be controlled by the type and quantity of the propellant or pyrotechnic charge.

Special care is required in the design of the mounting of piston devices. It is advisable not to apply mounting clamps to the body of the device, because they can cause deformation of the body which would result in binding of the piston. An ideal mounting arrangement is a cavity into which the device fits.

In another type of explosive-actuated devices the bellows principle is employed. The bellows device has several advantages over conventional actuator types: It can provide long straight strokes, both ends of the bellows device are hermetically sealed, and, compared with piston devices, O-ring seals are not needed in bellows devices. A further advantage of bellows devices is that thrust can be directed in a curved pattern or around a corner. It is, however, necessary to guide the bellows throughout its stroke by walls on all sides. In some applications, surfaces of the part that is to be actuated may be utilized as guide walls.

In the design of the mounting of explosive-actuated devices, including piston and bellows devices, the influence of the heat generated by the device should be considered. There is no need for concern when the operation is instantaneous. However, when the device contains a time delay in the form of a column of slow-burning explosive, the generated heat may be sufficient to damage nearby located electronic components or plastic parts. In such cases, relocation of the device or an insulation is required. It can be expected, however, that the burning rate of a time delay column will be affected by a thermal insulation of the device. It is, therefore recommended in such cases to test the device including the insulation under the expected operating conditions and to recaliber the time delay column as required. The burning rate of a time delay column in a device will also be affected if the device is mounted adjacent to a heat sink. The best solution in such a situation is to relocate the device.

A. *Actuators*

Actuators are devices which transform an electrical signal into mechanical motion. Linear motion actuators that function on the piston principle, as described in the introduction to this chapter "Piston and Bellows Devices", are most commonly used. Usually, these actuators are designed to yield a predetermined motion, force or velocity upon initiation. They find ideal application in cases where weight and size are important factors, and in many cases they can be used to replace solenoids, especially where electrical power is limited.

Piston-type actuators are used in many different applications, for example in satellite release systems, for stage separation, thrust reversal, tie down, stores release, switch and valve operations, in harness release devices for seat ejection systems, and in numerous other applications. Conventional piston actuators deliver output forces of 8 to 500 pounds and have stroke lengths of 0.1 to 3.0 inches. The size of these actuators ranges from 0.34 inch in length and 0.072 inch diameter to about 5 inches in length and 1.25 inch diameter. The weight of these actuators is surprisingly low: It ranges from 0.115 grams to 700 grams.

In these simple actuators without a piston locking device, after firing, the piston remains extended under moderate loads. In applications where higher loads are expected, the use of an actuator with a piston locking device is recommended.

A cross-section through a conventional, normally retracted, piston actuator without a locking device is shown in Fig. 4:1.

A typical piston actuator with a length of 0.725 inch and a diameter of 0.130 inch delivers a thrust of 20 pounds over a stroke of 5/16 inch, and the weight of this device is only 1.0 gram. Fig. 4:2 shows this actuator before and after firing. A force-versus-time diagram of this piston actuator is shown in Fig. 4:3. This diagram shows that the force curve does not originate at zero because the actuator has a preload of 20 pounds. The initial time interval, shown by the flat portion of the curve, is the time required for the bridge wire to

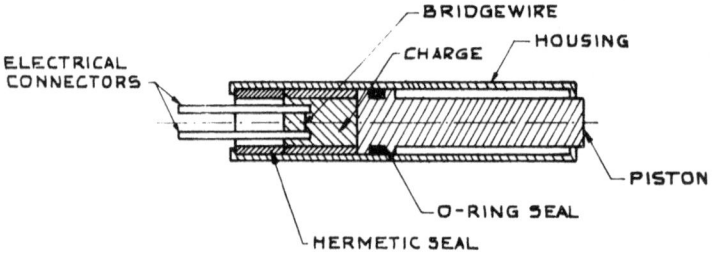

Fig. 4:1 A piston actuator.

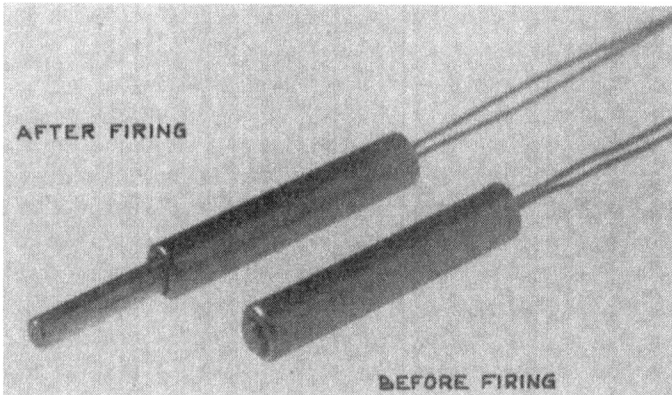

Fig. 4:2 Miniature actuator with extending piston (Courtesy of Atlas Chemical Industries, Inc.).

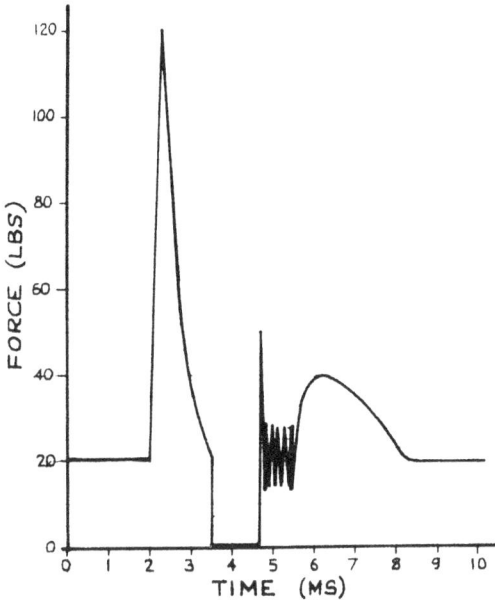

Fig. 4:3 Force vs. Time – a diagram of 20 lbs. Piston actuator with 5/16 inch stroke. (Courtesy of Atlas Chemical Industries, Inc.).

heat and ignite the powder. Peak force is produced during the first portion of the stroke and decreases as the piston moves out permitting gas pressure to diminish. After an elapsed time of 3.45 milliseconds, at the point at which the curve again crosses the 20-lb line, the stroke ends. The curve then shows a drop to zero force because the momentum causes the load to break contact with the actuator and continues its travel. The rising curve at 4.7 milliseconds indicates that the actuator again makes contact with the load. This irregular trace is the result of oscillation and of an attempt of the system to reach equilibrium.

These conventional actuators are available with various choices of bridgewires. The required firing current ranges from 0.30 amperes minimum all-fire and 0.04 amperes maximum no-fire current to 4.5 amperes minimum all-fire and 1.0 ampere maximum no-fire current.

These explosive-actuated mechanical actuators are ideally suited for applications in extreme environmental conditions. A typical actuator can withstand a one-half sine-wave pulse of 100 g for 3 milliseconds and 2,000 g for 1.5 milliseconds. Some actuators can stand up to shocks as high as 20,000 g in any plane. Most conventional actuators are capable of withstanding transportation and sinusoidal vibration of 5 − 2,000 − 5 Hz at 30 g or 0.34 inch double amplitude for 15 minutes. A typical acceleration resistance rating for actuators is 50 g in each of three mutually perpendicular planes for one minute.

In applications where only a very short stroke and a thrust of low magnitude are needed, and where extremely small actuator size and weight are a stringent requirement, miniature actuators can be used ideally. A typical miniature piston actuator, as shown in Fig. 4:4, has a piston stroke of 0.10 inch. a function time of 4 milliseconds, and delivers a thrust of 8 pounds.

A typical actuator with a retractable piston, before and after firing, is shown in Fig. 4:5. An internal thread in the piston end provides for attachment of parts or subassemblies to be retracted. When fired, the piston retracts into the housing. An internal locking device holds the piston in the retracted position under load. This actuator delivers a thurst of 25 pounds and a stroke of 5/16 inch. It has a length of only 1.0 inch, a diameter of 0.375 inch, and it weighs only 12 grams.

In applications where the stroke must be directed around a curve or a radius, bellows actuators are ideally used. They produce linear or non-linear motion with a long stroke. One of their advantages is that they do not require any O-ring seals, because they do not contain any

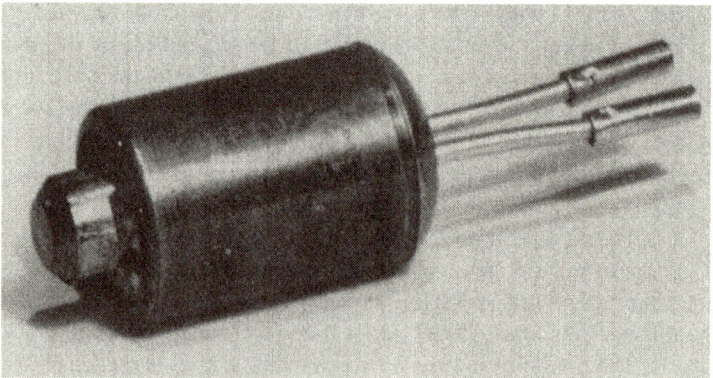

Fig. 4:4 Miniature piston actuator (Courtesy of Atlas Chemical Industries, Inc.).

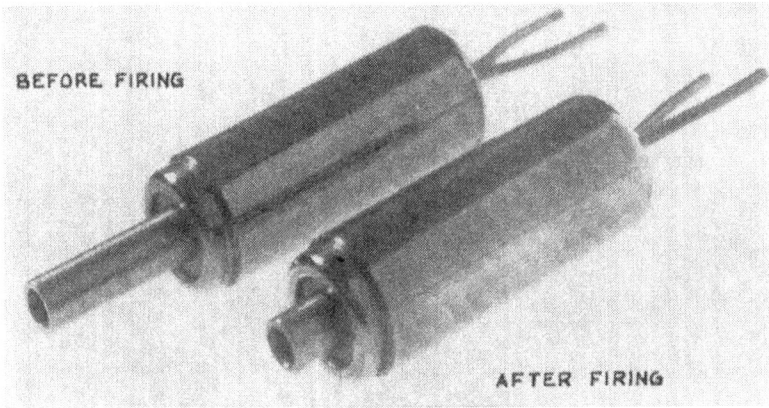

Fig. 4:5 Miniature retracting actuator (Courtesy of Atlas Chemical Industries, Inc.).

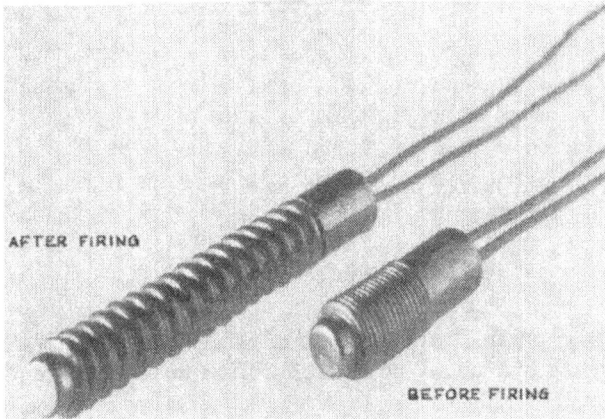

Fig. 4:6 Bellows actuator (Courtesy of Atlas Chemical Industries, Inc.).

sliding parts. When fired, caused by the developed gas pressure, the bellows expands and exerts the output force. A typical bellows actuator, as shown in Fig. 4:6, delivers a thrust of 10 pounds over a stroke length of 1.0 inch. Before firing, this actuator has a total length of 1.0 inch, a diameter of 5/16 inch, and a weight of 3 grams.

Lead styphnate/LMNR or KDNBF are commonly used as charge material in these actuators.

Another type of miniature actuator for applications where only a very short stroke and a very low thrust output are needed, is the dimple actuator. Fig. 4:7 shows a typical dimple actuator before and

87

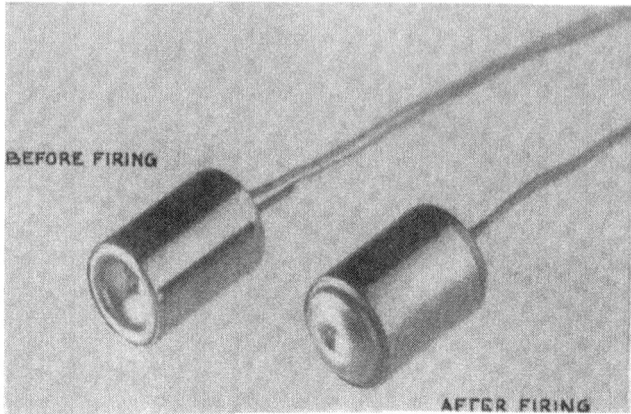

Fig. 4:7 Dimple actuator (Courtesy of Atlas Chemical Industries, Inc.).

after firing. Simplicity is the main advantage of these small actuators which do not contain any sliding parts. When fired, a metal diaphragm at the output end of the device is flexed out to provide the required stroke. The typical dimple actuator, as shown, delivers a thrust of 8 pounds over a stroke length of 0.10 inch, and its weight is only 3 grams. This miniature device has a length of 0.518 inch and a diameter of 0.295 inch.

Special types of piston actuators, which have the feature of a controlled velocity piston stroke, are used in aircraft seat ejection systems and in similar applications. This special feature is obtained by metering hydraulic fluid through orifices in the main piston, which results in a low "g"-output. A typical knee elevating ballistic actuator has a piston stroke of 5.25 inches and a piston velocity of 1.4 to 5.0 feet per second with a dynamic output of 800 pounds acting against a mass of 56.4 pounds. This actuator can withstand a compression load of 800 pounds and a tension load of 100 pounds. The weight of this special actuator is only 1.35 pounds.

B. *Thrusters*

Thrusters are propellant-actuated piston devices designed to produce a stroke of a specified force and velocity. They are used to operate switches, valves, release mechanisms, safe/arm devices, to separate structures and fastener components, and to eject capsule covers, doors, fairings, stores, parachute canisters, and similar equipment. Thrusters consist of a cylindrical housing, a piston which extends in the form of a push rod through one end of the housing, and a gas-generating electro-explosive power cartridge mounted at the opposite end of the housing behind the piston.

Two major types of thrusters are

1. Gas retaining thrusters having a captive piston,
2. Thrusters having a free piston.

In a gas-retaining thruster, the piston locks in the housing, retaining the piston and sealing in the gases. In some thrusters, the thrust of the piston releases a load-mounting insert.

In free-piston type thrusters, the piston end is used as a load mount. When the thruster is activated, the piston is forced out of the housing, releasing the load and the gases.

In both types of thrusters, the piston must be retained until commanded to activate, and for this reason, a simple and reliable retainer and release device must be provided. Two basically different types of release systems are generally used in thrusters. A shear pin of low strength is commonly used as a simple release system in thrusters designed for tensile loads of up to 5,000 pounds, whereas a ball detent release mechanism is frequently used in thrusters for tensile loads exceeding 5,000 pounds.

Fig. 4:8 Thruster (Courtesy of Hi-Shear Corp., Ordnance Div.).

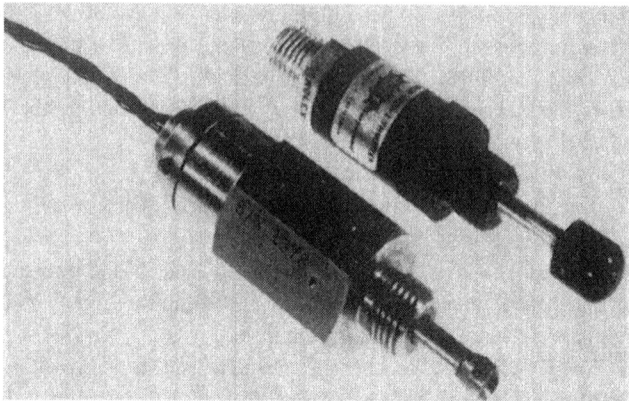

Fig. 4:9 Thruster (Courtesy of Holex, Inc.).

Thrusters have been developed for thrust peak loads from 20 to 13,000 pounds. Various types of mounting of the thrusters are possible by flanges and threads provided on the thruster housings, as shown in Fig. 4:8 and Fig. 4:9.

A cross-section through a typical thruster with a ball retent release mechanism and with two cartridges is shown in Fig. 4:10.

Thrusters are a good example of high power-to-weight ratio devices. A thruster for a thrust peak load of 1,200 to 2,500 pounds has a size of only 5/8 inch diameter, is 1.84 inches long, and weighs only 0.15 pounds.

In a special type of thruster which provides a stroke of about 2.5 inches and initial thrust of 3,000 pounds, as required for extension of stabilization fins on a high-speed ejection seat, equalized deceleration within a few milliseconds is achieved by a self-contained snubbing

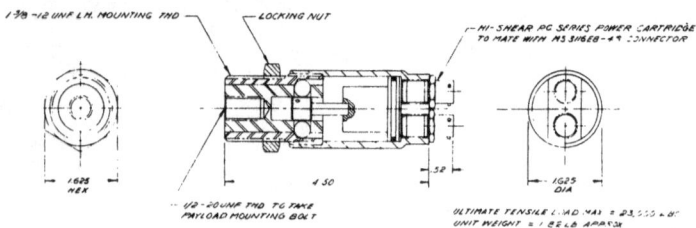

Fig. 4:10 Thruster with a ball release (Courtesy of Hi-Shear Corp., Ordnance Div.).

device which reacts against the piston head. This snubbing device consists of a set of snap rings mounted in a groove on the piston near its end. At the end of the stroke, these snap rings engage in a ring-shaped groove in the cylindrical housing and then lock the piston in the extended position.

For applications where delay sequencing is required, for example for ejection of two seats from a supersonic aircraft, delay sequencing actuators have been developed. A typical delay sequencing thruster is mechanically initiated and is equipped with dual cartridges and firing pins for redundancy. Either cartridge is capable of producing a mini-mum output of 150 pounds. A sear pin pull of 15 to 30 pounds is required to fire this thruster which produces a stroke of 0.875 inch. The weight of this unit is only 18 ounces.

Special types of thrusters or linear actuators are the ejectors and removers as used in aircraft, in aircraft canopy, and seat ejection systems. Ejectors are large two or three-tube telescoping thrusters which are mainly used as catapults for ejection seats. A typical ejector has a length of 39 inches and a diameter of 2.3 inches, a stroke of 66 inches, a weight of 8.2 pounds, and it is designed to eject a weight of 300 pounds at a minimum velocity of 60 feet per second and with an acceleration of 20 G's. The stroke time of this ejector is 0.22 second. As a firing method, gas actuation is utilized. On initiation, the gas developed by the initiator, which is attached by a flexible hose, exerts a force against a firing pin. Subsequently, the firing pin is propelled forward and, by striking, detonates the primer and thereby ignites the black powder and propellant charge in the cartridge. During the for-ward movement of the firing pin, the block assembly and inside tube are being unlocked, and the inside tube and telescoping tubes are forced by the expanding gases to move simultaneously until the shoulder of the telescoping tube comes in contact with the trunnion. The outside tube and the telescoping tubes remain with the aircraft, whereas the continuous moving block assembly and the inside tube are ejected with the seat.

Removers, designed for aircraft canopy ejection, generally function on the same principle as ejectors. They differ mainly in their acceleration, velocity, thrust and physical dimensions. A typical re-mover has a length of 16 inches, a diameter of approximately 2 inches, a stroke of 23 inches, a weight of 2.1 pounds and is capable of propelling a weight of 300 pounds at a velocity of 20 feet per second.

It produces a thrust of 2800 pounds, and its stroke time is 0.135 second. The firing method for this remover, utilizing gas actuation, is similar to the firing method used in the ejector. In some other types of removers, mechanical actuation is used. In such a mechanical system, first a safety locking pin is removed, and a sear pin is manually rotated through a pulley system which is connected to a lever on the pilot's seat. By rotating the sear pin, a spring-loading firing pin is released, thus unlocking the inside tube. The function of the firing pin in this system is the same as in the ejector, as described above.

C. *Pin Pullers*

Pyrotechnic pin pullers are used mainly as release mechanisms for instant separation of external stores, structural members, packages, equipment, emergency hatches, rods, cables or parachutes. Pin pullers are designed as locking devices which carry full shear or tension loads. They function in a similar way as the thrusters, with the main difference being that the piston pulls the extending shaft end inward. Pin pullers release by controlled low-pressure gas energy sufficient to sever a low-strength shear pin and to overcome static friction in the device and in the loaded pin end which results in a retraction of the piston and a release of the load. The shear pin holds the piston and pin shaft in the assembly in place prior to actuation. In many conventional types of pin pullers, after operation, a certain length of the piston shaft protrudes out of the pin puller housing. Sufficient clearance must be provided for this protruding end in the vehicle or assembly when using this type of pin puller.

In many pin pullers, the power cartridge is mounted on the end of the housing in order to keep the diameter of the device small. Where minimum length is a requirement, the cartridge is located at the side of the housing. For applications which require redundancy, two cartridges are mounted on either the side or the end of the pin puller housing. A pin puller with a dual cartridge arrangement on the sides of the housing is shown in Fig. 4:11A. A pin puller with the cartridge mounted at the end is shown in Fig. 4:11B, and a pin puller having the cartridge mounted at one side is shown in Fig. 4:11C. All three types of pin pullers are provided with an external thread on the housing next to the pin for ideal mounting in an assembly.

An unusual type of pin puller, in which the pin extends from the

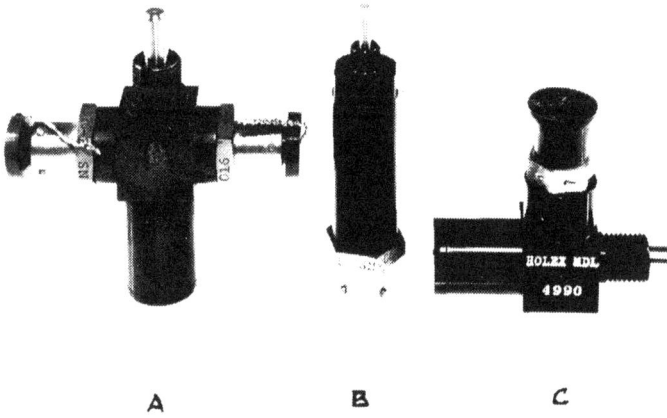

Fig. 4:11 Pin pullers (Courtesy of Holex Inc.).

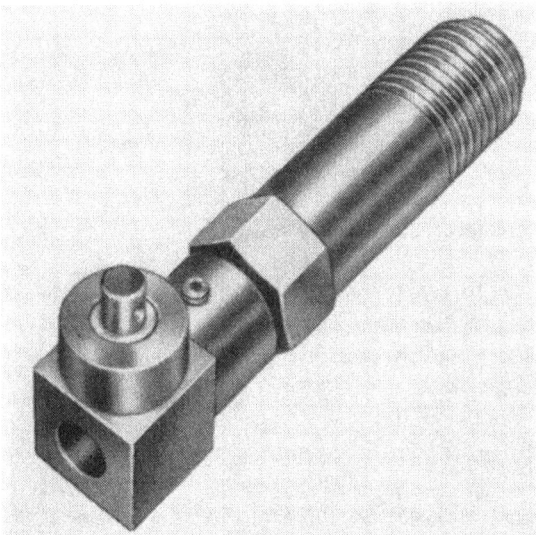

Fig. 4:12 Pin puller (Courtesy of Hi-Shear Corp., Ordnance Div.).

side of a long housing, is shown in Fig. 4:12. This pin puller was developed as a cutter release mechanism for a special application.

Conventional pin pullers have been developed for operating side loads in the range from 100 to over 18,000 pounds. The weight of a

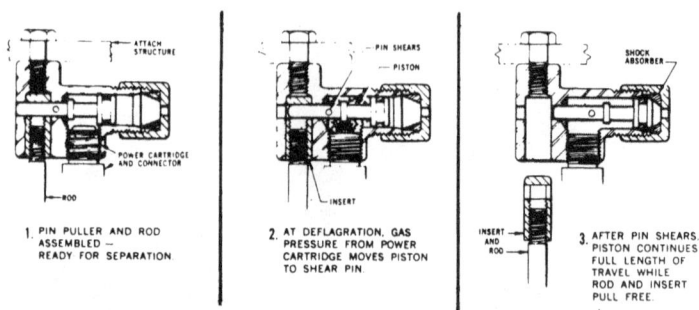

Fig. 4:13 Functioning of pin puller (Courtesy of Hi-Shear Corp., Ordnance Div.).

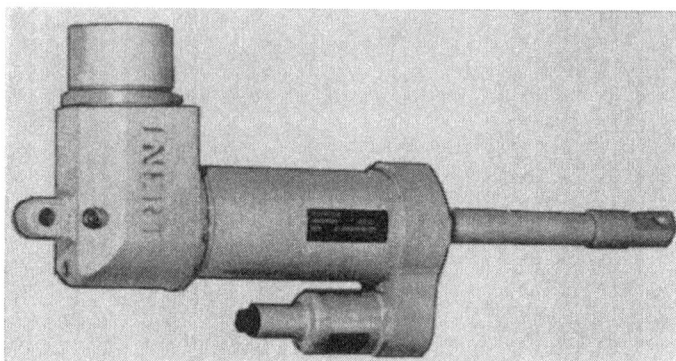

Fig. 4:14 Ballistic puller (Courtesy of Talley Industries).

pin puller for a side load of 18,200 pounds is surprisingly low — only 0.72 pounds. These pin pullers operate without fragmentation, retaining all parts and the gases. With the best suitable power cartridge for a given application, release times of 5 milliseconds or less are obtained with these pin pullers.

The functioning of a typical pin puller is shown in three phases in Fig. 4:13.

A special type of pyrotechnic puller, a ballistic puller, as designed for pre-ejection positioning of personnel or equipment in aircraft seat ejection systems, is shown in Fig. 4:14. This device has the special feature that its operating time, with no load to maximum load condition, varies very little so that physiological limits of the pilot or structural strength limits of the equipment are not exceeded under extreme operating conditions. A hydraulic buffer is incorporated as additional stroke control to match the normal ballistic stroke time. Ideal actuation over a wide load range with complete safety is provided by this dual control stroke system. A performance diagram of this ballistic puller is shown in Fig. 4:15.

PERFORMANCE CURVE

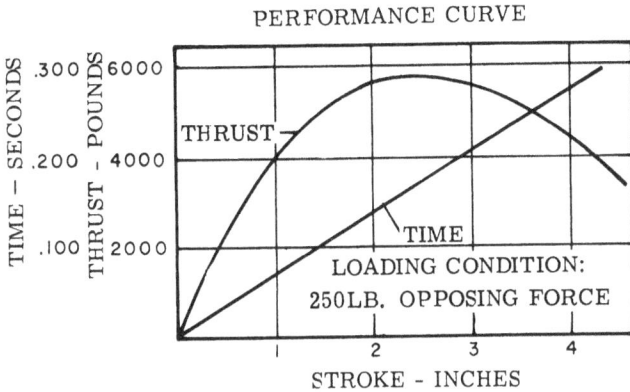

Fig. 4:15 Performance diagram of a ballistic puller.

D. *Cable and Hose Cutters*

Cartridge-actuated cutters are used to sever aircraft steel cable, electrical cable, pressure hose, fuel lines, tubing, rod and any type of line of high tensile strength. Advantageous features of these cutters are their small size, low weight and high reliability. Their power-to-weight ratio is higher than that of any other type of mechanical cutter. Usually, cartridge-actuated cutters are mechanical piston devices whose function is similar to that of thrusters.

Fig. 4:16 shows a cable and wire cutter which is designed to cut steel cables of up to 3/16 inch diameter. A dimensioned sketch and a cross-section through this cutter are shown in Fig. 4:17. The cutter consists of a cylindrical housing from steel, a piston cutter blade, an anvil, an end cap and a cartridge. Two cutouts opposite to one another are provided near the anvil end in the housing for holding the cable to be cut. The piston blade is retained by a shear pin until, after firing within milliseconds, sufficient pressure develops to shear the pin. At this instant, the blade strikes against the anvil and severs the cable. The device is sealed to prevent gas or flame leakage from caus-

95

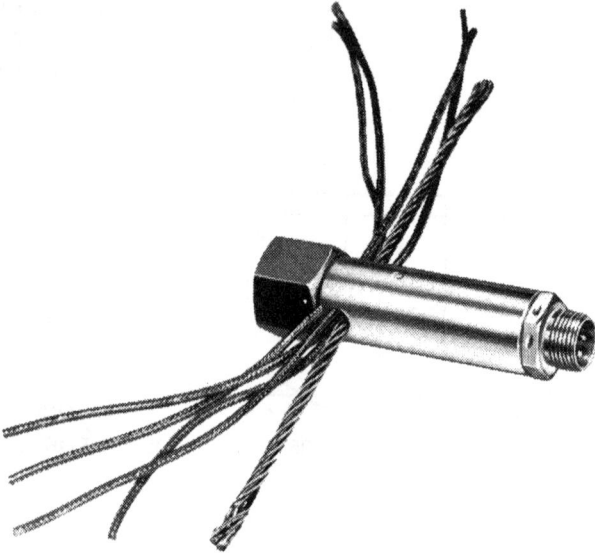

Fig. 4:16 Cable and wire cutter (Courtesy of Hi-Shear Corp., Ordnance Div.).

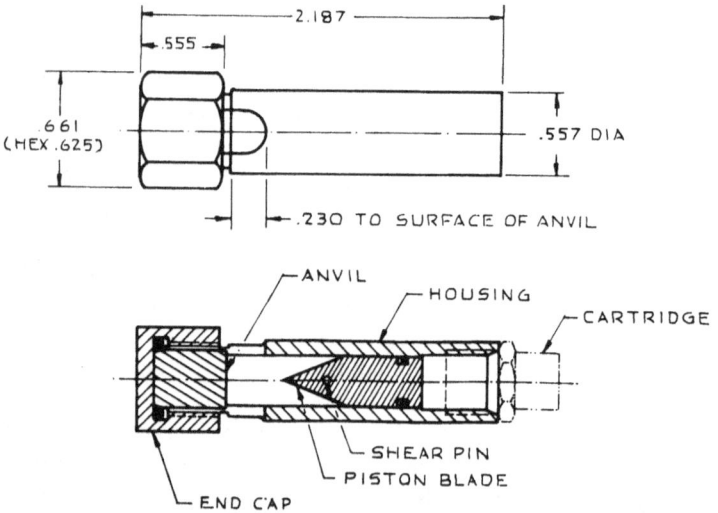

Fig. 4:17 Cable and wire cutter (Courtesy of Hi-Shear Corp., Ordnance Div.).

ing damage to adjacent components. After firing, this type of cutter can be re-used by replacing the piston blade, anvil and cartridge. The functioning time of this cutter is 5 milliseconds maximum. What is surprising is its low weight of only 0.10 pound.

Fig. 4:18 shows a wire bundle cutter of similar simple design as the cutter described above. This cutter is designed to sever a rubber-sheathed electric cable of 16 strands of No. 21 plastic-coated stranded wire of 3/16 inch maximum diameter. This device has a length of 2.375 inch, a diameter of 0.875 inch and weighs without power cartridge 0.12 pound. An external thread on the housing can be provided for easy mounting. Cutters of this type have been developed for wire bundle sizes of up to 1.0 inch diameter.

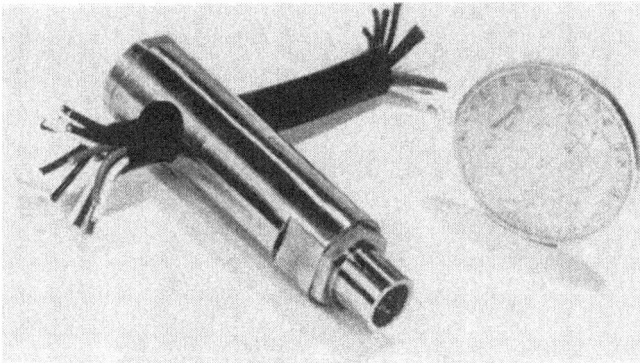

Fig. 4:18 Wire bundle cutter (Courtesy of Hi-Shear Corp., Ordnance Div.).

Several wire bundle and line cutters are shown in Fig. 4:19. The wire bundle cutter "A" is capable of cutting a 0.75 inch diameter bundle of 46 electrical conductors, whereas the largest cutter, device "B", is designed to sever a 2.10 inch diameter bundle of electrical wires. The advantageous possibility of utilizing various types of firing mechanisms is demonstrated in these large cutters, which can be equipped with either electrical, mechanical or gas firing mechanisms and can be provided with initiator cartridges for either instant or time delay firing.

A special type of cable cutter, which is equipped with redundant pressure cartridges and a safe/arm valve mechanism, is shown in Fig. 4:20. This cutter is designed to sever a 3/8 inch diameter 7 x 19

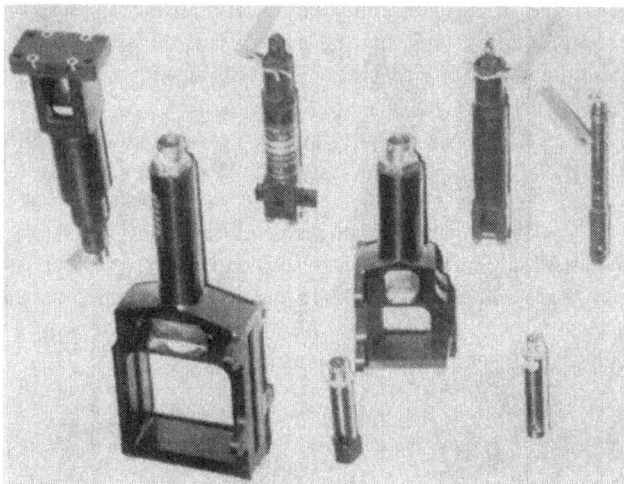

Fig. 4:19 Wire bundle and line cutters (Courtesy of Atlas Chemical Industries, Inc.).

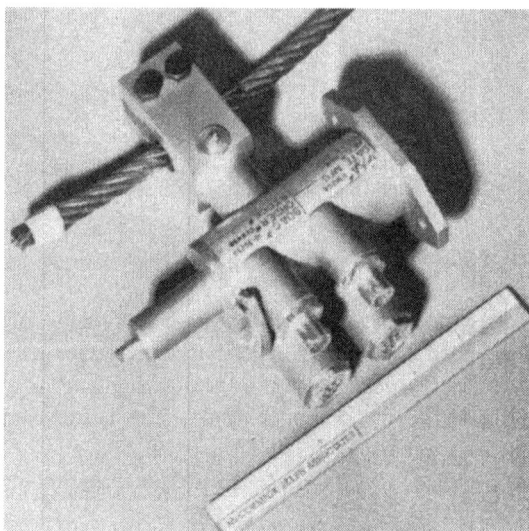

Fig. 4:20 Cable cutter with "safe/arm" valve mechanism (Courtesy of McCormick Selph).

strand stainless steel cable. Additional to the conventional piston blade and anvil, the housing of this special cutter contains a "safe/arm" valve mechanism and flame arresters. By rotating the valve mechanism 90 degrees, the device can be set in either "safe" or "armed" position. With the valve in the "safe" position, inadvertent firing of the cartridges results in venting of the gases through the flame arresters into the atmosphere. With the valve in the "armed" position, the gases are directed against the back surface of the piston, driving the blade forward and cutting the cable. A special feature of this cutter is its capability of operating safely in an explosive atmosphere, made possible by the built-in flame arresters.

In a special type of pipe cutter, as used, for example for emergency cutting of primary reactor coolant pipes and for similar applications, the part of the piston shaft directly above the forward end of the blade or punch is designed as a swaging die half. The piston is guided in a long slot in the cylinder wall, so that the shear punch will line up with a clearance slot which is cut across the center of the anvil. To both sides of the slot, the anvil is shaped to a swaging die half. When fired, the pipe is severed and both ends adjacent to the cut are closed by swaging, which is a desirable feature in many pipe cutting applications.

An unusual type of cutter, different from the previously described piston-type devices, is shown in Fig. 4:21. This cable cutter, which is initiated by a confined detonating fuse (CDF), is designed as a ring-shaped device. It is capable of severing a two-conductor shielded and jacketed 17/32-inch diameter stranded wire bundle, held together with heat-shrinkable tubing. A shaped-charge type initiator is used in this device. The advantages of this unconventional cutter are its small size and low weight as compared to piston-type cutters.

For puncturing diaphragms, and for similar functions, a special type of cutter has been developed. A typical diaphragm cutter, designed on the piston principle, is shown in Fig. 4:22. This cutter punches a 0.160-inch diameter disc in a 0.035-inch thick steel container diaphragm and locks itself into clear-flow position, allowing the release of the pressurized gas from the container. As tests showed, this device is capable of punching holes into twice as thick diaphragms. A unique feature of this cutter is the positive insurance of high post-fire residual resistance to prevent current drain on the battery. A typical diaphragm cutter has a length of 2.15 inches, a diameter of 0.63 inch,

Fig. 4:21 Cable cutter with CDF initiator fuse (Courtesy of McCormick Selph).

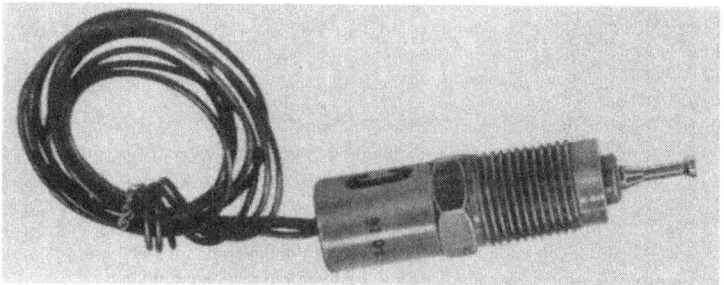

Fig. 4:22 Diaphragm cutter (Courtesy of Conax Corp.).

and its operating time is 7 milliseconds. It functions reliably in a temperature range from $-54°C$ to $+57°C$. Its operating pressure is 6,800 psi, its proof pressure is 9,000 psi, and its burst pressure is 17,000 psi.

E. *Line Cutters*

Line cutters are mainly used for severing the reefing lines of parachutes and similar nylon line, cord and small-gage electrical wires.

A parachute is reefed to reduce the projected diameter of the parachute canopy temporarily by a reefing line laid around the skirt of the canopy, and thus, the opening shock and the forces acting on the parachute and on the attached payload are greatly reduced. By severing the reefing line, the canopy skirt restriction is removed and the canopy opens to the full projected diameter. A reefing line cutter is usually mounted in a pocket or at a bracket provided at the canopy skirt. Line cutters for severing reefing lines with a tensile strength of up to 2000 pounds are used in recovery and cargo delivery systems.

Two different types of line cutters are generally used:

1. Cartridge-actuated line cutters, and
2. Lanyard-actuated line cutters.

Both types of cutters are used with and without time delay.

Cartridge-actuated line cutters are in their design similar to piston-type cable cutters. Their main difference is in the shape of the line aperture which must have smoothly rounded edges in order to avoid premature severing of the line which could be caused by rubbing on too sharp edges. Another difference of some line cutters as compared to cable cutters is in the design of the cutting blade guide inside the housing. The blade duct is not cut through to the line aperture, but a thin metal wall is provided between blade and aperture to protect the line prior to severing. The blade must cut through this wall

Fig. 4:23 Cartridge-actuated line cutter (Courtesy of Hi-Shear Corp., Ordance Div.).

first before cutting the line. Instead of such a wall, a shear pin in the side of the piston blade is often used as a retaining device.

A typical cartridge-actuated line cutter of small size, which has no built-in time delay, is shown in Fig. 4:23. This cutter is designed to sever a reefing line of 750-pound tensile strength. For redundancy, a dual-bridgewire gas-generating integrated power cartridge is used in this line cutter. The length of this cutter is 1.312 inches, the diameter 0.50 inch, and its weight is less than 0.04 pound.

Lanyard-actuated line cutters are used where electric current is not available. Usually, they are equipped with built-in time delays. They are actuated by mechanical pull, utilizing the opening force of the parachute. For mounting at the parachute skirt, these cutters are often equipped with a mounting lug at the anvil end. A firing pin is released by the pull having a magnitude of 15 to 35 pounds. The pin strikes a percussion primer which actuates a delay train. During actuation, the piston blade, guided in the cylindrical housing, is thrust against the line and cuts it. Time delays of up to 12 seconds can be obtained in presently available line cutters.

A typical lanyard-actuated line cutter, as shown in Fig. 4:24, has the special feature of being equipped with a flexible steel cable and

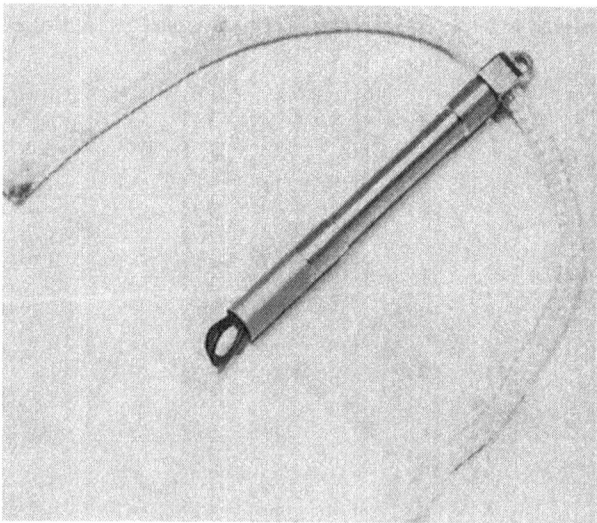

Fig. 4:24 Lanyard-actuated line cutter with time delay (Courtesy of Hi-Shear Corp., Ordnance Div.).

102

swivel, which are connected to the detent and which provide a capability of angular pulls of up to 90 degrees without impairing the operation of this line cutter. This cutter, which can be provided with a time delay of 0–10 seconds, has a length of approximately 5 inches, a maximum diameter of 0.50 inch, and a weight of only 2.1 ounces.

Fig. 4:25 shows a simpler and smaller type of a line cutter for a 750-pound nylon line. In this mechanically actuated device, an aluminum sear release is utilized. The lanyard pull direction of this cutter must stay within 15 degrees from the center line. For mounting this type of cutter, a circumferential groove for facilitating a snap ring in a

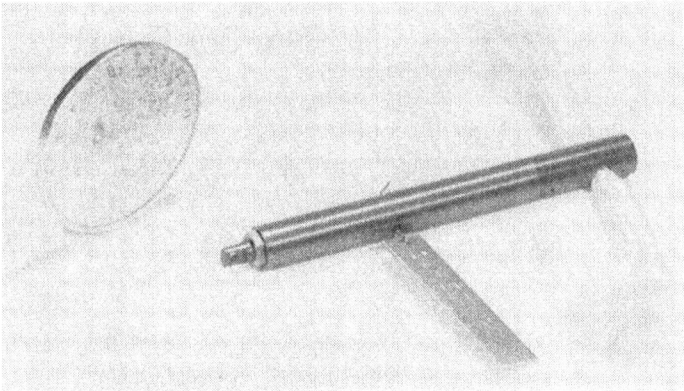

Fig. 4:25 Lanyard-actuated line cutter with sear release (Courtesy of Hi-Shear Corp., Ordnance Div.).

Fig. 4:26 Lanyard-actuated cutter with time delay (Courtesy of Holex Inc.).

small bracket is often provided at about half of the housing length. The dimensions of the line cutter, as shown, are: 3.56 inches long, and 0.321 inch diameter, and its weight is 0.38 ounces.

Fig. 4:26 shows another typical cutter, which has four mounting brackets and which is somewhat bulkier than the previously described cutters. This cutter, which is used in an aircraft seat ejection system, is designed to cut four wires of 0.150 inch diameter encased in a shrink tubing, and it has a built-in time delay of one second. An advantageous feature of this cutter is that it can be reused by replacing the cartridge, cutter, and, if necessary, the anvil.

Some basic rules for the design of line cutters are presented in the following:

1. The length of the cutter blade must be greater than the diameter of the line aperture to prevent the high-pressure gases from leaking out prematurely.

2. In lanyard-actuated cutters, the propellant charge must be blocked off from the firing mechanism.

3. Lanyard-actuated cutters should be so designed that the snatch force of the parachute or decelerator acts in the same direction as the firing pin, thus assisting the firing pin's striking force.

F. *Valves*

Cartridge-actuated valves for gas and fluid systems are usually one-shot devices. Their advantageous features as compared to solenoid valves or mechanically-actuated valves are their short operation time and the small size and weight of the power cartridge. A great variety of types, sizes and shapes of normally open and normally closed valves with two or more ports for tube or manifold connection have been developed. The size range of typical cartridge-actuated valves is demonstrated by the two devices as shown in Fig. 4:27.

The function principle of cartridge-actuated valves is clearly presented in Fig. 4:28, which shows the cross-section through a normally-open gas valve after firing. The gas inlet is at the vertically positioned tapped end of the body. The outlet duct is machined to a slightly conical shape into which the conical ram end fits tightly. In open position, before actuation, the flat piston end is located close to the cartridge. When fired, the conical ram end slams into the seat, thus closing the valve against a 1200 psi gas pressure. The gases and frag-

Fig. 4:27 Cartridge-actuated valves (Courtesy of Conax Corp.).

Fig. 4:28 Normally open cartridge-actuated gas valve after firing.

ments from firing the cartridge are fully·contained to prevent contamination of the working gas system.

A normally open cartridge-actuated valve before and after actuation is shown in Fig. 4:29. When fired, the ram with its semicircular-
shaped cutout at the forward end forces a tubular part, which is
clamped between both port nipples, out of the duct and finally closes
the valve. The desirable feature of zero leakage is achieved by the tight
fit of the ram end in the duct, which is similar to a cold-weld joint.

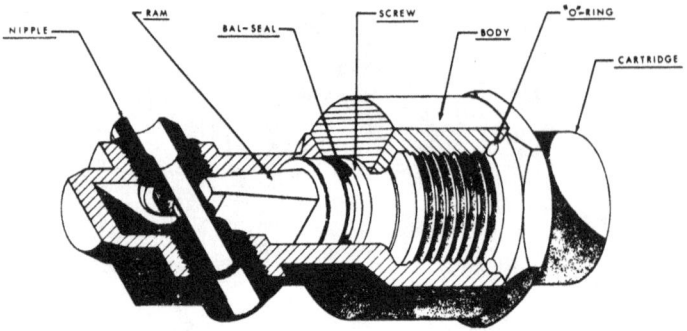

BEFORE ACTUATION

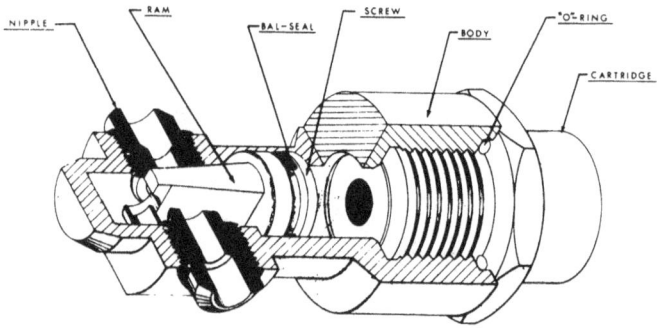

AFTER ACTUATION

Fig. 4:29 Explosive actuated valve normally open (Courtesy of Pyronetics Inc., Subsidiary of Cosmodyne).

In a normally closed valve, as shown in Fig. 4:30, before actuation, the duct is closed by a plug at the end of each port nipple. During actuation, these plugs are sheared off by the forward end of the ram, and a circular opening in the ram lines up with the open duct. By providing a cold-weld type tight fit of the sides of the ram with the ends of the port nipples the feature of zero leakage is also attained in this valve.

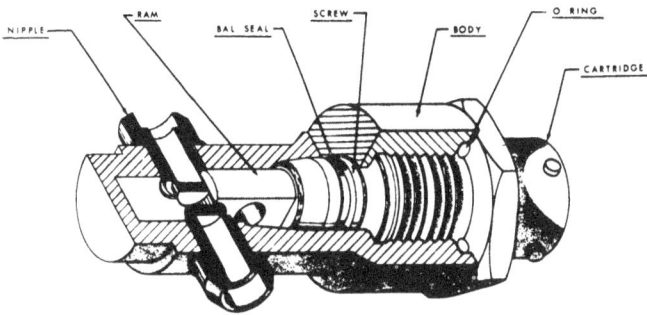

BEFORE ACTUATION

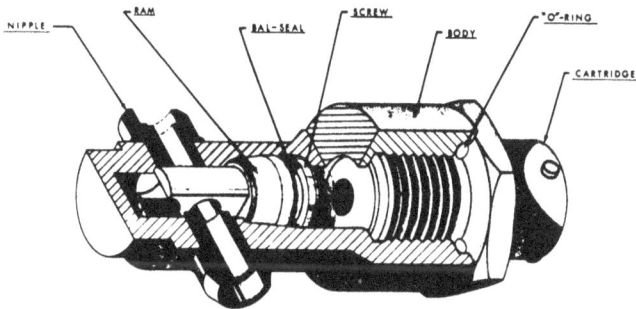

AFTER ACTUATION

Fig. 4:30 Explosive actuated valve normally closed (Courtesy of Pyronetics Inc., Subsidiary of Cosmodyne).

A three-way cartridge-actuated valve, as used for an aircraft emergency hydraulic system, is shown in Fig. 4:31. In the unactuated normal position, the hydraulic fluid flows from the main supply to the aircraft control system, while the secondary hydraulic supply is isolated by a shear nipple. By firing, the main supply is closed, and the control system is connected to the secondary hydraulic supply. This valve is designed for an operating pressure of 3000 psi, and it has a

Fig. 4:31 Three-way cartridge-actuated valve (Courtesy of Pyrodyne, Div. of William Wahl Corp.).

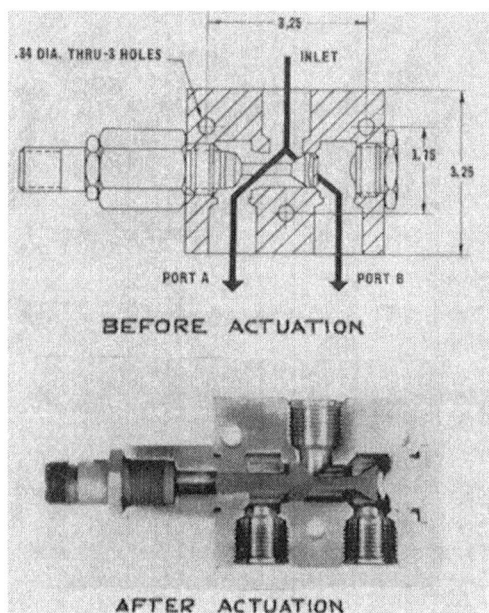

Fig. 4:32 Cartridge-actuated combination valve (Courtesy of Conax Corp.).

response time of less than 3 milliseconds. The valve body is made from an aluminum alloy forging and the internal parts from corrosion-resistant steel.

Another combination valve, as shown in Fig. 4:32, having three ports, redirects the flow of gas, fluid or vapor. Before actuation, the flow from the inlet is unrestricted to port A, while port B is closed. When fired, the diaphragm at port B is sheared, opening the flow to port B and closing port A. The ram end is nose-shaped to provide a tight closure of port A. This valve is designed for an operating pressure of 2000 psi, a proof pressure of 3000 psi, and a burst pressure of 5000 psi. The weight of this valve is 1.97 lbs.

A different type of a combination valve, which provides dual-function flow control, is shown in Fig. 4:33. This device combines both a normally closed and a normally open valve and is equipped with two cartridges. In the pre-actuation condition, flow is restricted

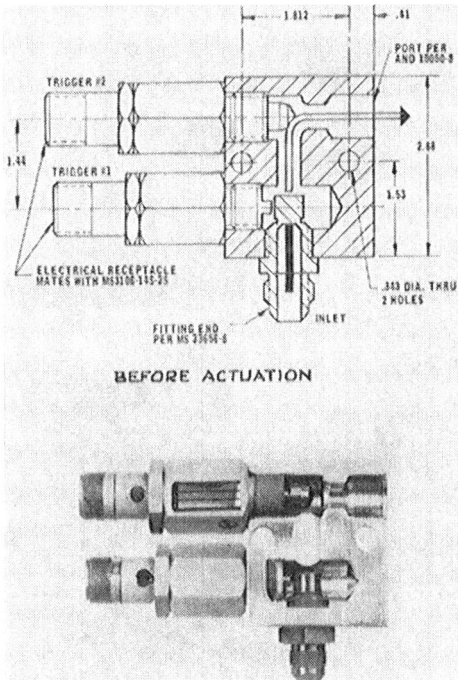

Fig. 4:33 Cartridge-actuated dual-function valve (Courtesy of Conax Corp.).

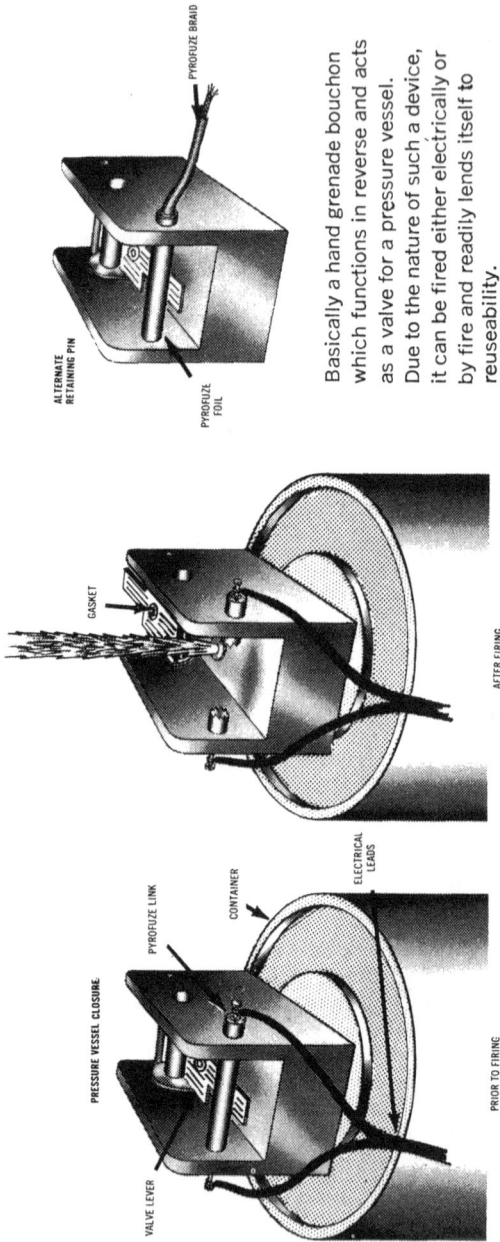

PYROFUZE BRAID

ALTERNATE RETAINING PIN

PYROFUZE FOIL

Basically a hand grenade bouchon which functions in reverse and acts as a valve for a pressure vessel. Due to the nature of such a device, it can be fired either electrically or by fire and readily lends itself to reuseability.

GASKET

AFTER FIRING

PRESSURE VESSEL CLOSURE

PYROFUZE LINK

CONTAINER

ELECTRICAL LEADS

VALVE LEVER

PRIOR TO FIRING

Fig. 4:34 Utilization of "Pyrofuze" materials in normally closed valve device (Courtesy of Pyrofuze Corp.).

in the dead-end passage of the inlet fitting. Firing of trigger cartridge
No. 1 shears off the cap of the inlet fitting, thus allowing flow to the
outlet port. Firing of trigger cartridge No. 2 drives the ram into the
passage opening, thus shutting off the flow.

Cartridge-actuated valves similar to these valves as described and
various unconventional types of valves have been developed for pressures of up to 10,000 psi, and for operation under extreme environmental conditions.

An unusual device which can be used in place of a normally
closed valve, is shown in Fig. 4:34. The valve outlet or container
opening is held closed by a valve lever, supported by Pyrofuze links.
When fired, the link disintegrates, thus releasing the valve lever and
allowing the fluid to flow out of the outlet.

G. *Switches*

Explosive-actuated electric switches find ideal applications for triggering, stage separation, firing delay, safe and arm systems and many
other uses in spacecraft, missiles and underwater vehicles. Most commonly used are miniature switches for closing or opening of circuits
instantly or with predetermined time delays. A typical miniature
switch for four circuits is shown in Fig. 4:35. This switch has a length
of 1.4 inches and a housing width and height of 0.38 inch each, and
its weight is only 20 grams.

Fig. 4:35 Miniature switch for four circuits.

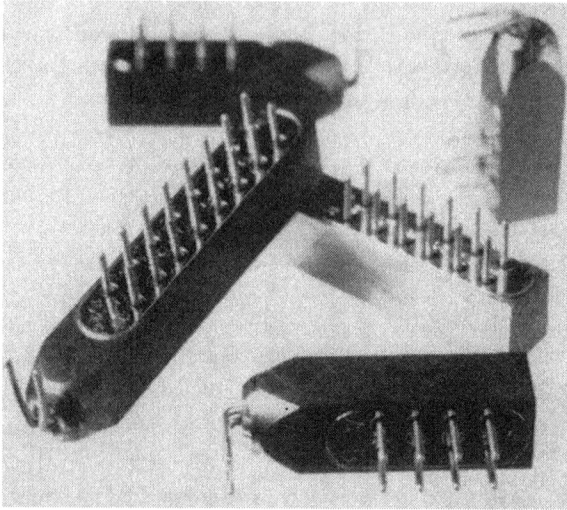

Fig. 4:36 Miniature switches for two, four, six and eight circuits (Courtesy of Atlas Chemical Industries, Inc.).

Miniature switches for two, four, six and eight circuits are shown in Fig. 4:36.

These switches are hermetically sealed in a bronze or aluminum housing. A bridgewire is used to ignite the charge. The expanding gases generated by the ignited charge drive a piston and contact slider forward, thus actuating the normally open or normally closed circuits. Immediately after actuation, a locking recess or an interference fit prevents the slider from moving backwards out of the switching position. This provision results in the capability of the switch to withstand a shock of 2000 G's. The ignition charge in commonly used miniature switches consists of 50 milligrams of LMNR or KDNBF. The function time of these switches is approximately two milliseconds. A graph showing the relationship of function time versus firing current for a miniature switch is shown in Fig. 4:37.

A variety of bridgewires for different all-fire and no-fire currents can be used in these switches. For example, with a bridgewire of 0.33 ohm, an all-fire current of 4.5 amperes and a no-fire current of 1.0 ampere can be used. With a bridgewire of 25.0 ohms, a suitable all fire current is 0.10 ampere and the no-fire current is 0.01 ampere. Insulation resistance ratings of conventional explosive-actuated switches,

KDNBF IGNITION
-54° C TO +71°C

LIMITS FOR .9999 RELIABILITY
AT 90% CONFIDENCE

FUNCTION TIME (MS)
FIRING CURRENT (AMPERES)

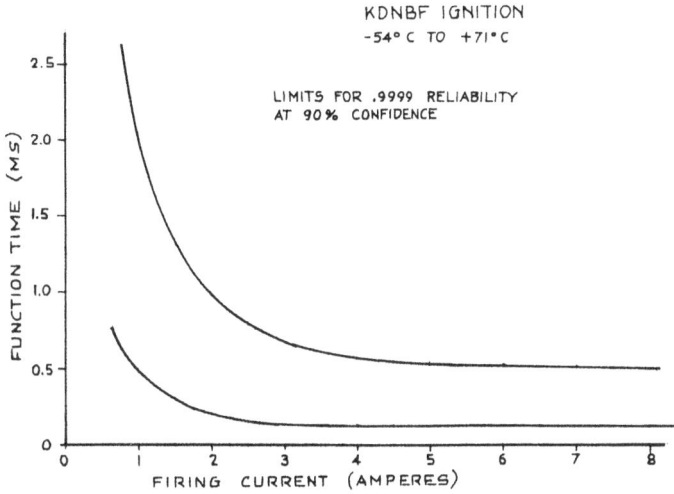

Fig. 4:37 Relationship of function time vs. firing current for miniature switch (Courtesy of Atlas Chemical Industries, Inc.).

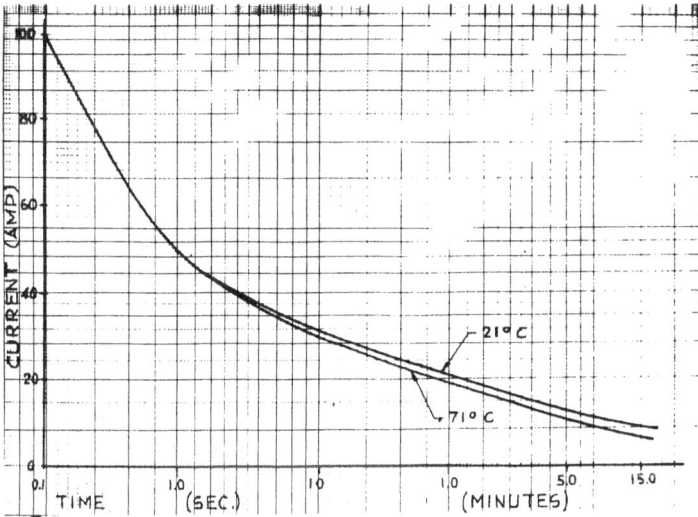

CURRENT (AMP)

21° C

71° C

TIME (SEC.) (MINUTES)

Fig. 4:38 Relationship of current capacity vs. time for miniature switch (Courtesy of Atlas Chemical Industries, Inc.).

113

measured at 500 vdc, are: 500 megohm minimum before firing, 10 megohms minimum after firing, between contact heads and housing, and 10,000 ohms minimum after firing. Conventional miniature switches are designed for a closed contact current capacity of 10 amperes for a time maximum of 6 hours. The relationship of current capacity versus time for a miniature switch is shown in the graph in Fig. 4:38.

Time delays of up to 30 seconds can be obtained in miniature switches by inserting pyrotechnic delay trains between the ignition wire and the charge. For some special applications, miniature switches with built-in time delays of up to 60 seconds have been developed. The length of the pyrotechnic delay train must be sized according to the burning rate of the selected composition. Consequently, time delay switches require a longer housing than instant switches.

Time delays can also be obtained by a separate electronic timer connected to the switch. Usually, a pyrotechnic time delay train is more rugged, whereas an electronic timer is more accurate.

For applications where more than eight circuits must be closed or opened, and where space limitations prohibit the use of an extremely long switch, a switch unit with a housing and contact pin arrangement, as shown in the upper left corner of Fig. 4:39, can be used. The ignition charge in this high-current switch consists of approximately

Fig. 4:39 High-current switches (Courtesy of Atlas Chemical Industries, Inc.).

60 milligrams of LMNR. The length of this switch is 2.25 inches, the width is 1.60 inches, and the height of the housing is 0.51 inch. These high-current switches have a contact current capacity of 12 amperes and of 200 amperes for 100 milliseconds at a temperature of 71°C.

The smallest and lightest explosive-actuated devices are the micro-miniature switches. They are one-shot devices of low power requirement and high reliability. A typical micro-miniature switch, as shown in Fig. 4:40, needs only a 2 millisecond pulse of 1 ampere for actuation, and it has a size of 0.14 inch diameter and 0.56 inch long. Its weight is only 0.6 gram. These micro-miniature switches are ideally used in sequencing and programming devices.

Fig. 4:40 Micro miniature switch.

Programming switches must perform sequential switching operations at predetermined intervals. A great number of poles can be integrated into a switching unit, and very short as well as long time intervals between the first and the last operation, for example, from 50 milliseconds to several minutes, can be attained.

These programming switches are usually custom-designed to specific requirements. Since they are hermetically sealed, combustion products are contained. A typical programming switch, as shown in Fig. 4:41, which has four switching ports, consists of a housing in which lengthwise a delay column containing a powder charge of a constant burning rate is packed. For special requirements, the column may consist of several zones of powder with different burning rates to obtain longer or shorter timing intervals, thus using the available space most efficiently. The squib leads at the right end of the unit shown in Fig. 4:41 connect to a bridgewire which ignites the delay column. On the side of the housing, perpendicular to the delay column, switching ports are provided. A micro-miniature switch is pressed into each port.

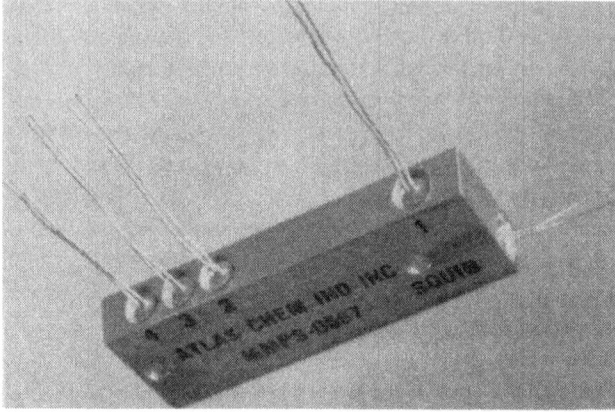

Fig. 4:41 Programming switch. (Courtesy of Atlas Chemical Industries, Inc.).

After firing, as the burning front advances along the delay column, it fires each switch as it passes its lower end. These switches, which are capable of handling 2 amperes, are usually single-pole, normally-open or normally closed switches. In applications where more than one circuit must be switched at a given instant, or where more than 2 amperes must be handled, the programming switch can be used to fire a suitable conventional miniature switch, as described above. Conventional miniature switches have a capacity of 10 amperes, and they are equipped with up to eight poles.

The average time interval of programming switches can be determined as follows:

$$T_\theta = T_{nom} \, 1 - 0.0007 \, (\theta - 70)$$

where: T_θ = average time interval at ambient temperature θ

T_{nom} = average time at $21°C$

Time spread at any temperature θ = $\pm10\%$ of T_θ

The time sequence may begin less than 10 milliseconds after current is applied to the bridgewire. The first pole, which is often a normally-closed contact and which opens at T_θ, is used to disconnect the bridgewire in order to prevent unnecessary drain on the power supply. A current of 1.0 ampere for only 10 milliseconds is required to initiate such a programming switch.

Mounting of explosive-actuated switches usually does not present problems. In applications where switches must be mounted on printed-circuit boards, the contact pins can be passed through holes in the board and are then soldered to the proper electrical wires. Thus, the solder serves not only the purpose of connecting the wires electrically, but also secures the switch to the board.

In some applications, where a time delay switch must be mounted close to a heat source or a heat sink, the device may have to be thermally insulated and calibrated with the insulation. In other cases, it may be necessary to shield the switch from radio-frequency interference.

For emergency systems, such as aircraft seat ejection systems, where electric current for initiation is not available, mechanically-fired ballistic switches have been developed. A typical lanyard-actuated switch has a cylindrical housing which contains a percussion primer, a delay element, a propellant change and a piston. A spring-loaded sear and a firing pin are mounted in the initiator end of the unit. Initiation is accomplished by pulling the lanyard, which releases the firing pin which, in turn, hits the percussion primer. At the end of the delay time, closing of the switch is caused by the generated gas pressure. Generally, delay times from 0.25 seconds to 10 seconds or more can be obtained with these ballistic switches. The weight of a typical ballistic switch is only 4.5 ounces.

5   Explosive Bolts

Explosive bolts are reliable and efficient mechanical fastening devices having the special feature of a built-in release. They are ideally used in spacecraft, missiles, aircraft and underwater vehicle systems, for example for launcher operation, stage separation, release of external tanks, rocket sled release, thrust termination and many other applications.

Generally, an explosive bolt consists of a bolt with a cavity containing a permanent explosive charge or a removable cartridge. At the length where the bolt is intended to break and separate when fired, either a notch is machined outside around the bolt shaft, or the cavity is undercut to weaken it at the predetermined length.

Two basically different types of explosive bolts are commonly used:

1.   Explosive bolts in which the shock wave generated by the detonation of the high explosive is utilized to break and separate the bolt by exceeding its ultimate tensile strength.
2.   Explosive bolts in which the high pressure generated in the cavity of the bolt and acting against the end of the bore is utilized to break and separate the bolt.

The high-explosive type bolt is more generally used than the pressure-type bolt, because it has a higher reliability. Since the breaking characteristics of high-explosive bolts can vary because of the shock wave type of mechanism, depending on the surrounding structure, these bolts are to be tested in fixtures and under load conditions identical to those of the actual application to assure the highest reliability. ·

The pressure-type bolt is less sensitive to the surroundings with regard to changes in break as a function of the applied stresses than the high-explosive type bolt. Other special features of pressure-type bolts which can be utilized in certain applications are minimum swelling, and generation of high ejective forces.

Numerous different shapes and sizes of explosive bolts have been developed for a great variety of applications. In some cases, it is advantageous to use an explosive bolt with a separate cartridge, rather than with a built-in charge. The use of a separate cartridge offers logistic advantage because in an early stage of assembly, the bolt without cartridge can be installed to serve the purpose of connecting the components of the structure, and the cartridge can be stored and handled independently, and it can be installed a short time prior to firing.

Explosive bolts are highly resistant to environmental conditions. Since they have no moving parts, prior to operation, they can withstand high vibration and shock loads without experiencing accidental detonation. Some simple types of explosive bolts are fragmenting when actuated. They can only be considered for crude applications, where thrown-off fragments are not expected to cause damage to structures of equipment. For spacecraft applications, non-fragmenting explosive bolts are generally used to exclude any possibility of damage to the vehicle and equipment. The extent of fragmentation depends to a great extent on the type and degree of confinement surrounding the bolt shaft and on the location of the break line. Properly designed non-fragmenting explosive bolts fired in the totally unconfined condition will produce a "banana-peel" effect when fracturing.

For the firing of explosive bolts in crude applications, "hot-wire" ignition can be used, whereas in most spacecraft applications, the conventional 1-amp 1-watt bridgewire ignition system is used.

Explosive bolts for tensile loads from 1,200 pounds to over 240,000 pounds have been developed. For proper design of explosive bolts, it is important to select the best suited material and a design configuration which will provide the most reliable functioning, preferably without fragmentation. Hardened steel, such as 4340 steel, rather than mild steel, is recommended as material for explosive bolts because it has the advantageous characteristic that it fractures cleanly. However, the degree of steel hardness must be in balance with the required tensile strength and fatigue strength. Generally, a heat treatment to the ultimate tensile strength requires an extremely fine surface finish. At the same time, a high heat treatment can cause fine cracks in the surface which may lead to premature tensile failures of the bolts.

At the desired break area, an undercut or a notch is machined either inside or outside the bolt shaft, as has been mentioned in the introduction to this chapter. To prevent failures, it is recommended to provide a radius in the base of the notch or undercut, rather than sharp corners which would be undesirable stress raisers.

An explosive which has a high detonation pressure and which is to be detonated by a booster is ideally suited as base charge material in explosive bolts. A high detonation pressure of the main charge is obtained by a high pressed density of the charge material. The shape of the detonation wave caused by the charge is greatly affected by the area of initiation. A height-to-diameter ratio of 1:10 of the base charge is recommended for highest effectiveness.

Very careful control is required during the loading of the explosive charge. A too large quantity of explosive material can cause an excessive amount of swelling and "banana peeling" of the bolt shaft at actuation, and a too small quantity of explosive material can result in no separation at all. The ideal quantity will produce a clean break, a minimum of fragmentation and no swelling of the bolt shaft. To determine the optimum charge weight of explosive material for specific explosive bolts, it is recommended to conduct a theoretical analysis using the thermo-hydrodynamic

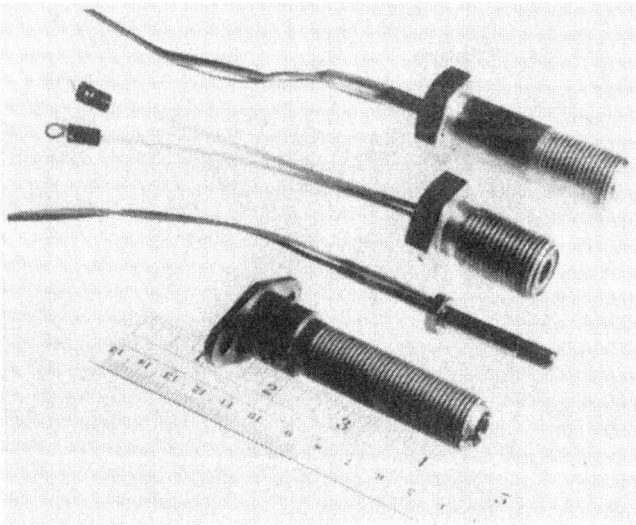

Fig. 5:1 Non-fragmenting explosive bolts for release and separation applications (Courtesy of Holex Inc.).

theory of explosives in conjunction with detonation velocities and Maxwell's theorem for the velocity of molecules.

The explosive bolts shown in Fig. 5:1, which are designed for aircraft and missile release and separation applications, are of the integral charge high-explosive type and utilize the high detonation velocity of an explosive train to generate a shock wave of such a magnitude as to cause the bolt to separate at the predetermined notched or undercut break line. These bolts separate cleanly and produce no fragments.

The firing characteristics of these bolts are: no-fire current = 1.0 ampere, all-fire current = 2.0 ampere, and recommended fire current = 10 or more amperes for each bridgewire. The bridgewires in these explosive bolts have a resistance of 0.18 ± 0.03 ohms, and 0.24 ± 0.03 ohms, respectively. The all-fire and no-fire values represent 99.9% reliability at a 95% confidence level.

The sketches in Fig. 5:2 show some typical applications of this type of non-fragmenting explosive bolts. The dotted line across each bolt shaft indicates the predetermined break line.

A functioning time-versus-firing current curve for explosive bolts of the type, as described above, is presented in Fig. 5:3.

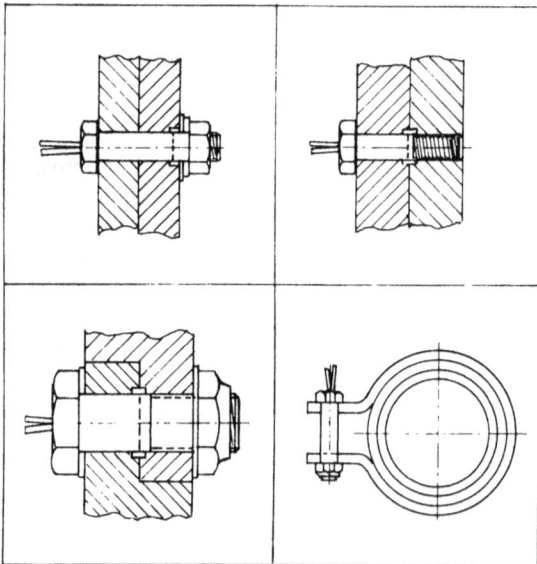

Fig. 5:2 Typical applications of explosive bolts (Courtesy of Holex Inc.).

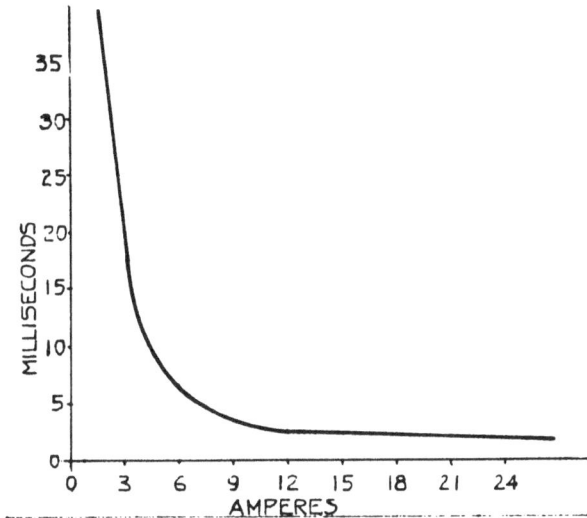

Fig. 5:3 Functioning time vs. firing current for explosive bolt (Courtesy of Holex Inc.).

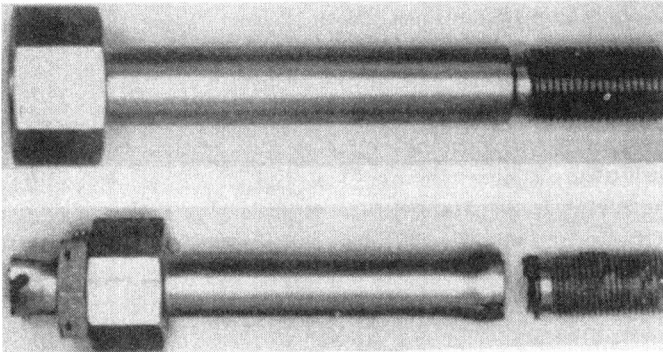

Fig. 5:4 Non-fragmenting explosive bolt before and after firing (Courtesy of McCormick-Selph).

A non-fragmenting explosive bolt before and after firing is shown in Fig. 5:4. A minor swelling and longitudinal cracks near the clean separation are clearly visible in this illustration.

A special type of explosive bolt designed to separate in two pre-determined areas, as used in the launch system of the Titan II launch vehicle, is shown in Fig. 5:5. Both bolt ends up to each notch were

123

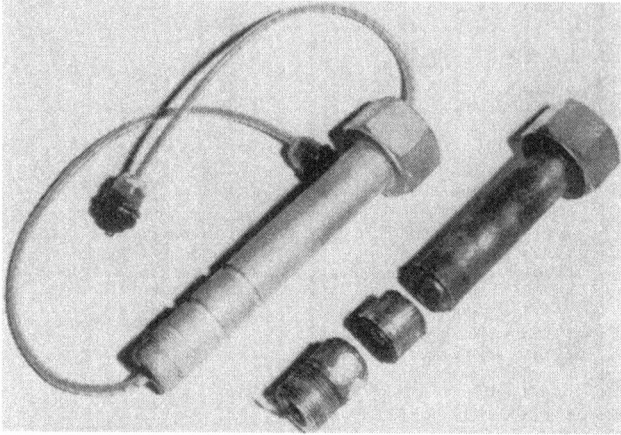

Fig. 5:5 Explosive bolt with two separations (Courtesy of Bermite Powder Co., ORDCO).

machined hollow and were equipped with a charge, whereas the shaft portion between both notches remained solid. Four of these bolts, which are rated at 180,000 pounds tension each, were used as the final link in the sequence of events to the launching of the vehicle. They were used to hold the launch vehicle after engine ignition, while an internal system check was conducted. If all systems were "go", the bolts were fired to separate the vehicle from the launch pad. If the systems were "no-go", the rocket engines were shut down.

A small non-fragmenting explosive bolt, developed for a parachute release system, is shown in Fig. 5:6. The separate cartridge offers the logistic advantage that the bolt can be installed in the assembly without the cartridge, and that the cartridge can be installed a short time prior to firing. As is clearly visible in the drawing of this bolt, the hollow area in the shaft ends at the beginning of the notch, which is the predetermined break area. The photograph of this bolt shows the clean break after firing. The tensile strength of this specific explosive bolt is 30,000 pounds, and the function time is 15 milliseconds.

An unconventional type of explosive bolt is the explosive shear bolt, as shown in Fig. 5:7. This bolt consists basically of two major parts which are held together by a shear pin. The bolt head is larger than a standard bolt head to accommodate the cartridge. Unlike conventional explosive bolts, the explosive shear bolt with its low-strength shear pin does not require the brute force of an explosion to fail high-strength bolt shaft

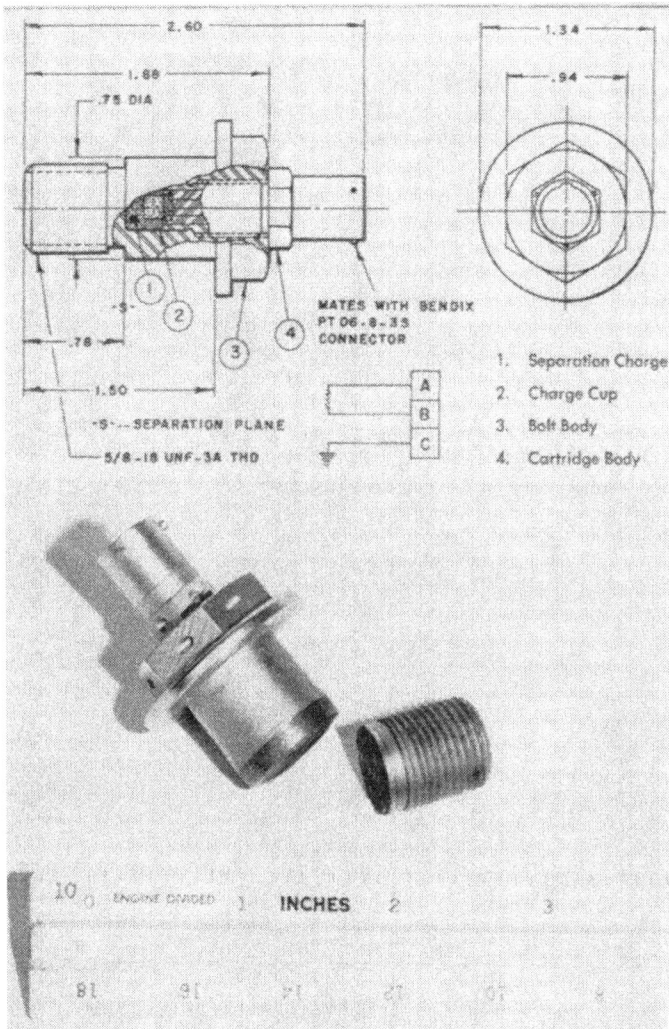

Fig. 5:6 Non-fragmenting explosive bolt with separate cartridge (Courtesy of Bermite Powder Co., ORDCO).

materials in order to cause separation. In the explosive shear bolts, the
separation is accomplished by gas pressure generated by the deflagration of
the cartridge which is mounted in the bolt head.

The explosive shear bolt offers the advantage that deformation of the

125

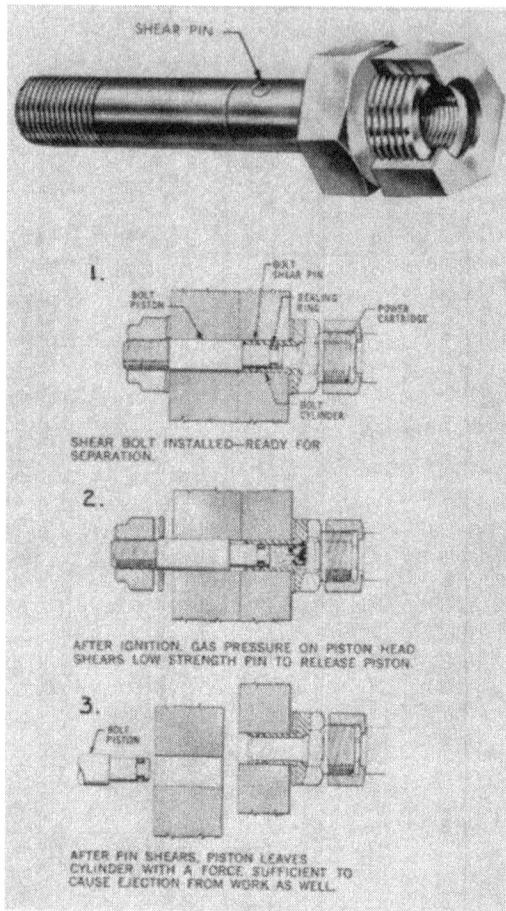

Fig. 5:7 Explosive shear bolt (Courtesy of Hi-Shear Corp., Ordnance Div.).

structure upon separation is prevented because of the lack of explosive force. The cartridge can be removed and reinstalled to facilitate periodic inspections and to prevent possible deterioration of the cartridge. This type of explosive bolt is non-fragmenting, but it separates into two parts, one of which is ejected from the structure.

For special applications, where electric current is not available or where electric initiation cannot be considered, pneumatically-actuated and mechanically-actuated explosive bolts have been developed. In both types,

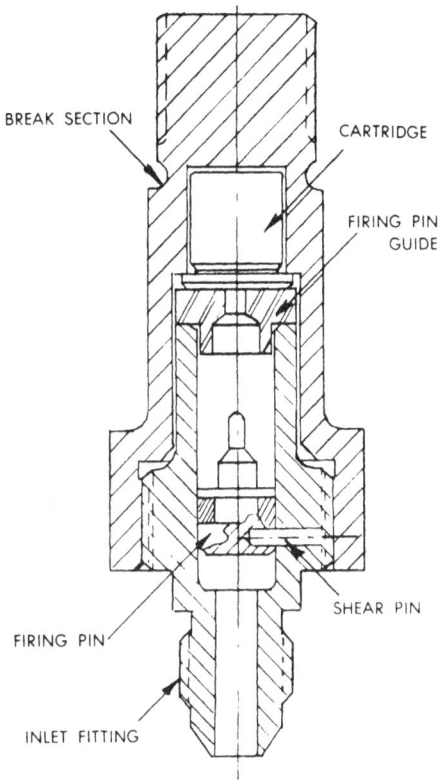

Fig. 5:8 Pneumatic-actuated explosive bolt (Courtesy of Cartridge Actuated Devices, Inc.).

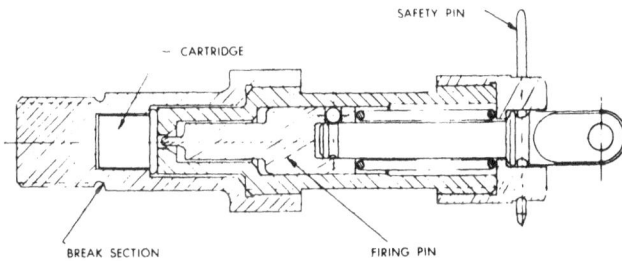

Fig. 5:9 Mechanically-actuated explosive bolt. (Courtesy of Cartridge Actuated Devices, Inc.).

a firing pin mechanism is used to ignite the charge, and a shear pin secures the firing pin in place and thus prevents premature firing.

A cross-section through a pneumatically-actuated explosive bolt is shown in Fig. 5:8, and a cross-section through a mechanically actuated explosive bolt is presented in Fig. 5:9.

6   Explosive Nuts

In many applications it is advantageous to use explosive separation nuts rather than explosive bolts as unfastening devices. Explosive nuts are also known under the term "bolt release mechanisms". Various types of explosive nuts, mainly captive and non-captive nuts have been developed. Non-captive explosive nuts are only recommended for applications where released parts are not a design consideration.

An explosive separation nut usually consists of a cylindrical housing containing a number of thread segments supported by the inner wall of the housing, and a gas-generating "hot-wire" or explosive bridgewire (EBW) cartridge mounted in the closed free end of the housing. In cases where redundancy is required, two cartridges can be mounted in the end of the device. On captive separation nuts, the housing is usually equipped with a flange for mounting the device on one half of the structure to be separated, whereas non-captive nuts have no provisions for flanged mounting. In most explosive nuts, the thread segments made from stainless steel provide the capability of carrying full torque and tension loads of wrenching bolts heat treated to 160,000 − 180,000 psi ultimate tensile strength.

Major advantages of explosive separation bolts are: Positive bolt ejection, short release time, low power requirement, low reaction force affecting the structure, and containment of thread segments and gas. Separation and release are accomplished with the deflagration of a rather small propellant charge, unlike some explosive bolts that require an impact-type high-explosive charge to fail the fastener material. Explosive separation nuts eliminate the need for custom-developed devices, such as explosive bolts with specific grip lengths, large bolt heads and enlarged shaft diameters, notches or other "weak links" machined into the shaft, and special hole preparation requirements. In some explosive nuts, a Teflon or Nylon insert provides self-locking feature of the thread and a gas-tight seal between thread segments, bolt and housing. The average release time for explosive nuts is 3 milliseconds. With some separation nuts a release time of 1.6 milliseconds has been achieved.

In early types of captive explosive separation nuts, which retained the thread segments, a cam action at separation was employed to disengage the nut thread segments from the bolt, and a shroud was provided to capture the threaded segments after separation. The shroud made the device very bulky, and because of the long movement, an undesirably high impulse was transmitted to the structure.

In a greatly improved type of explosive separation nut, a system of sliding pistons and cylinders is utilized to disengage the thread segments, which requires a very short movement of the internal parts. Thus, the energy requirement and the release time are substantially reduced, and minimum reaction impulses are transmitted to the structure. A reduction of the required energy and of the release time of over 50% has been achieved in these improved explosive nuts as compared to the early conventional devices.

In general, two basic types of flanged captive explosive nuts are used:
  a.   Bolt-torquing nuts, having an integral flange base,
  b.   Nut-torquing explosive nuts, having a removable flange base.
Non-captive explosive separation nuts are available as bolt or nut-torquing types.

In special applications, explosive nuts can be made integral parts of the structure. After having been actuated, some types of explosive nuts can be reconditioned for reuse, mainly for ground testing.

The scheme of operation of an explosive-actuated nut separating from a standard bolt is shown in Fig. 6:1. A very short release time and reusability are advantageous features of this separation nut. The release time of this device is two milliseconds. It is available in sizes to accomodate bolt threads from 1/4 inch to 1 1/2 inches, and it functions reliably in temperature ranges from $-62°C$ to $+150°C$.

Non-captive explosive separation nuts and a high-strength internal wrenching bolt, as used with these nuts, are shown in Fig. 6:2.

The cross-sectional illustration in Fig. 6:3 explains the functioning sequence of a captive explosive nut.

In order to retain the bolt after separation and to prevent potential damage to the structure or equipment by ejected bolts, bolt catchers are

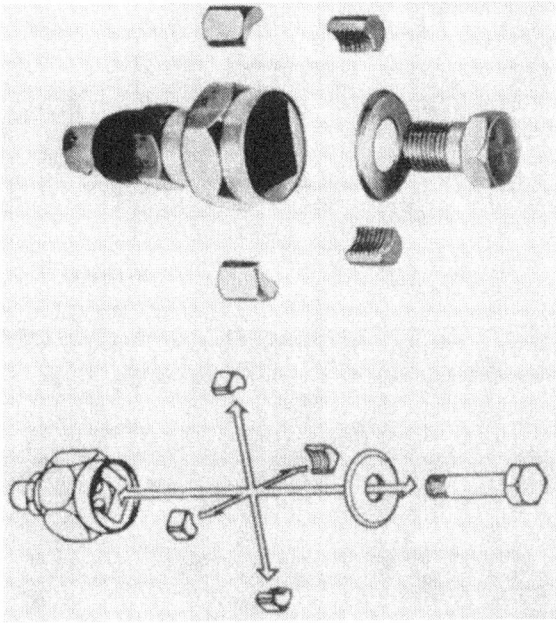

Fig. 6:1 Operation scheme of an explosive-actuated nut (Courtesy of Conax Corp.).

Fig. 6:2 Non-captive explosive nuts and an internal wrenching bolt (Courtesy of Hi-Shear Corp.).

131

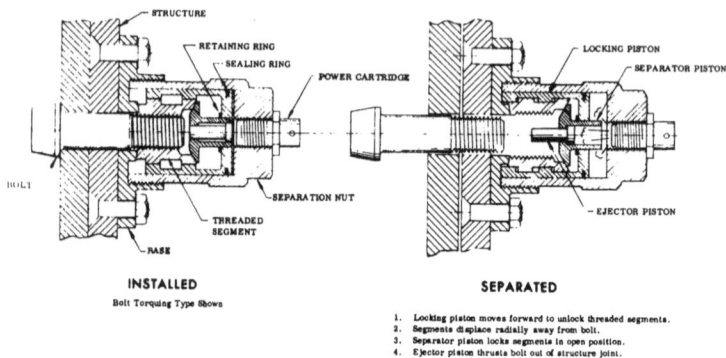

Fig. 6:3 Functioning sequence of a captive explosive nut (Courtesy of Hi-Shear Corp.).

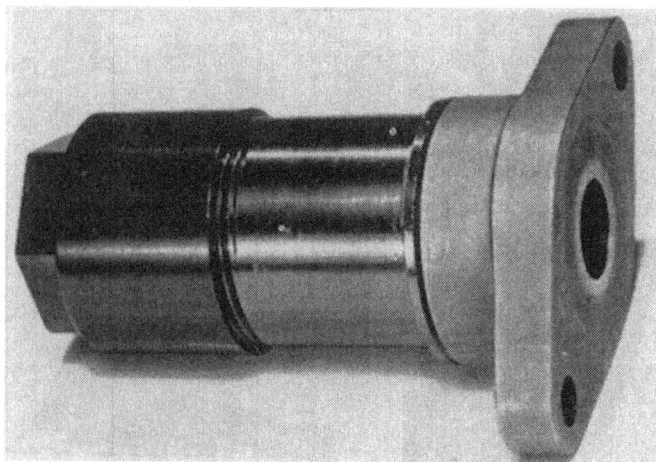

Fig. 6:4 Bolt catcher (Courtesy of Hi-Shear Corp.).

used in conjunction with explosive separation nuts and separation bolts. The bolt catcher is mounted at the bolt-head side of the structure to be separated. A set of leaf springs mounted inside the bolt catcher holds the ejected bolt in its end position, absorbs the shock imparted by the bolt and prevents the bolt from rebounding into the hole and from becoming trapped by structural parts. The bolt catcher can be refurbished for reuse by replacing the cap and sleeve. A typical bolt catcher, as described, is shown in Fig. 6:4. Access for refurbishing of this bolt catcher is provided by a threaded cap.

The functioning sequence of a structure separation utilizing an explosive separation nut and a bolt catcher is shown in Fig. 6:5.

A separation nut and bolt catcher application as used for fairing separation of the Orbiting Astronomical Observatory (OAO) is presented in Fig. 6:6. Twenty-two separation nuts and bolt catchers were arranged peripherally and longitudinally along the edges of the fairing halves to provide separation and bolt containment as the fairings were jettisoned from the spacecraft.

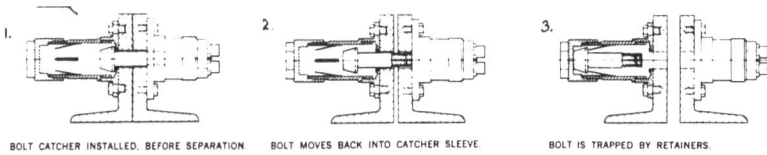

BOLT CATCHER INSTALLED, BEFORE SEPARATION.   BOLT MOVES BACK INTO CATCHER SLEEVE.   BOLT IS TRAPPED BY RETAINERS.

Fig. 6:5 Separation sequence utilizing explosive nut and bolt catcher (Courtesy of Hi-Shear Corp.).

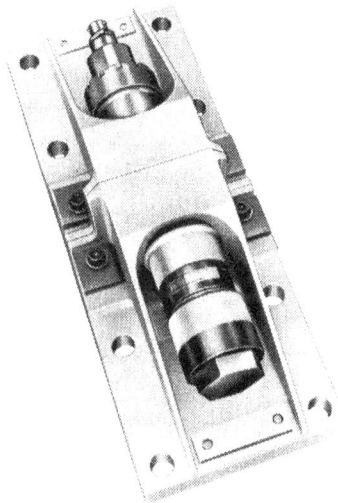

Fig. 6:6 Explosive nut and bolt catcher for fairing separation (Courtesy of Hi-Shear Corp.).

133

7   Release Mechanisms

Pyrotechnics technology is ideally employed in a great variety of release mechanisms in aircraft, spacecraft, missile and underwater applications. As in most other devices described in this book, of greatest advantage in these release mechanisms are: high power-to-weight ratio, small size, low power requirements and high reliability of pyrotechnic devices.

Pyrotechnic release mechanisms find wide use in the form of clamp separators, bolt releases, rod and cable separators, latch separators, and explosive links. Numerous unconventional types of release mechanisms have been developed for special applications. The most important conventional release mechanisms and some unusual separation devices are described in this chapter.

A. *Clamp Separators*

Supply lines, umbilicals, container covers, tanks and similar equipment can easily be opened, separated and released by clamp separators. Until actuated, in closed position, a clamp separator functions as a fastener. Gas pressure generated by a power cartridge is utilized for actuation in most conventional clamp separators, in which the sliding piston principle is employed to release the tension on the clamp. When actuated, the clamp separator imparts a separation impulse to the band or clamp to release it from the structure.

According to the tension load rating, various types of tension release methods are used in these clamp separators: shear pin release, separation groove release, and ball detent release. The shear pin release is generally used in clamp separators which are rated below 3,500 pounds tension load, while the ball detent or separation groove release system is employed in clamp separators rated above 3,500 pounds tension load.

For applications where containment of the combustion products is required, the piston is retained in the clamp separator body, thus containing the combustion products, while for application, where release of the combustion products can be tolerated, a clamp separator which releases the piston can be used. The release schematics of contained-piston type and free-piston type clamp separators are shown in Fig. 7:1.

It is an advantageous feature of the clamp separator, as shown, that it can be installed without the cartridge to reduce hazardous conditions. To facilitate periodic inspections and to prevent possible deterioration, the cartridge can be removed and reinstalled at any time. A normal release time of these clamp separators is 5 to 10 milliseconds.

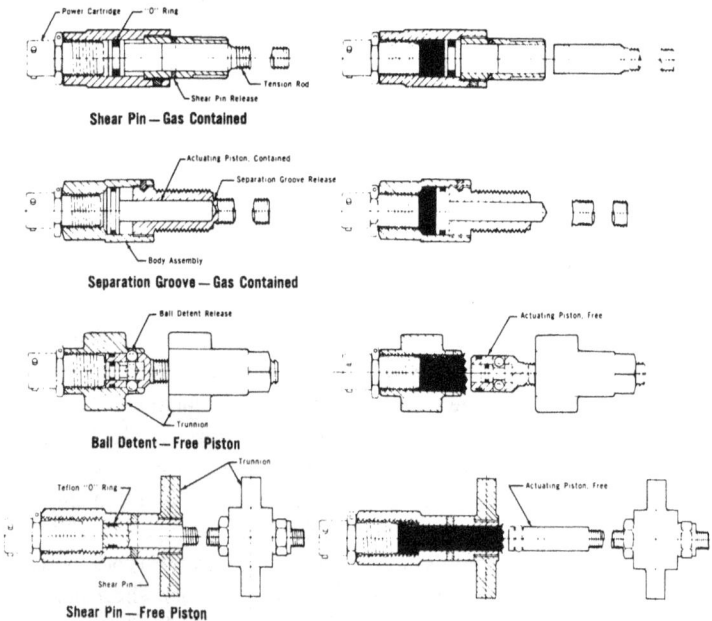

Shear Pin — Gas Contained

Separation Groove — Gas Contained

Ball Detent — Free Piston

Shear Pin — Free Piston

Fig. 7:1 Release schematics of clamp separators (Courtesy of Hi-Shear, Corp.).

A vent line coupling with an explosive latch for emergency venting of missile tanks, as used in the Titan II ground equipment, utilizing two clamp separators, is shown in Fig. 7:2.

An unconventional type of clamp separator, as shown in Fig. 7:3,

Fig. 7:2 Explosive latch (Courtesy of Aeroquip Corp., Marman Div.).

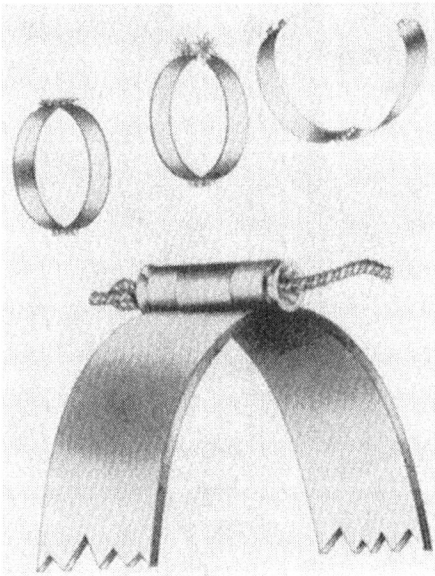

Fig. 7:3 Clamp separator with "Pyrofuze" braid bolt (Courtesy of Pyrofuze Corp.).

utilizes a Pyrofuze braid in a hollow bolt, and a Pyrofuze braid initiator as a release device. The concept presented in Fig. 7:4 shows the feasibility of using a double-shear bolt consisting of a high-strength aluminum alloy case containing a scroll of Pyrofuze foil and Pyrofuze filament, which passes from one bolt end to the other to ignite the scroll.

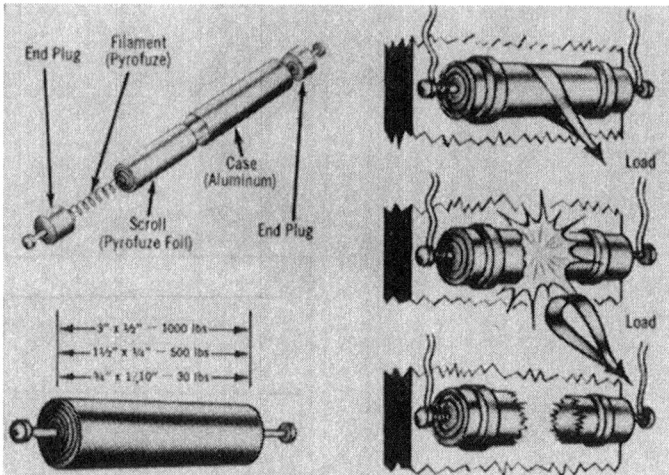

Fig. 7:4 "Pyrofuze" bolt for clamp separators and similar devices (Courtesy of Pyrofuze Corp.).

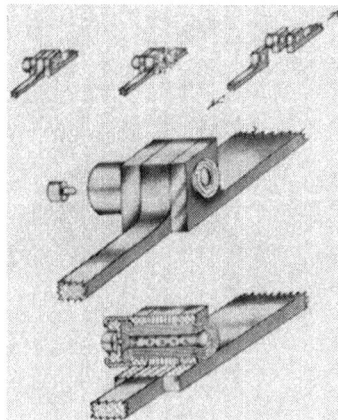

Fig. 7:5 "Pyrofuze" clamp bolt with percussion primer (Courtesy of Pyrofuze Corp.).

138

A different type of Pyrofuze clamp bolt is shown in Fig. 7:5. The bolt consists of a scroll of Pyrofuze foil which is initiated by a percussion primer. All these Pyrofuze clamp separation devices can only be considered for applications, where complete containment of gases and other combustion products is not required.

B. *Rod Separators and Tension Release Devices*

For applications, where a structural component, such as a rod, a cable, or a link under tension must be separated, a variety of release devices has been developed. A simple and small tension release device is the explosive stud, as shown in Fig. 7:6. The threaded ends of the stud are used to join two structural components, which are to be separated later. When actuated, the stud separates in the recessed area, releasing the solid stud end which does not contain electrical leads. The explosive stud, as shown, has a tensile strength of 10,000 pounds.

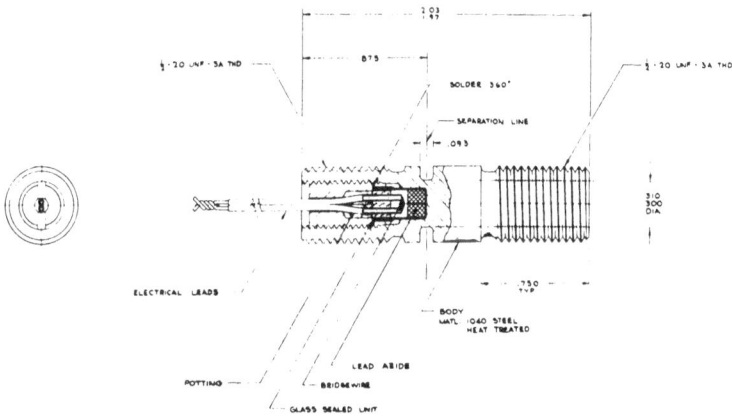

Fig. 7:6 Explosive stud (Courtesy of SDI Special Devices Inc.).

Another small tension release device, the explosive link, as shown in Fig. 7:7, has the advantage of simplicity in design and in application. Two sturdy lugs are provided for connecting the structural components to be separated. Such explosive links are frequently used for parachute release, door release and similar applications.

The tension release device, as shown in Fig. 7:8, was developed for spacecraft applications and is equipped with two cartridges to provide redundancy. It is designed to contain the combustion pro-

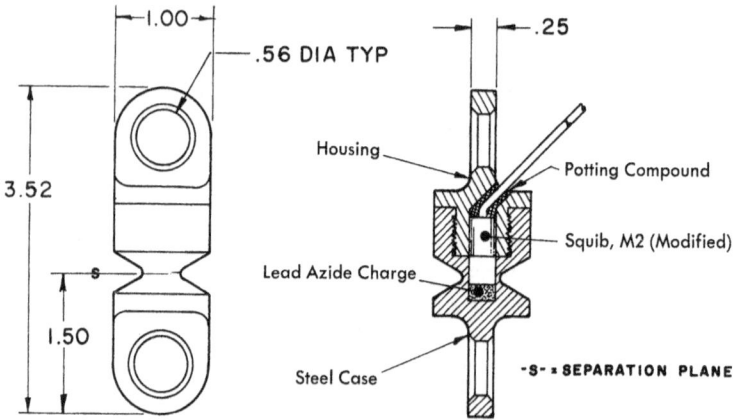

Fig. 7:7 Explosive link (Courtesy of Bermite Powder Co., ORDCO Div.).

Fig. 7:8 Tension release device with two cartridges (Courtesy of Harvey Engineering Laboratories).

ducts and fragments, and it has the advantageous feature of being reusable by replacing the expendable crushed chuck and the cartridge. More than ten firings can be expected with one release device of this type. The unit shown has a 3/8 inch diameter, a length of approximately 2.75 inches, and a load capacity of 8,000 pounds. Its weight excluding the cartridge is 0.25 pound.

Fig. 7:9 Release bolt for recovery canister (Courtesy of Holex Inc.).

An unconventional explosive-actuated separation bolt, as developed for release of the recovery canister from the Gemini spacecraft, is shown in Fig. 7:9. The bulky portion of the housing contains a piston which, after firing of the cartridge, is thrust forward toward the threaded end of the device. This causes a shearing-out of the bottom of the hollow cylinder at the recess, thus separating the bolt from the cylinder portion of the device. Several of these separation bolts of 3/8 inch diameter were used to mount the recovery canister on the forward bulkhead ring of the spacecraft. All separation bolts were actuated simultaneously. Separation occurred in the recessed area. The small bolts remained with the separated canister, while the cylinder portions stayed on the bulkhead ring on the spacecraft. These bolts were designed for a tensile strength of 7,100 pounds and for a shear strength of 6,000 pounds. The length of this bolt is 3.78 inches, and its maximum diameter is 1.25 inches.

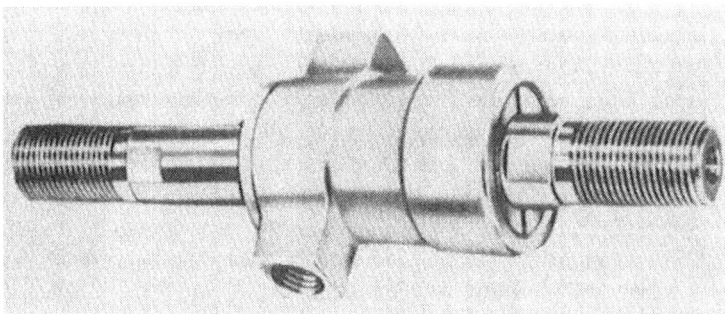

Fig. 7:10 Tension rod separator (Courtesy of Hi-Shear Corp.).

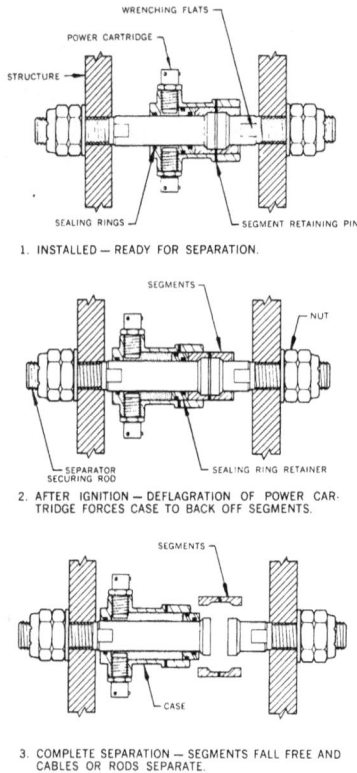

1. INSTALLED — READY FOR SEPARATION.

2. AFTER IGNITION — DEFLAGRATION OF POWER CAR-TRIDGE FORCES CASE TO BACK OFF SEGMENTS.

3. COMPLETE SEPARATION — SEGMENTS FALL FREE AND CABLES OR RODS SEPARATE.

Fig. 7:11 Release sequence of tension rod separator (Courtesy of Hi-Shear Corp.).

For applications, where a rod or cable under tension must be separated, special tension rod separators, designed to be an integral part of the rod or cable assembly to be separated, have been developed. Such separators are available for a tension load range from 5,000 to 55,600 pounds and in thread sizes from 1/4 inch to 3/4 inch. A typical tension rod separator equipped with two cartridge ports is shown in Fig. 7:10. For actuation, comparatively low pressure gas generating cartridges are used. The weight of these tension rod separators is reasonably low. The unit weight excluding the cartridge of a 1/4-inch rod separator is only 0.20 pound, while the weight of a 3/4-inch device is 1.20 pounds. A release sequence of a rod separator is shown in Fig. 7:11.

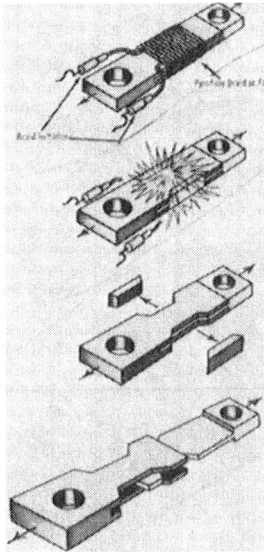

Fig. 7:12 Separation sequence of "Pyrofuze" release link (Courtesy of Pyrofuze Corp.).

An unusual tension release device, as shown in Fig. 7:12, utilizes Pyrofuze braid or foil and Pyrofuze initiators. The slotted joint is held in closed condition by two cam lock blocks, which are restrained by windings of Pyrofuze braid. When ignited, the restraining windings of braid burn off rapidly and release the cam lock blocks, thus separating the joint. A release device, as shown, with a cross-sectional size of 1.0 × 1.125 inch was tested under a tension load of 3,500 pounds. During destructive testing, failure occurred at the lug ends, and not in the Pyrofuze section. Joints of this type for a restraining capability of more than 90,000 pounds have been developed.

Pyrofuze release devices of this type were used to retain intact a multi-segmented re-entry shape during boosted re-entry from space. Upon actuation during re-entry, all units had to function reliably and expose a final fixed surface.

C. *Ball Release Mechanism*

In various separation devices, the ball release system is utilized advantageously. A typical ball release mechanism, as presented in Fig. 7:13, does not fragment during separation, and retains all combustion pro-

143

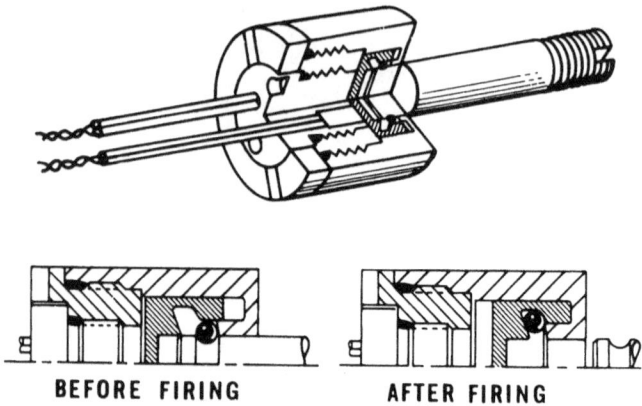

BEFORE FIRING   **AFTER FIRING**

Fig. 7:13 Ball release mechanism (Courtesy of Conax Corp.).

ducts so that it can be used near delicate equipment and in an explosive atmosphere.

As the cross-section of this device shows, before firing, the separable rod is locked in the housing of the device by means of steel balls engaged in a semi-spherical groove near the end of the rod, being held in this position by a small diameter bore of a piston. When actuated, the generated gas pressure drives the piston toward the rod, allowing the steel balls to engage in a large diameter groove in the piston, thus unlocking the rod and separating it.

A normal functioning time of this ball release mechanism is 5 milliseconds. According to requirements, these devices can be equipped with single or dual primers and with either lead wires or standard connector terminations. Ball release devices of the type as described are available in bolt sizes from 1/4 inch to 3/4 inch. The operating loads of these release mechanisms are equivalent to the rating of standard AN bolts.

D. *Parachute Release Mechanisms*

Since parachutes used for aerial delivery, final-stage recovery and personnel descent remain inflated at low wind velocities and tend to drag and overturn the load or the jumper after ground impact, the parachute canopy must be separated from the load by means of suitable release devices. In aircraft deceleration and drone recovery parachute systems, the parachute canopy is separated from the vehicle after the landing deceleration or recovery operation has been accomplished.

For personnel parachutes, manually-operated release devices are used, while squib-actuated release devices are mainly used in aerial delivery, aircraft deceleration and stage recovery systems, because squib actuation is highly reliable and the best suited method for accomplishing the release. In parachute release systems which must be actuated at touchdown, the squibs are fired by actuating an electrical switch through the push-action of a feeler wire, which projects a short distance below the bottom of the vehicle. In parachute release systems which must be operated during descent, barometric pressure switches or timers are used to actuate the electrical-explosive or mechanical release device.

The most important types of squib-actuated parachute release devices are: Latch-type, hook-type, and swivel-type disconnect devices.

. Latch-type disconnect devices

Most latch-type disconnect devices consist of a bracket in which a swing-arm type latch is mounted in lugs and held closed by a piston or plunger. One half of the piston length is engaged in a cylindrical cavity in the bracket and one half in a similar, but longer cavity in the swing arm. The latch bracket is attached to the load, while the swing arm is attached to the parachute. When the cartridge or squib is actuated, a shear pin locking the piston in the housing is sheared, and the piston or plunger is forced by the generated high-pressure gases into the cylindrical cavity in the swing arm, disengaging from the cavity in the bracket, and thus releasing the parachute. A simple latch-type parachute disconnect device, which is designed on this principle, is shown in Fig. 7:14.

A different, more complex parachute latch release device, designed on the locking hook principle, is presented in Fig. 7:15. The latch is locked in closed position by a rotating hook and secured by a shear pin across the upper forward hook end and the yoke-shaped bracket. When actuated, the vertically-located piston is driven downward by the generated high-pressure gases, breaking the shear pin, and subsequently forcing the hook to rotate into the open position, thus releasing the latch.

145

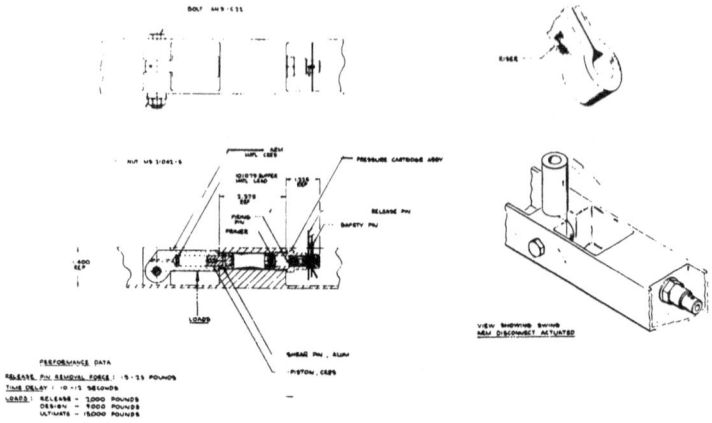

Fig. 7:14 Simple latch-type parachute disconnect device (Courtesy of SDI Special Devices Inc.).

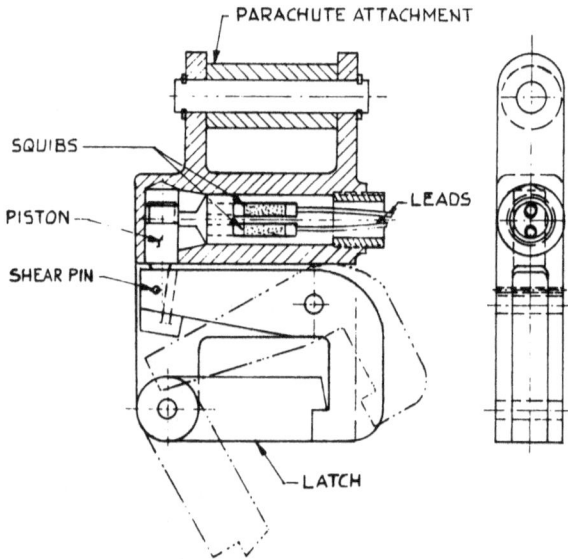

Fig. 7:15 A latch-type parachute release device.

2. Hook-type disconnect devices

Generally, hook-type parachute release devices consist of a lug-shaped bracket with a rounded hook mounted so as to allow rotation in the lower portion of the bracket. A compression spring mounted in the bracket holds the hook in closed position by exerting pressure against a cam extension on the upper part of the hook. When actuated, the high-pressure gases force a piston to move upward against the spring, which, in turn, releases the cam from the locked position, allowing the hook to rotate to the open position and to release the ring which is used to attach the load to the hook.

3. Swivel-type disconnect devices

    For applications where either the parachute canopy or the load must be allowed to rotate, a swivel-type disconnect device is ideally used. A small swivel disconnect device in which the ball release principle is employed, is shown in Fig. 7:16. In closed position, both halves of the

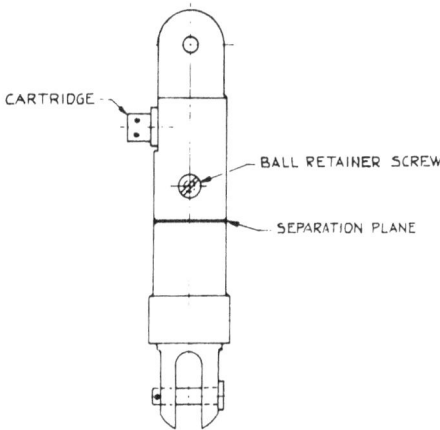

Fig. 7:16 Swivel-type parachute disconnect device.

device are held together by a number of steel balls located in holes in the wall of the upper half. The balls are held in locked position against the lower half of the device by a piston which is secured by a shear pin. When actuated, the high-pressure gases move the piston, breaking the shear pin and reaching a position in which a portion of the piston having a reduced diameter allows the steel balls to move inward, thus releasing the lower half of the device with either the parachute canopy or the load.

For applications where axial separation of a load must be ac-

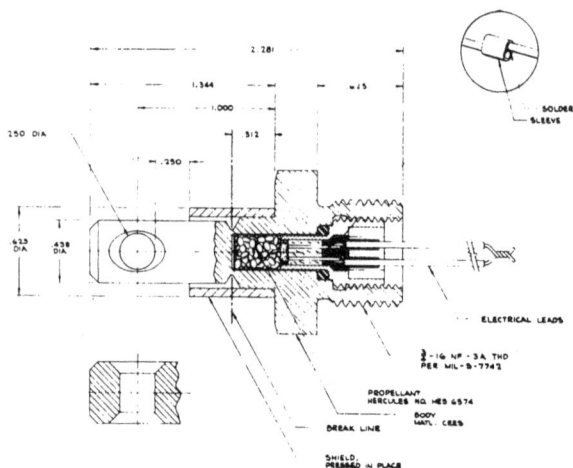

Fig. 7:17 An axial release device for a parachute or ballast (Courtesy of SDI Special Devices Inc.).

complished, but where provision for swivel movement is no requirement, simple axial release devices have been developed. A cross-section through a small and simple axial release device for parachute or ballast disconnect is shown in Fig. 7:17. Its design is based on the principle as used in various explosive bolts. When actuated, the cylindrical lug end of the device is separated from the body, breaking off in a recess machined around the lug near the bottom of the cavity provided for the charge. A tubular shield for protection from fragments and combustion products is attached to the body around the break area. This release device is designed for a load of 500 pounds.

In air-drop systems, mechanical and explosive-actuated release devices are used. Major requirements for disconnect devices for these systems are a high degree of reliability of release at ground impact and avoidance of premature mid-air release during parachute deployment and descent. To prevent premature load release, integral time-delay devices are frequently used. Various cargo-parachute release systems have been developed, based on the load-stress relaxation principle. In most of these systems, pyrotechnic delay devices have been employed to prevent premature release during parachute deployment. In some air-drop release systems, reefing line cutters are utilized as time-delay devices. In these cutters, the pyrotechnic delay trains are initiated by lanyards attached to the firing wires at the start of deployment. After the delay of several seconds, the cutters fire, severing holding cords and allowing the cutter housings to drop out of their sockets, thus freeing the release arms of the disconnects. The loads are held in place by their own weight exerted on the attachments to the release devices. At touchdown, as this force is relaxed, the cargo suspension sling attachments are freed from the release device, thereby separating the parachute canopies from the load.

8   Gas Generators

Gas generators are pyrotechnic devices which produce gases at high pressure for a specific duration and of a predetermined volume, as needed to energize various control, servo, mechanical and electrical systems and devices, such as gyroscopes, hydraulic pumps and turbine starters, to inflate flotation bags, balloons, emergency escape slides and ejection devices, to pressurize hydraulic accumulators, liquid-propellant systems and fire extinguishers, to ignite liquid or solid propellants, and to energize electrical batteries.

Generally, a gas generator consists of a pressure chamber, a gas-producing propellant charge, an ignition system, a nozzle and auxiliary components as required for the desired output of work. The gas-producing charge can have the form of a billet or grain of cast or extruded propellant. The igniter charge usually consists of an ignition primer and a booster. When the ignition primer is fired, it ignites the booster charge which causes the pressure and temperature inside the gas generator to rise to the operating level of the gas-producing propellant, resulting in the ignition of the propellant charge, which then burns at its characteristic rate and temperature, and thus, generates the required volume of gas at the predetermined pressure.

Gas generators offer the advantage over heavy and bulky stored pressurized gas systems, gas compressors or liquid hydraulic systems, in that they are of very small size and light weight, and that they function reliably, which are very important characteristics in aircraft, spacecraft, missile and underwater vehicle applications. Another considerable advantage of gas generators, as compared with stored pressurized gas systems, is the greater safety of gas generators, because their pressure, which is developed only during firing, is usually lower than that of stored gas systems.

Gas generators have the minor disadvantage that they can be used for only one shot. However, most gas generators can be refurbished easily by cleaning and replacing the propellant charge and the igniter.

Two basically different types of solid-propellant gas generators are used: hot-gas generators and cool gas generators. Both types are classified according to their operating gas temperature. The temperature range of hot-gas generators is from 980°C to 2760°C, while the gas temperature of cool-gas generators ranges from −37°C to 980°C.

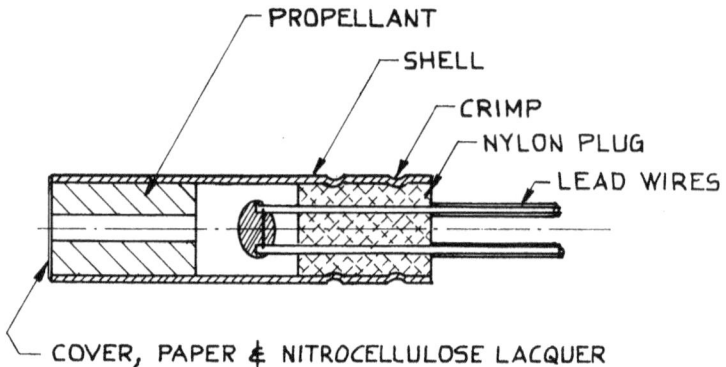

Fig. 8:1 A simple hot-gas generator (Courtesy of Atlas Chemical Industries, Inc.).

A simple hot-gas generator of very small size is shown in Fig. 8:1. It consists of a cylindrical shell which contains a grain of propellant and an ignition device that ignites the propellant when electrically initiated. Gas generators of this type are mainly used for activation of batteries and for inflation of flotation and ejection devices and balloons, and for mechanical actuation. The gas generator, as shown, is designed to displace 100 cc of water or similar liquid in 2.5 seconds. It has a length of 0.80 inch, a diameter of 0.217 inch, and its weight is only 4.4 grams.

Besides electrical initiation, manual firing mechanisms of the lanyard type and pneumatic firing mechanisms are sometimes used, especially in applications where electrical current is not available.

Various types of propellants, such as single-base, double-base and composite double-base propellants in precise formulations are used in gas generators to develop hot gases at predictable and reproducible rates.

An extremely high thermal stability is required for gas generator charges as used in jet turbine starters, where external skin temperatures of about 260°C are experienced. A special type of charge material, the HES type propellants, has been developed for these special applications. They are capable of withstanding environmental temperatures of 177°C to 260°C for four hours. These propellants are available in a castable form, which is ideal for use in gas generators. The thermodynamic properties of these HES propellants are listed in Table 8:1.

Table 8.1

HES Propellants
Thermodynamic Properties

Calculated Properties	HES Propellant Number			
	5250	5808	6574	6478
Specific Impulse I_{sp} (1000 to 14.7 psi) lb-sec/lb	235	231	185	226
Chamber Temperature T_c, °K	2679	2775	2671	220
Moles gas/g, n_g	0.0403	0.0347	0.0229	0.0474

The high efficiency of a typical hot-gas generator, as used in a turbine starter, is illustrated by the following example: The starter consists of a solid-propellant, double-end burner cartridge which fuels a gas generator used to produce the required pressure. The gas generator has a diameter of about 1.9 inches and a length of about 7.7 inches, including the 2.2-inch long nozzle, and it weighs 1.77 pounds, of which the propellant material is 0.273 pound. The device containing this small amount of propellant delivers 133.7 gas horsepower, produces 14 cubic feet of gas at a temperature of 21°C and 1 atm pressure, and it develops a maximum chamber pressure of 2100 psi at 73°C ambient temperature, 1600 psi at 21°C, and 1000 psi at −54°C.

A larger hot-gas generator, equipped with a removable initiator cartridge, is presented in Fig. 8:2. The cross-section shows clearly the arrangement of the igniter and propellant charges and of the nozzle.

Cool-gas generators are mainly used in applications where the gas temperature must be compatible with the materials used in the systems, as

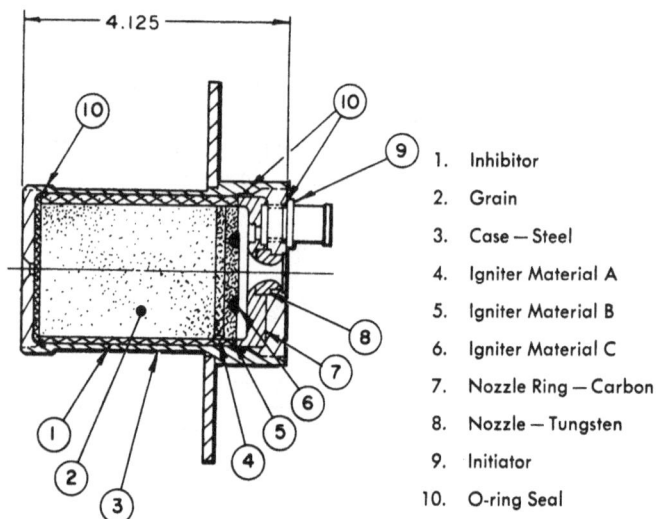

1.	Inhibitor
2.	Grain
3.	Case — Steel
4.	Igniter Material A
5.	Igniter Material B
6.	Igniter Material C
7.	Nozzle Ring — Carbon
8.	Nozzle — Tungsten
9.	Initiator
10.	O-ring Seal

Fig. 8:2 A hot-gas generator (Courtesy of Bermite Powder Co., ORDCO Div.).

for example in a flotation system consisting of rubberized fabric which can withstand only a rather low maximum temperature.

A cool-gas generator consists of a shell containing solid propellant for gas generation and pressurization, a coolant for gas temperature control, and an electrical initiator. After initiating, the burning propellant develops hot gas which is mixed with the coolant after rupture of a diaphragm that isolates the stored propellant and coolant.

A cool gas generator, as used to inflate a flotation system, is shown in a cutaway view in Fig. 8:3. The device consists of a large sphere which contains a main chamber charged with carbon dioxide (CO_2) and a small quantity of ethyl alcohol, and a smaller chamber filled with a clean-burning solid propellant. The propellant chamber is sealed off from the main chamber by a propellant burst disc. Burning of the propellant is initiated by the electrically-actuated igniter. Subsequently, the pressure built up by the burning propellant causes the propellant burst disc to rupture, which occurs at a predetermined pressure and time based on selection and design of the propellant grain. The hot propellant gas passing through the ruptured disc into the main chamber is directed against the curved surface of a deflector, which causes a thorough mixing of the hot gas with the stored carbon dioxide-alcohol mixture. During this mixing operation, the hot

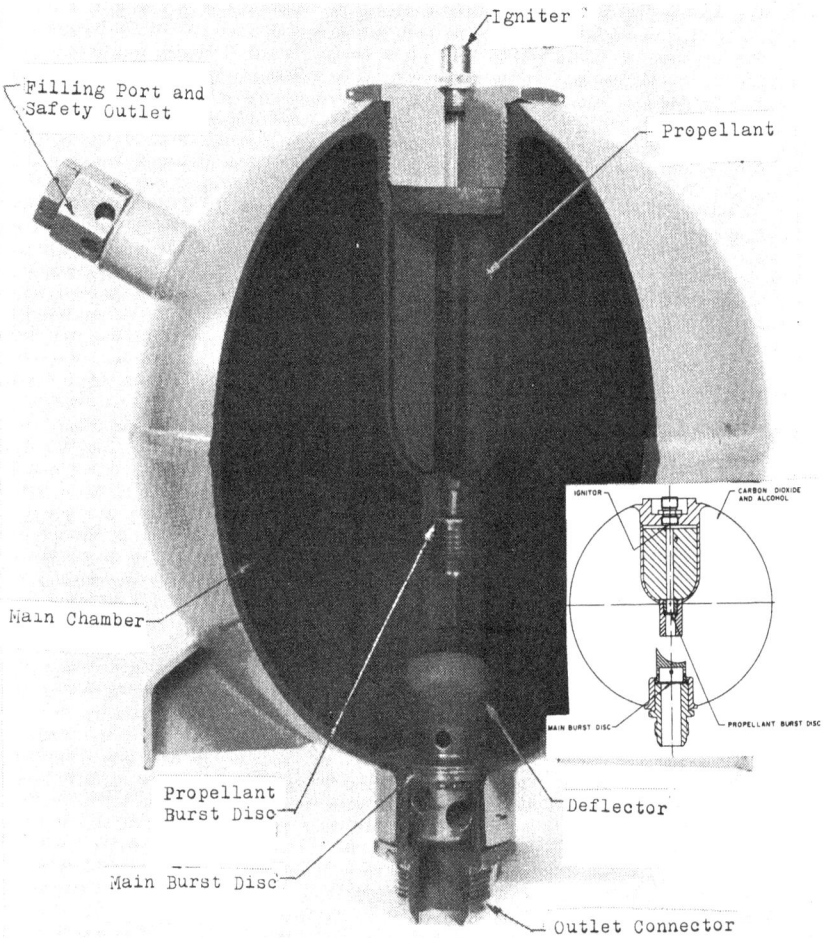

Fig. 8:3 A cool gas generator (Courtesy of Walter Kidde & Co., Inc.).

propellant gas causes an increase of the temperature in the main chamber, which results in a rise of the pressure of the mixture in the main chamber to a predetermined level, at which the main burst disc ruptures. To prevent possible damage of the flotation device, the fragments of the burst disc are captured in the outlet fitting. The generated cool gas flows through the outlet into the inflatable flotation device at nearly ambient temperature.

155

Carbon dioxide is used mainly because of its advantageous features of high storage density and low pressure. The presence of the alcohol minimizes the effect of high temperatures on the carbon dioxide. When the temperature rises, the alcohol absorbs a large amount of heat by vaporizing, while the alcohol remains a liquid at low temperatures. By adding heat to the carbon dioxide, the propellant gases speed up the generation of carbon dioxide pressure and thus prevent the formation of solid carbon dioxide due to gaseous expansion during the discharge.

A combined filler port and safety outlet is provided on the upper half of the sphere. The safety outlet contains a burst disc designed to rupture at 4800 to 5600 psi in case of overpressurization within the gas generator which might be caused by high ambient temperature.

The cool gas generator, as shown, which has a diameter of 12 inches and a weight of about 30 pounds, is designed to inflate a flotation bag of 80 cubic feet volume in less than 5 seconds. Two of such bags are used in a helicopter flotation system. The inflation pressure is 2 psi. The operating temperature of this gas generator ranges from −29°C to +52°C.

Another similar cool gas generator, designed to inflate a one-man life raft of a volume of 4.5 cubic feet, weighs only 2.5 pounds and operates at a temperature range from −54° to +71°C.

It is very important to keep the temperature of the generated gas used for inflation of flotation devices, emergency slides, balloons and similar devices at an acceptable level, because a too cold gas can damage the delicate device resulting in cold cracking, while a too hot gas can weaken or damage the device, and as the gas cools down to ambient temperature, a pressure decrease will be the result.

The temperature-entropy relationship for carbon dioxide is briefly described in the following paragraph: The main chamber of a cool gas generator is usually filled at a specific volume of 0.024 cubic foot per pound or to a density of 68 percent of the weight of water that would fill the volume. The internal energy of the stored gas at the lowest temperature of −29°C is approximately 153 Btu/lb. The desired point to reach, 15 psia at −29°C, indicates an internal energy of approximately 275 Btu/lb. The energy required from the propellant is the difference of both internal energy levels, which amounts to 275 − 153 = 122 Btu/lb. At an initial temperature of about 52°C, the internal energy of the stored gas is approximately 221 Btu/lb. When the same amount of energy from the propellant is added, a pressure at the same volume (which is 7.1 cubic feet) of approximately 28 psia, an internal energy of 343 Btu/lb. and a temperature of about 185°C is obtained. Since pressure and temperature would be too high, a small amount of ethyl alcohol, which has a heat of vaporization of 360 Btu/lb., is added to the carbon dioxide to absorb the excess energy and to limit the internal energy of the carbon dioxide to 298 Btu/lb, corresponding to a temperature of +52°C and a pressure of 20 psia. The ethyl alcohol absorbs 45 Btu/lb of carbon dioxide. Under these temperature and pressure conditions, the alcohol vapor generates approximately 2 psi pressure. At lower temperatures, only a portion of the alcohol vaporizes, and a large portion of the heat generated by the deflagrating propellant is utilized for vaporizing the carbon dioxide. This results in an acceptable high temperature performance with little adverse effect on the low temperature performance.

Table 8.2

Properties of Cool-Burning Propellants

Characteristic	Performance Data of Propellant	
	PHM	PKN
Flame temperature, °C measured calculated	1065 1038	927 891
Ballistics, reproducible plateau burning rate, in/sec	0.135	0.123
Ignition, reproducibly ignitable between	−54°C and +71°C to more than 1000 psia	−54°C and +71°C
Stability, Taliani slope at 100 mm	0.85	0.94
Auto-ignition test temperature, °C	Can withstand 1 hour at 149°C	—
Exhaust	Clean, smokeless	Clean smokeless

(Ref.: No. 94)

Functioning of a cool gas generator is not affected by its external environment as long as a pressure differential of 25 psia exists between the chamber pressure and the exhaust pressure to provide sonic flow of the generated gas.

A solid propellant ideally suited for cool gas generators has the following properties:

Molecular weight	22.79
Flame temperature at constant chamber pressure	1360°C
Burn rate at constant chamber pressure	0.08 in/sec
Characteristic velocity	3876 ft/sec
Mole of gas/100 gms	4.48

Properties of the cool-burning propellants PHM and PKN, suitable for use in cool gas generators, are listed in Table 8:2.

A cool gas generator of a different shape and size than described before, as used for inflation of a flotation bag for a helicopter, is shown in Fig. 8:4. The small inserted illustration demonstrates the difference between the sizes of the gas generator and the flotation bag. This cool gas generator, which functions on the same principle as described before, has an envelope volume of approximately 300 cubic inches and weighs 15.2 pounds. It inflates a flotation bag of 56 cubic feet volume to a pressure of

Fig. 8:4 A cool gas generator for inflation of a flotation bag (Courtesy of McCormick-Selph).

0.5 psig to 6 psig in less than 4 seconds over a temperature range of -7°C to $+52^{\circ}$C.

When comparing cool gas generators with stored pressurized gas systems, it is found that with an increase of the volume delivered, the cool gas generator becomes much more efficient compared to bottled gas systems. The advantage of cool gas generators is shown by the following example: A high-pressure stored gas system for the same application would have a volume of about 2000 cubic inches, it would take 10 seconds to inflate the flotation bag, and the weight of the system would be about 25 pounds, as compared to 15.2 pounds weight of the cool gas generator, as described.

A. Gas Cartridge Actuators

For applications where a substantially smaller volume of cool gas is required than is produced by cool gas generators, gas cartridge actuators are ideally used. They consist of a small container filled with pressurized gas and a normally closed cartridge-actuated valve which is mounted on the open port of the gas container. When actuated, the high-pressure gases generated by the cartridge drive a piston in the valve forward. A punch provided at the forward end of the piston cuts a sealing diaphragm in the outlet port, thus releasing the pressurized gas.

Gas cartridge actuators are used for a variety of applications, for example, to inflate small devices, to move a piston, to pressurize or purge a system, to transfer liquids, to spin a gyro or a small turbine, to cool a component and for numerous other applications.

The main advantages of gas cartridge actuators are their long-time storage capability, low weight, small size, fast operation, high reliability and reusability. The advantageous small size of these devices may be demonstrated by the following example: For a free gas volume of 1000 cc to 1500 cc and a stored gas pressure of 1500 psi, the volume of the gas container would be between 10 and 15 cc.

Gas cartridge actuators can be refurbished for continuous use by refilling the gas container and by replacing the sealing diaphragm and the cartridge. Commercially available gas cartridge actuators have gas container volumes ranging from 2.6 cubic inches to 30 cubic inches, and are designed for filling pressures from 838 psi to 5000 psi. The sizes of these actuators range from 1.0 inch diameter and 3.3 inches long to 2.0 inches diameter and about 15 inches long. Carbon dioxide

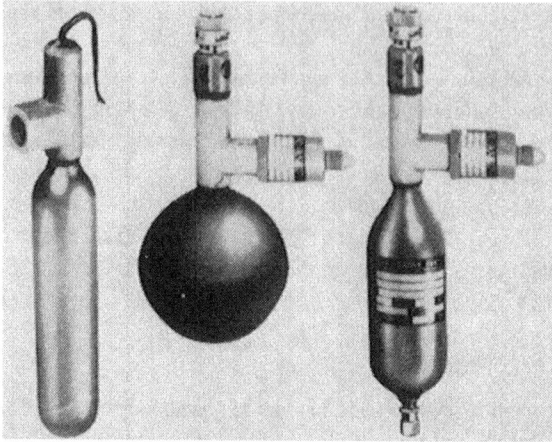

Fig. 8:5 Gas cartridge actuators (Courtesy of Conax Corp.).

and nitrogen are mostly used in these devices, but also helium, argon, hydrogen, methane and oxygen are used as suitable gases.

Typical gas cartridge actuators with gas containers of various shapes and sizes are shown in Fig. 8:5.

In a combination of a gas cartridge actuator with a long piston actuator, as shown in Fig. 8:6, the released gas pressure of the gas cartridge actuator is utilized to drive the piston.

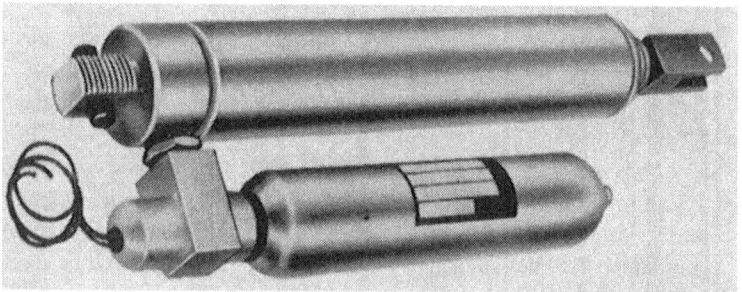

Fig. 8:6 A gas cartridge actuator mounted on a piston actuator (Courtesy of Conax Corp.).

9   Location Aids

In most reentry operations, it cannot be expected that the vehicle, in spite of the bright glow from the ionized vapor trail or from glowing heat shield particles, can be visibly observed and accurately located without special devices, mainly because of the very limited time of visibility and because of the difficulty of instantaneous orientation. To ensure recovery of reentry vehicles, spacecraft components, or missiles, effective and reliable methods for their speedy detection of location are a prerequisite. Visual and electronic methods are used in order to accurately locate them on sea and land. Since the visual and the electronic methods have certain limitations, both methods are employed concurrently. During adverse weather conditions, the effectiveness of visual detection and location methods is often reduced to a low grade.

As visual location devices, flares, smoke bombs, dye marker, photoflash cartridges, and electric flashing lights are used, while homing radio beacons, Sofar bombs and radar are employed as electronic location devices. According to the objectives of this book, only pyrotechnic location devices, such as flares, smoke bombs, photoflash cartridges and Sofar bombs are described in this chapter.

A. *Flares*

Flares are pyrotechnic light producers which are used extensively as signalling, illumination and warning devices, e.g. aircraft, marine and railway signals, tracking and location aid for missiles and spacecraft, torches, road emergency flares, and lighting devices for military targets and aerial photography.

According to their functioning characteristics, flares are categorized as a slow-burning type, a fast-burning type, and an incandescent type. The burning time of slow-burning flares is up to 2.5 minutes, while fast-burning flares usually have a burning time of only a few seconds.

The pyrotechnic materials to be used in flares are carefully evaluated and selected for their optimum light-producing and radiation properties, ignitibility, burning rate and stability. White light producing materials are used in flares mainly for lighting of targets and aerial photography, while colored light producing materials are used in flares for signalling. For target illumination and similar purposes, two basic types of flares are used: ground-to-ground or air-to-ground flares. Ground-to-ground flares can be hand-held or deployed ballistically by launchers, guns or similar means, while air-to-ground flares are usually deployed by aircraft or missiles. Depending on their specific application, flares can be of extremely simple or of a more complex design. Usually, a flare consists of a cylindrical housing containing the pyrotechnic light-producing material and an igniter or a fuse. For the ignition of simple ground-to-ground flares, a mild-detonating fuse (MDF) is frequently used.

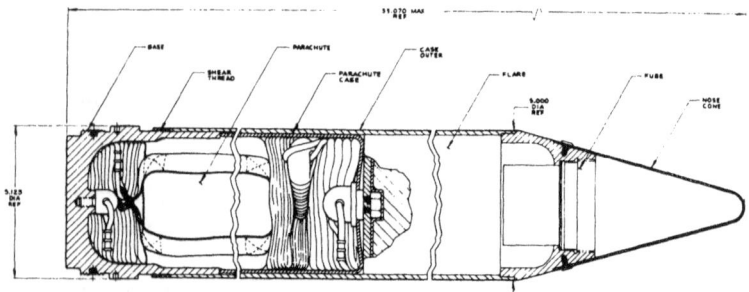

Fig. 9:1 A parachute flare (Courtesy of SDI Special Devices Inc.).

Air-to-ground flares and high-performance ground-to-ground flares are often equipped with a parachute which is stored in a separate compartment in the cylindrical housing. A typical parachute flare is shown in Fig. 9:1. At a predetermined altitude, a built-in timing mechanism fires the propellant charge which shears the thread on the base. Subsequently, the outer case falls away, and simultaneously, the hot gases ignite the pyrotechnic light-producing composition. The parachute deploys, and the flare, producing 1,000,000 candle power, illuminates the area underneath. The operating time of this flare at sea level is 2.5 minutes. Parachutes used for flares must provide a slow descent rate to keep the flare air-borne for the duration of the burn.

A special type of parachute flare is equipped with a built-in launch rocket and stabilizing fins. It is man-rated and fired from the ground. The cap of the rocket barrel is equipped with a firing pin. To launch this flare, the firing cap is removed and placed on the bottom end of the rocket barrel, and the system is activated by striking the firing cap with the palm of the hand, thus energizing a primer placed in the bottom of the rocket barrel. This device which may carry a payload consisting of white, red, or green clustered-star burning illuminating materials, white or red parachute-supported illuminating compositions, or red or green parachute-supported smoke-producing materials, reaches an altitude of 700 to 800 feet. The parachute-supported illuminating flares of this type have a burning time of 25 to 50 seconds. The burning time of the smoke-producing signal device, which is also parachute supported, is from 6 to 22 seconds, while the not parachute-supported illuminating stars have a burning time of only 4 to 8 seconds.

A recently developed air-to-ground flare which produces 5,000,000 candle power and which has a burn time of 5 minutes descends at a rate of 5 feet per second, suspended from a specially developed hot air balloon. The flare is released by the pilot with external rack initiation. The system is equipped with a timer which is preset on the ground and which has settings from 2 to 25 seconds. When the flare is released from the aircraft, the timer is activated by an integral arming wire. After run-out of the timer, the nose cone is separated from the housing by releasing a tie band, and the cone is ejected by a compression spring. Simultaneously, a drogue chute is deployed, and its opening shock is utilized to initiate a cable cutter with a built-in time delay to cut a retainer cable, and thus permits the chute to pull the storage bag off the packed balloon. As the bag is being pulled off, the balloon is deployed. Inflation of the balloon is accomplished by ram air forced through one-way scoops provided near the balloon's equator. When the balloon is about 90% inflated, the flare ignition sequence is mechanically initiated by means of a bridle which is attached to the inside of the balloon. Atop of the flare candle in line with the bottom opening of the balloon, a heat generator consisting of a housing from phenolic-asbestos, containing a solid propellant, is provided. The light-producing composition used in this flare consists of a magnesium-sodium-Laminac mixture. To give the pilot a burn-out warning, a small quantity of a red composition is pressed into the candle, causing a color change during the last seconds of the total burning time. When initiated, both the heat generator and the candle are ignited simultaneously by an intense flame of short duration. During burning of the flare, a separation at the housing is melted, thus separating the candle from the heat generator, letting the candle tip over and drop to a distance of 20 feet below the bottom of the balloon, arrested by suspension cables.

The descent rate of this system is greatly affected by the weight decrease caused by the burning of the flare and by the decrease of the balloon's buoyancy. Since the weight decrease is greater than the buoyancy decrease, at a predetermined time, 260 seconds after deployment of the drogue chute, a vent is cut in the balloon top to maintain the descent rate of 5 feet per second. After a time span of 60 seconds, a large opening is cut in the balloon by a self-destruct system to release the entrapped air and to let the system descend at about 20 feet per second, thus clearing the airspace of the spent system.

A. *Light-producing Materials and Compositions*

Ideal materials for brilliant white-light production are magnesium, aluminum, titanium and zirconium. Because of their high cost, titanium and zirconium are not as commonly used in flares as magnesium and aluminum. Both magnesium and aluminum are used in form of powder of flakes. Light-producing compositions usually consist of a mixture of about 50% of these metal powders or flakes, about 40% oxidizer, and about 10% binder. Suitable oxidizers are potassium perchlorate and nitrates of potassium, sodium, barium and strontium. In flares for military applications, sodium nitrate is used preferably, because it is less sensitive to detonation by impact than chlorates.

Frequently used binder materials for flare compositions are: oils, asphaltum, resins and shellac. The volume percentage of binders for light-producing compositions depends on the design type of the flare. For a flare composition, which is kept in compressed condition in a cylindrical housing, a rather small amount of binder material is required, while a larger amount of binder must be used in flares which are completely exposed and are not equipped with a housing.

As mentioned above, magnesium and aluminum are most frequently used for white-light production. Magnesium, Mg, is a silver-white lustrous, malleable metal. It has a specific gravity of 1.74, a melting point of 650°C, and a boiling point of 1120°C. In powdered or flake form it burns readily with a brilliant bluish-white light and with evolution of substantial heat of combustion. Magnesium powder liberates hydrogen when in contact with water.

In some white-light producing compositions, magnesium powder is used in combination with aluminum powder, while in other compositions, a small amount of sulfur is used in combination with aluminum powder.

The compositions used for the production of colored light are basically similar to the compositions used for white-light production. The most commonly used colors for colored-light flares are: red, green and yellow. They are obtained by adding certain elements to the composition. Strontium nitrate or strontium oxalate are frequently used for producing bright red light. An amount of about 30% strontium nitrate can be used in a mixture of 35 to 40% magnesium and 20 to 30% potassium perchlorate plus a binder.

For producing bright green light, an amount of about 40 to 60% of barium nitrate can be used in a mixture of 15 to 30% magnesium, about 20% hexachlorobenzene or a similar amount of potassium perchlorate, plus a binder. Copper compounds, such as copper powder or cupric oxide are often used to intensify the green light.

For producing yellow light, an amount of about 12 to 20% sodium oxalate can be used in a mixture of 10 to 30% magnesium and 25 to 50% potassium perchlorate plus a binder.

The properties of strontium nitrate and strontium oxalate, which are used for red-light production, are briefly explained in the following. Strontium nitrate, $Sr(NO_3)_2$, is available in the form of white granules of powder. It has a specific gravity of 2.98 and a melting point of 570°C. When heated with combustible materials it releases oxygen and imparts a red color to the flame. Strontium oxalate, $SrC_2O_4 \cdot H_2O$, is available in the form of a white crystalline powder. It decomposes on heating and burns with a red light.

A different type of light-producing devices other than the flares, as described, are the flash cartridges which are mainly used for the tracking of missiles. At certain intervals, flash cartridges are released from the missile. The clearly visible trace, which is left by the flashes, is recorded by a camera.

A flash cartridge usually consists of a metal housing containing loose metal powder and an explosive ignition charge. The metal powder is dispersed without an oxidizer by the bursting charge and burns in the ambient air. Binders are not used in these flash cartridges. Only in some special flash cartridge types, an oxidizer, such as potassium perchlorate is used.

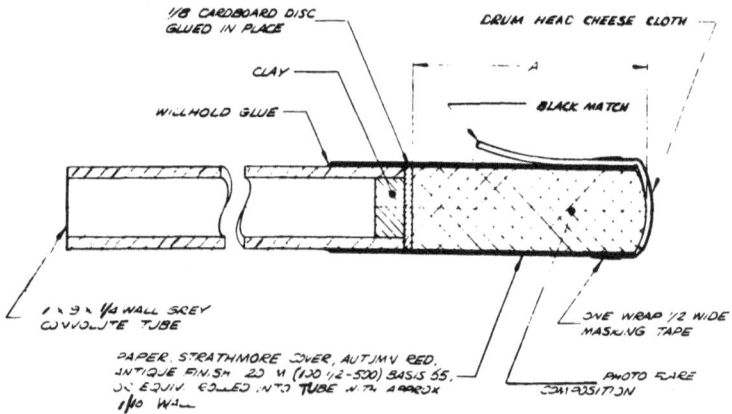

Fig. 9:2 A photo flare (Courtesy of SDI Special Devices Inc.).

A very simple type of a flare, a photo flare, which is similar in design to a torch, is shown in Fig. 9:2. In commercially available photo flares, the length of the flare composition, indicated as dimension "A", ranges from 4 inches to 23 inches. The main difference of this photo flare from a conventional torch is in the composition and in the free end or handle. In the photo flare, as shown, the free end consists of a convolute tube, while in a torch, a wooden handle is used. The burning time of these photo flares ranges from 1 minute to 5 minutes.

Various other types of flares, for example ion flares, infrared flares, and underwater flares have been developed. They are usually electrically ignited. In design and function they are similar to the flares as described; the main difference is in the pyrotechnic composition used in these special devices.

B. *Smoke Generators*

As flares are used as visual location devices mainly in darkness, smoke produced by smoke generators finds extensive use for tracking and signalling at daylight. On auxiliary airfields, smoke is often used as ground wind direction indicator. Another application of smoke is that as an effective camouflage device in military operations and training. Ships and troop movements can be hidden by dense smoke clouds. While smoke as visual location aid is mainly used at daylight, it is being utilized for camouflaging at both day and night: white smoke at daylight and black or dark gray smoke at night.

Smoke generators are also widely used as special effect devices in the motion picture industry to simulate gun fire, bullet hits and other battle effects, explosions, rocket launches and similar events. For these various applications, smoke generators in the form of smoke cartridges, smoke pots, smoke puffs and pellets have been developed to produce smoke of a great variety of colors, as for example: white, black, grey, red, green, blue, yellow, orange, purple and pink. Depending on the application, smoke generators are ignited either by a fuze or by an electric current.

The most commonly used smoke-generating materials and compositions and their properties are described in detail in Chapter 1, Section H of this book.

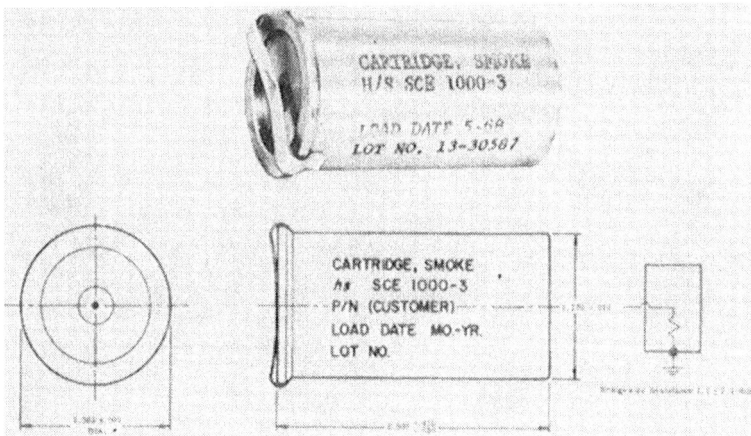

Fig. 9:3 A smoke cartridge (Courtesy of Hi-Shear Corp.).

The smoke cartridge, as shown in Fig. 9:3, was designed as an effective device for spotting drones or other high-flying vehicles travelling at high speed, including supersonic speed. This bridge-wire-equipped cartridge is attached to the drone, and it is initiated on receipt of a radio impulse. A current of 5 amperes is applied to the bridgewire which results in a function time of 2 milliseconds. If the minimum all-fire current of 3.5 amperes would be applied, the average function time would be about 5 milliseconds. When the cartridge is fired, an internal smoke package is ejected producing a flash and a puff of white smoke which appears 0.5 second after ejection. Both the flash and the smoke are visible to the naked eye in clear atmosphere at a distance of more than 12,000 yards, in slant range.

Conventional smoke pots, as used for effects, are initiated either by fuze or by electric current. Smoke pots range in size from 0.5 inch diameter and 4 inches long to 1.0 inch diameter to 12 inches long. A cross-section through a simple fuze-initiated smoke pot is shown in Fig. 9:4.

Another frequently used type of smoke generator is the smoke puff. It consists of a small polyethylene bag, 0.001 to 0.003 inch thick, filled with a smoke-generating composition. Ignition is accomplished by a built-in electric squib. The squib lead wires extend through the open end of the bag which is closed with a tie string. A

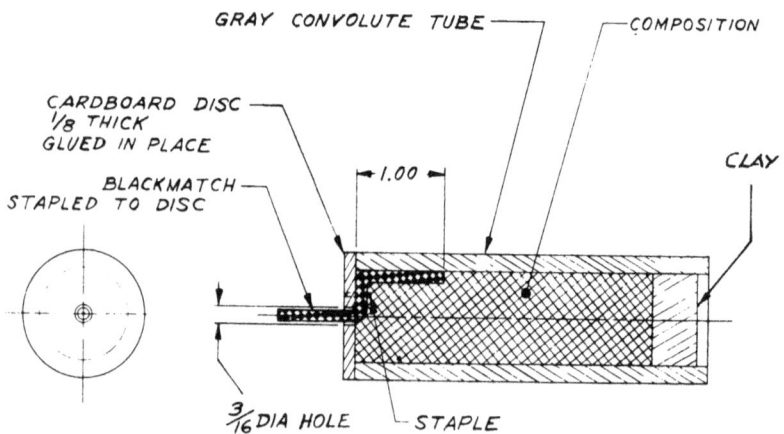

Fig. 9:4 A smoke pot (Courtesy of SDI Special Devices Inc.).

typical smoke puff contains a quantity of 3 grams of smoke-generating material.

Smoke pellets, which are used for motion picture battle effects, usually consist of a hard-pressed smoke-generating composition, formed in cylindrical shape in conventional sizes from 0.5 inch to 1.0 inch diameter and about 2 inches long.

The smallest smoke generators, as used to simulate bullet hit effects, consist of a 1/16 inch thick disc ranging in diameter from 0.218 to 0.515 inch. The disc contains a small amount of smoke generating material and is equipped with a bridgewire for ignition. Lead styphnate is often used as initiating material in these smoke generators. A different type of bullet hit-simulating smoke generators is made in a cylindrical shape of 1/4 inch diameter and a length ranging from 1/4 inch to 5/8 inch. They are also bridgewire-ignited.

A more complex smoke-generating device, as used to simulate a cannon shot, is shown in cross-section in Fig. 9:5. This effect device contains three different smoke-generating materials in separate compartments. The red fire material and the black powder are ignited by separate squibs. The charcoal, arranged between both squib-ignited materials, is ignited by the heat from both materials.

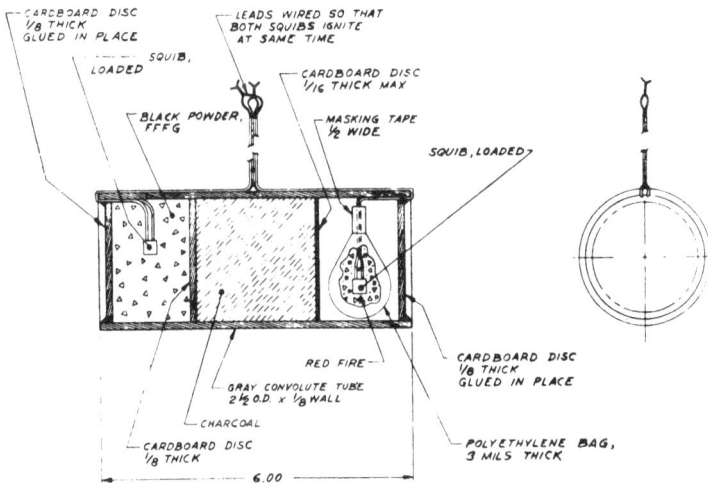

Fig. 9:5 A smoke generator for simulation of cannon shot (Courtesy of SDI Special Devices Inc.).

C. SOFAR Bombs

SOFAR bombs are used as effective location aids for spacecraft, missiles and ships. These devices are no bombs, but more accurately, they are signal generators. The term "SOFAR" is an abbreviation for "Sound Fixing and ranging". SOFAR bombs are hydrostatically operated controlled-depth explosive devices, which prior to water landing or impact are ejected from the spacecraft or missile or from the ship in distress. As the SOFAR bomb sinks in the water, at a pressure corresponding to the selected depth, the device is armed and then fires. This selected depth is called the "deep sound channel".

At this specific ocean depth, which is about 3000 feet in the Pacific Ocean and about 4000 feet in the Atlantic Ocean, sound energy is transmitted for thousands of miles without great loss in intensity. These "deep sound channels" in the oceans are the basis for the sound fixing and ranging system (SOFAR), which is being used successfully for exact determination of the point of water impact of a spacecraft or missile or of the location of survivors of a ship catastrophe. Electronic sensing stations, equipped with hydrophones and other suitable devices have been established throughout the world. The difference in time of the received signals at various sensing stations are used to determine the exact point of impact on the ocean by constructing the hyperbolas that intersect at the detonation point

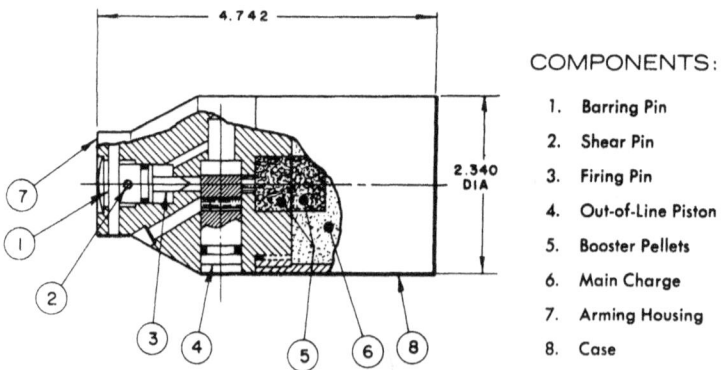

COMPONENTS:

1. Barring Pin
2. Shear Pin
3. Firing Pin
4. Out-of-Line Piston
5. Booster Pellets
6. Main Charge
7. Arming Housing
8. Case

Fig. 9:6a A SOFAR bomb (Courtesy of Bermite, North Hollywood Branch, Div. of the Whittaker Corp.).

of the SOFAR bombs which were ejected prior to water landing or impact.

A typical SOFAR bomb, which is designed to locate the impact point of missiles, is shown in Fig. 9:6A. This device contains an explosive charge of 0.5 pound, and it has a gross weight of 2.0 pounds. It is equipped with an out-of-line arming mechanism. Arming and firing is accomplished by hydrostatic pressure.

A similar type of SOFAR bombs for missile application, designed to fire at a sea depth of 4200 feet, is shown in Fig. 9:6B.

Fig. 9:6b A Production line of SOFAR bombs (Courtesy of Bermite, North Hollywood Branch, Div. of the Whittaker Corp.).

10   Special Devices

The great advantages offered by pyrotechnics, such as high power-to-weight ratio, small size and low weight of pyrotechnic devices, and the high reliability in operation are the main reason for the utilization of pyrotechnics for a rapidly increasing number of special devices needed for various functions. The one disadvantage that most of the pyrotechnic devices can be used only as "one-shot" items is in most cases outweighed by the great advantages as listed above.

Only a few special pyrotechnic devices, such as heaters, earth anchors, and underwater devices are described in this chapter, to demonstrate the great potentials of pyrotechnics for a variety of functions and applications.

A.   *Heaters*

Pyrotechnic heaters can be used as one-shot devices to raise the temperature of gases, liquids or solid materials. Usually, when a particular temperature must be maintained, the use of an electric heater with an appropriate control device is advisable. In cases, however, where for example, the temperature of a fluid must be increased and in cases, where electric current must be conserved, a pyrotechnic heater can be used ideally. Pyrotechnic heaters are often used for raising the temperature level in electrolytic batteries so that they reach their rated capacity.

    Pyrotechnic heaters have the advantage that they provide fast heating, they are of small size and low weight, they can be immersed in a fluid, they operate with a high reliability, and they have a low power consumption. Since they have no moving parts, they are relatively insensitive to shock, vibration and acceleration forces.

Various different types of pyrotechnic heaters are shown in Fig. 10:1.

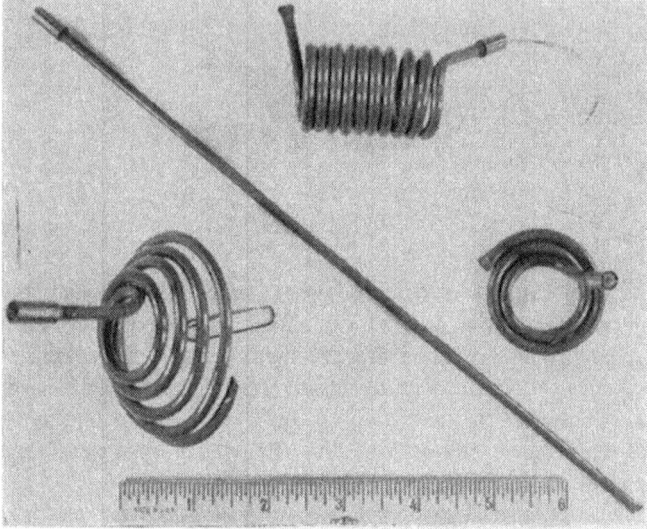

Fig. 10:1 Pyrotechnic heaters (Courtesy of Atlas Chemical Industries, Inc.).

A typical pyrotechnic heater consists of a cold-rolled steel tube which is loaded with a pyrotechnic composition having a predetermined total heat release. The tube is swayed down to a diameter of 0.152 inch. An electric igniter is screwed onto one end, and the free end of the tube is sealed. Heaters of this type in a length of up to 10 feet can be made. To obtain a greater heat output, several lengths of such type of heaters can be joined by using threaded coupling fittings. These tubular heaters can be coiled to a minimum inside diameter of 3/4 inch. Heaters in a variety of shapes, as shown in the illustration, can be made to suit various specific applications.

A pyrotechnic heater, as described, developing a surface temperature of $677°C$, can produce 0.625 Btu/in. with a burning rate of either 4 or 6 seconds per inch. The weight of these heater units is less than 2 grams per inch length.

To ignite the slow-burning heat-producing composition, passage of current through the bridgewire for approximately 10 milliseconds is required. An all-fire current of 1.0 ampere and a no-fire current of 0.1 ampere is used. Larger heaters than these, as described, require more current or may be set off by percussion caps.

The smallest pyrotechnic heater commercially available at present has a diameter of about 1.0 inch and a thickness of about 1/4 inch. It produces 5.5 Btu in 1 second and reaches a surface temperature of 650°C. The largest standard size pyrotechnic heater has a diameter of 2.5 inches and a thickness of 0.8 inch. It delivers 264 Btu in 3 seconds and reaches a surface temperature of 760°C.

Various methods of mounting, for example by using simple metal clamps, can be considered for these heaters. In cases where insulation is used over the heater, the insulation material can sometimes be utilized for securing the heater in place by means of tape and straps.

All these pyrotechnic heaters are sealed units, and they do not vent gases or other combustion products during or after operation. The distribution of heat may be controlled by bending the heater tube into almost any possible configuration.

B. *Earth Anchors*

In many areas on earth, the soil conditions make reliable anchoring of structures to the ground very difficult. Radio transmitter structures, high antenna masts, radar structures, utility poles, control towers, suspension bridges, retaining walls, aircraft catapult and arresting sys-

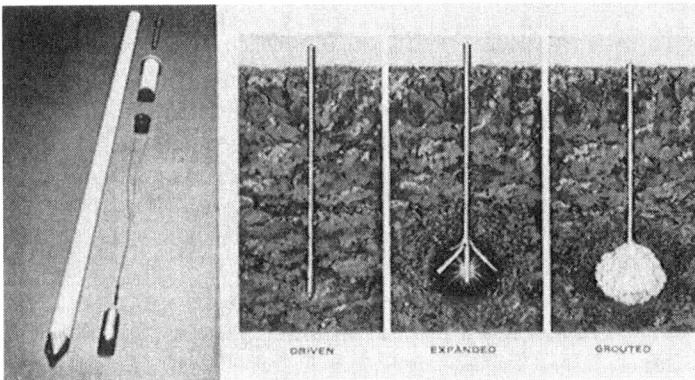

Fig. 10:2 An explosive earth anchor and its application (Courtesy of Harvey Engineering Laboratories, Div. of Harvey Aluminum, Inc.).

tems and portable buildings and shelters must be securely anchored by guy wires.

Explosive earth anchors have been developed as effective devices for securing structures to the ground. A typical explosive earth anchor, as shown in Fig. 10:2, utilizes a "virgin soil" technique, which has the advantage that it does not loosen the soil during installation of the anchor, but it rather compacts the soil, and thus achieves the rated 20,000-pound tension loading capacity for various soil types and conditions ranging from hard clay to water-saturated sand and silty soils.

Minimum effort and equipment are required to install the earth anchor. A two-inch diameter steel tube, which is equipped with a small explosive charge in the lower end, is driven into the ground. In cases where the soil is extremely hard, it may be required to pre-drill a 2-inch diameter hole for the earth anchor. After the tube has been driven into the ground and only a short end extends above the surface, the explosive charge is initiated by electric current. The explosion force forms the lower tube end into open umbrella-like "tines" and opens up a highly compacted underground cavity which is then filled through the tube with a quick-setting "grout" or cement-like material. The hardening time of a recommended type of grout (FLOROK) is approximately one hour, while a suitable cement (High Early Cement) has a cure time of up to six days. Using an explosive method for spreading out the lower tube end has the desirable effect of work hardening, resulting in an increased strength of the tines. The anchor tube with the tines bound into the hardened grout ball form the anchor. A lug-type attachment fitting or another suitable bracket can be screwed onto the upper threaded tube end for installing guy wires or similar structural devices.

Two main types of explosive earth anchors are used: Fixed-length anchors, and extendable anchors. The simple fixed-length anchor consists of a high-strength steel tube which, equipped with an ogive-shaped cap at the lower end, is inserted 6 to 12 feet into the ground. After firing the charge, the lower end is spread out and is anchored in the grout material.

The extendable earth anchors have a primary length of 6 feet which can be extended in increments of 2 feet to a total length of 12 feet. At the lower end of both types of anchors, six external 8-inch long grooves are machined lengthwise into the tube wall. The ends between these grooves form the tines after the firing of the charge.

Fig. 10:3 Explosive earth anchors installed (Courtesy of Harvey Engineering Laboratories, Div. of Harvey Aluminum, Inc.).

The explosive cartridge used in these earth anchors contains approximately 0.20 pound of desensitized RDX explosive.

Explosive earth anchors in three different stages of installation are shown in Fig. 10:3. In stage A, the tube driven into the ground is ready for firing. In stage B, the tines are spread out after firing, and in stage C, the grout has been filled in.

C. *Underwater Devices*

Parallel to the exploration of outer space, extensive work is being done in the field of oceanography and undersea research, which is of great importance, considering the fact that a much larger surface area of the earth consists of water than of land. To accomplish the required tasks in undersea research, numerous new devices had to be developed. It was found that pyrotechnic devices can be used to great advantage in many undersea applications. However, the different and severe environmental conditions that exist under the surface of the sea present unusual and difficult problems in the development of reliable pyrotechnic devices, which must be capable of functioning after a long exposure to high external pressure at depths of several thousand feet. Some of the devices, including electrical connectors and cartridges must be sealed watertight, and the materials used for these devices must be capable of withstanding the saltwater effects. For this reason, stainless steel is preferably used for these devices.

The capabilities of cartridge-actuated underwater devices range from powerful cable cutters to quick disconnect couplings capable of withstanding a tension load of more than 20,000 pounds at a depth of 7 miles.

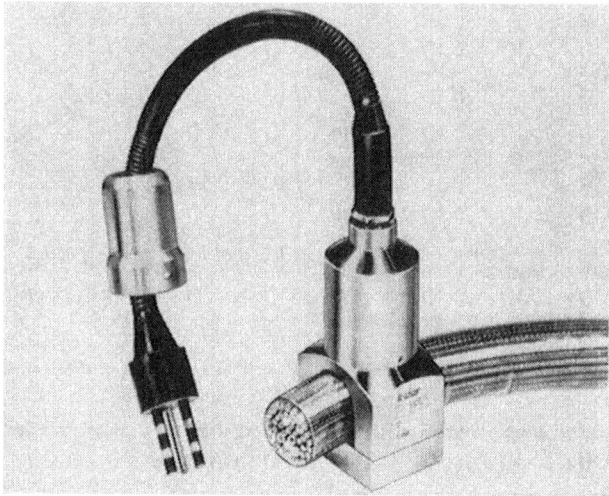

Fig. 10:4 Deep-sea cable cutter (Courtesy of Hi-Shear Corp.).

The deep-sea cable cutter, as shown in Fig. 10:4, is used to sever a heavily insulated, sheathed multistrand electrical cable of 1 inch diameter. This cutter is hermetically sealed. When actuated, the cutter blade ruptures the seal and cuts the cable. The power cartridge used in this cutter is a hot-wire igniter designed for marine ordnance application and is bonded to a 12-inch electrical cable. Similar cutters can be used to sever steel cables or anchor lines.

An underwater release device, which is designed to separate one submerged structure or component from another, is presented in Fig. 10:5. It can be used in primary or emergency systems, for example to separate buoyant structures from anchored structures or solid ballast from underwater vessels. This system can withstand external pressures

TYPICAL INSTALLATION

CONNECTOR

PC63 SERIES POWER CARTRIDGE

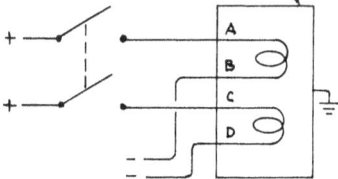

SEAL

LOCKWIRE PROVISION

CAPTIVE SEPARATION DEVICE

MOUNTING PLATE

SEALS

MOUNTING BOLT

STUD ASSEMBLY

SEAL PLATE LOCKNUT

PLATE OR UNIT TO BE SEPARATED

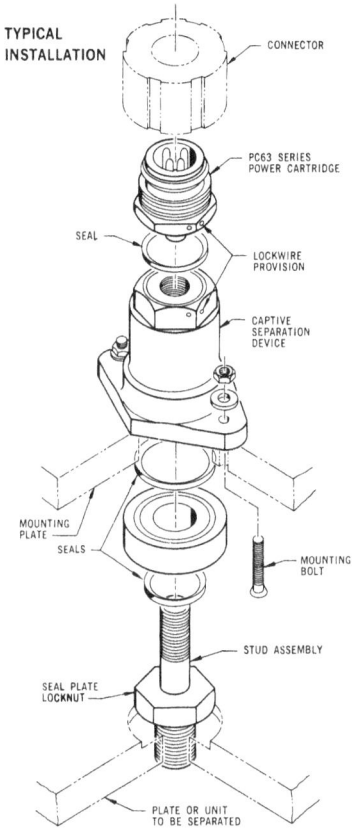

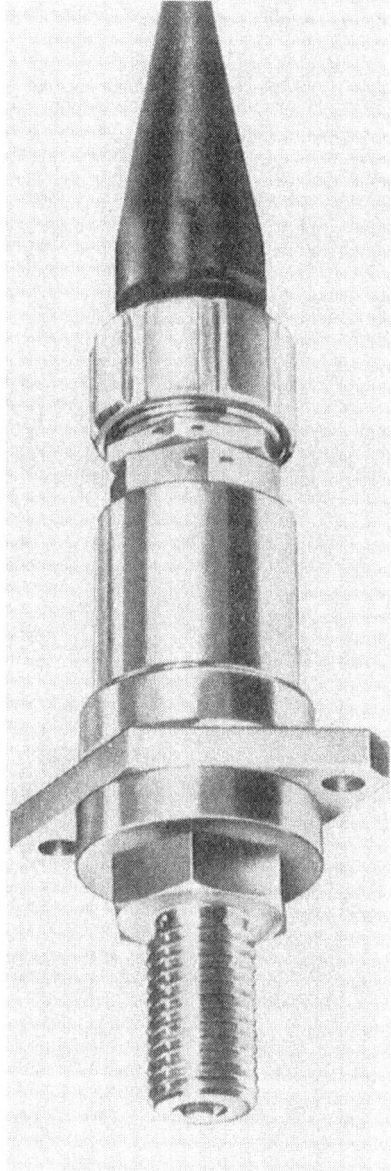

Power Cartridge

Current required to fire:
10-250 volts, AC or DC
5 amps per each of two bridgewires

ELECTRICAL SCHEMATIC

Fig. 10:5 Deep-submergence structure release device (Courtesy of Hi-Shear Corp.).

179

CYLINDRICAL
HOUSING

CLOCK
STARTER
SWITCH

CLOCK-TYPE
TIMER

HEX WRENCH
ADJUSTMENT
RECESS

COUNTER
WINDOW

FIRING UNIT

BATTERY

POWER
CARTRIDGE
CONNECTOR

FLOOD SWITCH
(HIDDEN)

POWER
CARTRIDGE

SEPARATION
DEVICE

CLEVIS
STUD

EXPENDABLE
END

SHACKLE,
THIMBLE,
AND CABLE
NOT FURNISHED

EXPLODED VIEW

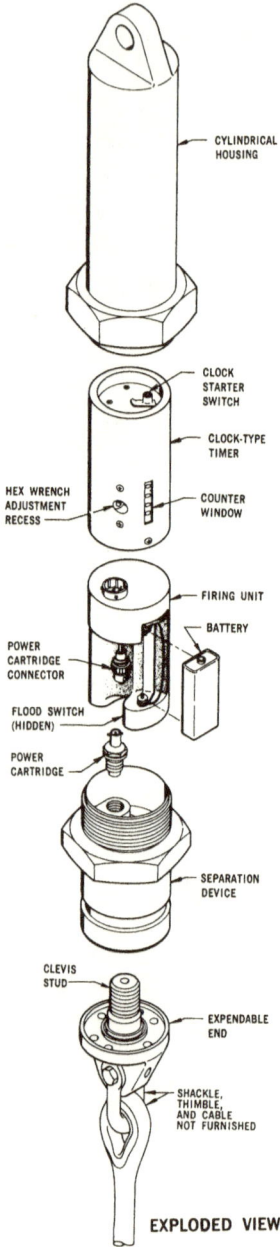

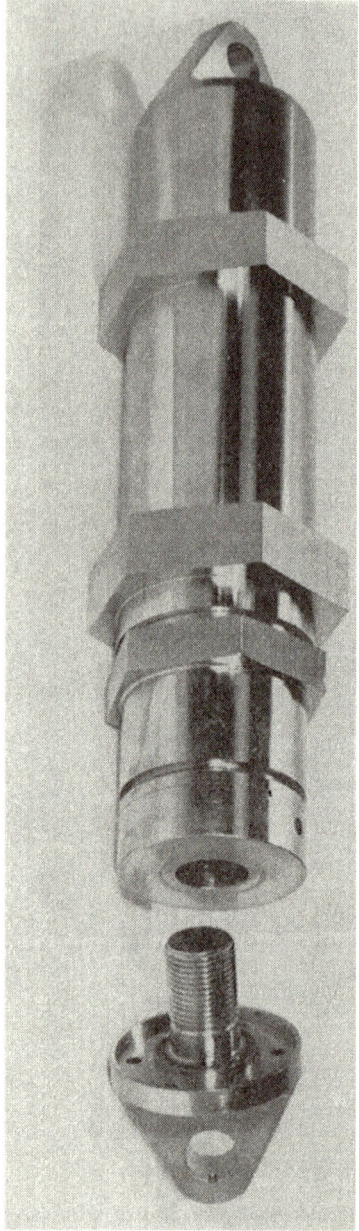

Fig. 10:6 Deep-submergence cable and buoy release device (Courtesy of Hi-Shear Corp.).

in excess of 5000 psi or a depth of 10,000 feet, and it will function after submergence for as long as 3 years.

This release system consists of three basic components: The captive separation device, a power cartridge, and a stud assembly. A positive lock is maintained by the bolt-mounted device until the cartridge is electrically commanded to fire, initiating a gas pressure to actuate the separation device, which releases and ejects the stud without fragmentation in less than 10 milliseconds. All parts are retained and locked into place in the device, and only a low reaction force is transmitted to the structure during release. The cartridge mates with a waterproof connector. To resist corrosion, all major parts are made from stainless steel, and the external surfaces are polished to mirror brightness to minimize marine growth. After firing, this release device can be refurbished to be reused.

For applications where a submerged cable or buoy must be released after a specified elapsed time, special timer-equipped release devices have been developed. The deep-submergence cable and buoy release, as shown in Fig. 10:6, is a self-contained, self-timing and self-activating tension cable separator that is recoverable and reusable. A special feature of this release device is that it has no external electronic or mechanical triggering mechanism or electric wires. A built-in timing device, which can be programmed for a maximum elapsed time of 1000 hours, will release the attached load consisting of a cable or buoy, after the preset timespan has elapsed. After ignition by the timing device, a gas-generating cartridge activates the release mechanism to release the load in less than 10 milliseconds after ignition. Both the timer and the firing unit are battery-operated.

This deep-submergence cable release mechanism consists of six main components. A cylindrical housing connects the release device with the expendable end. The timer and firing unit are mounted inside the housing, and the gas-generating cartridge connects the firing unit with the release device. The clock-type timing mechanism, which is mounted in a waterproof cylinder, is capable of 41 days continuous operation. It has a temperature range from 0°C to +50°C, and an accuracy of 0.5%.

The firing unit, which is mounted in a semi-cylindrical plastic housing, is contoured to fit between the timer and the separation device physically and electrically, and it contains the cartridge, the battery and a flood switch. In case there is a leak in the housing, the flood switch causes the cartridge to ignite and release the expendable end. The separation device is a gas-actuated system of sliding pistons and cylinders that disengage the threaded segments holding the clevis stud, to eject the expendable end.

This release mechanism has a clevis stud of 1.0 inch diameter having a strength of 60,000 pounds. The maximum operational depth of the mechanism is 35,000 feet. Having a length of 17 inches and a diameter of 2 inches, its weight is about 16.5 pounds.

In some deep-sea release devices used with oceanographic systems it is advantageous to use a ball-type swivel end for cable attachment. An undersea release device of this type is shown in Fig. 10:7. The housing of this free-flooding separation device contains a clam-shell joint half that grasps the ball-shaped end of the expendable cable swivel. When actuated, the clamshell half is forced open and the ball end of the swivel is released. For attachment of oceanographic equip-

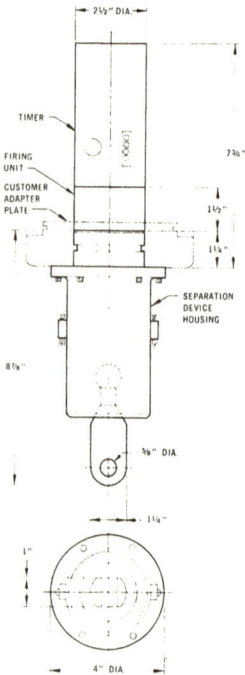

Fig. 10:7 Ball-type underwater release device (Courtesy of Hi-Shear Corp.).

ment, a mounting flange is provided at the upper end of this release device.

In numerous underwater applications which do not require highly sophisticated release equipment, especially developed explosive bolts are used as simple and reliable separation devices. It is a major requirement for these bolts that they are hermetically sealed and that they function without fragmentation.

An explosive bolt for deep-sea application, designed for mounting an external instrumentation package on the manned submersible "Deepstar", to be operated at a depth of 4000 feet, is shown in Fig. 10:8. Several of these bolts are used in conjunction with a tubular steel U-shaped hanger over the overhanging eyebrow of the submersible. In case the instrument package should become entangled on a rock or with undersea cables during a mission, the bolts are fired and the rack with the instruments is jettisoned.

In some undersea operations, after mission completion, deep-sea equipment must be inflated for buoyancy. A small-size system to be used for inflation, consisting of a pressure bottle and an explosively-actuated valve, is presented in Fig. 10:9. It is designed to operate at a depth of 8000 feet. The pressure bottle contains Nitrogen gas at 7500 psi pressure. Upon actuation of the valve, which has a 1/4-inch diameter orifice, the gas is released into the device to be inflated. Both the pressure bottle and the valve are made from stainless steel 17-4PH

Fig. 10:8 Explosive bolt for deep submergence release after firing (Courtesy of Bermite, North Hollywood Branch, Div. of the Whittaker Corp.).

Fig. 10:9 Deep-submergence nitrogen storage system (Courtesy of Pyrodyne, Inc.).

condition H 1025. The system has a total weight of only 3.12 pounds.

Numerous other functions than described above can be accomplished by utilizing cartridge-actuated devices in deep-sea operations. Major requirements are, that these devices are properly sealed and that the best suitable materials are selected to prevent corrosion and malfunction.

Part III

Pyrotechnic Systems

11   Aircraft Systems

Because of their high power-to-weight ratio, high reliability, low weight and small size, pyrotechnic systems are employed advantageously in both military and civilian aircraft operations. To provide for quick escape in case of emergency on the ground, the new generation of commercial jet aircraft is being equipped with large inflatable escape slides. At the same time, extensive tests are being conducted in the development of egress door blow-out systems for commercial aircraft to provide for rapid evacuation in emergency on the ground.

Pyrotechnics find also extensive use in seat ejection systems, in release systems for external fuel tanks and stores and as jet engine starters on military and experimental aircraft. On dangerous missions, small pyrotechnic destructors are carried by military aircraft to provide the pilot with highly-efficient means for destroying the aircraft in case he is being shot down by the enemy.

Drones used as targets and for surveillance are usually equipped with pyrotechnic systems for tracking and recovery. Flares are ejected for tracking the vehicle, and at the end of a drone mission, drogue chutes are ejected by pyrotechnic drogue gun systems for drone recovery.

For several years, feasibility studies have been conducted with the objective to increase the survivability of aircraft crews and passengers by jettisoning under exact control various aircraft parts, such as an engine, a large fuselage portion, or the rotor unit of a helicopter, if necessary. Large parachutes or flexible wings can be considered for a safe descent of the recoverable structure containing the passengers. These feasibility studies have lead to the conclusion that the development of pyrotechnics has reached the point that pyrotechnic systems and devices can safely be employed in such emergency rescue and recovery systems.

A. *Seat Ejection Systems*

To save the lives of military pilots, test pilots and crew members when they must give up their troubled aircraft in flight, they have to be equipped with a reliable escape system capable of ejecting them within a very short time and landing them safely on the ground. During World War II, the ejection seat was a proven device in the German Luftwaffe (= Air Force), with more than 60 reported successful ejections in combat missions, before it was brought to the United States for study. Since that time, through extensive testing and design work, several different seat ejection systems for subsonic and supersonic aircraft have been developed, and they have proved in many hundred cases of emergency to be reliable life saving devices. In addition to seat ejection systems, emergency escape capsules have been developed for use in highly-advanced supersonic aircraft.

Two basically different types of seat ejection systems, which are described in this chapter, are used:

a. Seats ejected by a pushing force, and

b. Seats ejected by a pulling force.

Both types of seats are ejected upward. A third type, a downward seat ejection system, is not described, because it does not provide escape possibility at low altitude.

In both seat ejection systems, which must fit into a limited space in the aircraft cockpit, the small size, high power-to-weight ratio, high reliability and low weight of a variety of pyrotechnic devices, such as initiators, drogue guns, ejectors or catapults, line cutters and mortars are ideally utilized.

Reliable seat ejection systems are needed at low altitudes and low speeds, where emergencies frequently occur, and at high altitudes and high speeds. At speeds which exceed the velocity of sound by factors of two or three, the use of open ejection seats is not possible, because the air around the aircraft is subjected to heating caused by the intense shock waves.

When actuated, an ejection seat of the first category, equipped with a drogue chute and parachute system, is ejected with the occupant by means of a catapult (also called "ejection gun"), which is fixed to the aircraft behind the seat. The catapult often consists of three telescoping tubes and is fired by an explosive cartridge. To increase the height during ejection, some seat systems, as for example the Martin-Baker seats, are additionally equipped with a rocket motor mounted underneath the seat. As the seat leaves the aircraft, ejected by the catapult, the rocket motor is ignited, and the combined force of the catapult and the rocket propels the seat and the occupant to a height of about 300 feet. This height is sufficient to ensure full para-chute deployment when the seat is ejected at ground level without forward speed. In this seat system equipped with the automatically sequencing rocket catapult propulsion unit, a manual override mechanism is provided.

A lift-off test with a Weber ejection seat is shown in Fig. 11.1, and a rocket motor unit as used on the Martin-Baker Mk. 9 ejection seats is presented in Fig. 11.2.

An explosive-actuated system is used in most military aircraft for jettisoning the cockpit canopy in an emergency. This canopy release

Fig. 11:1 Ejection seat lift-off test (Courtesy of Weber Aircraft Co.).

Fig. 11:2 Rocket motor of mk 9 ejection seat (Courtesy of Martin-Baker Aircraft Co. Ltd.).

system is usually linked to the ejection seat actuation system, requiring only one operation by the occupant to jettison the canopy and to eject himself with the seat. To ensure that the canopy is clear of the aircraft before the seat ejects, the system is equipped with suitable time-delay devices.

After the seat leaves the aircraft, the sequencing system activates a drogue gun to fire a drogue chute which is attached to the top of the seat. The deployed drogue chute stabilizes the seat and reduces its speed so that the occupant's parachute can be deployed without a too great opening shock.

When the seat's velocity has been reduced sufficiently by the drogue chute, a barostatically controlled time-release device, which is attached to the seat, releases the drogue chute from the top of the seat, thus transferring the drogue's pull to the canopy of the occupant's parachute and pulling it out of its pack. Simultaneously, the time-release device releases the occupant's safety harness from the seat, and the occupant is pulled from the seat by the opening parachute, while the seat falls free.

In case an ejection is made at an altitude exceeding 10,000 feet, the barostat control device delays the action of the time-release

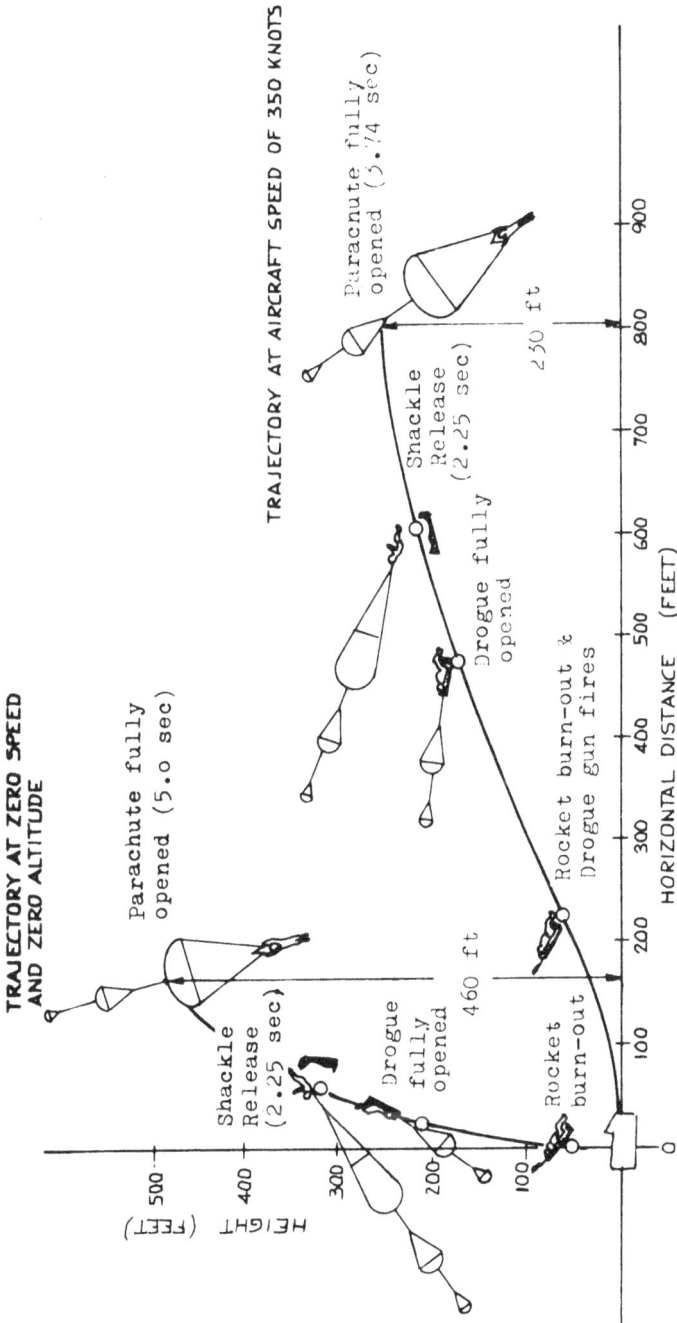

TRAJECTORY AT ZERO SPEED AND ZERO ALTITUDE

TRAJECTORY AT AIRCRAFT SPEED OF 350 KNOTS

Parachute fully opened (5.0 sec)

Shackle Release (2.25 sec)

Drogue fully opened

460 ft

Rocket burn-out

Paracnute fully opened (5.74 sec)

Shackle Release (2.25 sec)

Drogue fully opened

Rocket burn-out & Drogue gun fires

250 ft

HORIZONTAL DISTANCE (FEET)

HEIGHT (FEET)

Fig. 11:3 Trajectories of mk 9 ejection seat (Courtesy of Martin-Baker Aircraft Co. Ltd.).

191

device, until seat and occupant have descended to that altitude, which causes the occupant to move quickly down through the cold and rarified atmosphere on the seat stabilized by the drogue chute, and breathing oxygen which is contained on the seat and automatically turned on during ejection.

An important feature of all seat ejection systems is an automatic cartridge-actuated harness retraction system, which ensures that the occupant is pulled back in the seat in the most favorable position prior to ejection. Combined with the harness retraction system is an inertia lock system which permits the occupant full freedom of movement in the cockpit, but at the same time automatically restrains him against the normal G-forces encountered while flying, and automatically provides full restraint in the event of ejection or crash landing. A leg restraining system is incorporated in this harness system to prevent the occupant's legs from flailing when the ejection seat leaves the aircraft.

Trajectories of the Martin-Baker Mk 9 Ejection Seat at an aircraft speed of 350 knots and at zero speed and zero altitude are presented in Fig. 11:3. Seat ejection systems of this type, as described, have a speed range of up to 540 knots and an altitude range from 0 to 60,000 feet, and they deliver a maximum launch force of up to 17 G.

Greatly improved seat ejection systems have been developed during recent years with the major objective to limit ejection spinal loads to levels well below the limits of human tolerances and thus, to prevent serious injuries. The Douglas ESCAPAC Ejection Seat was developed to meet these requirements. To further improve its performance capabilities, a seat stabilization system was incorporated in this ejection seat, mainly to counteract the adverse effects of aerodynamic and dynamic center of gravity variations. Equipped with a suitable ejection sequence system, this seat is ideally used in tandem applications. Either crewman can accomplish initiation of the fully automatic emergency escape sequence by pulling one of the seat firing control handles.

This advanced stabilized ejection seat is presented in Fig. 11:4, and the backrest of this seat with the parachute installation and the mortar and retraction inertia reels is shown in Fig. 11:5. The drogue gun developed for this seat system is shown in Fig. 11:6. The slug is attached by means of a withdrawal cable to the latch pin of the drogue container lid and the extraction deployment bag. The breech

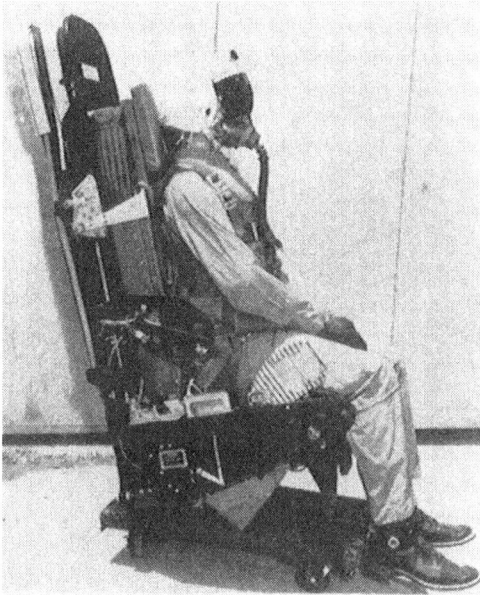

Fig. 11:4 Advanced stabilized ejection seat (Courtesy of Douglas Aircraft Co.).

of the drogue gun housing accommodates a removable mechanical firing mechanism and the cartridge. The details of the parachute mortar, which is shown in installed condition in Fig. 11:5, are clearly shown in Fig. 11:7.

The retraction inertia reels, which are located in the back of the seat, serve the purpose of automatically retracting the harness straps for prepositioning of the crewman prior to ejection. These retraction inertia reels are self-contained units consisting of a gear reduction assembly with an integral pneumatically fired gas generator, adapted to the standard dual-strap type inertia reel.

The escape system in tandem arrangement operates as described in the following:

The automatic emergency escape sequence is initiated by the crewman pulling either his face curtain firing control or the alternate firing control located on the front center of his seat bucket.

Ejection initiation by the rear crewman first jettisons the canopy, and power retracts his shoulder harness. After an interval of 0.4

Fig. 11:5 Backrest with parachute installation (Courtesy of Douglas Aircraft Co.).

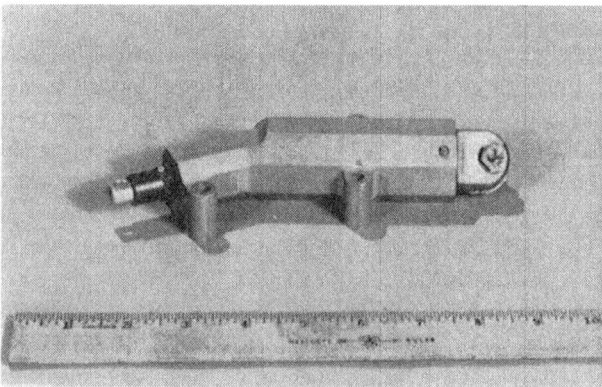

Fig. 11:6 Drougue gun for an ejection seat (Courtesy of Douglas Aircraft Co.).

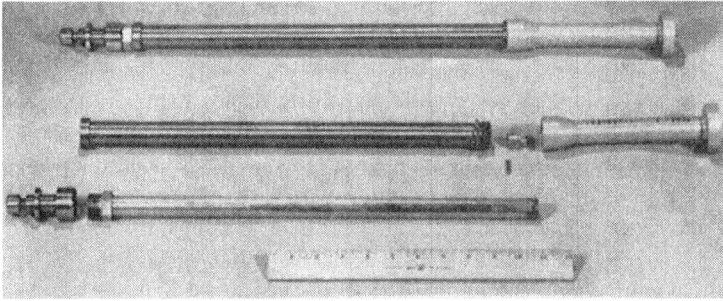

Fig. 11:7 Parachute mortar for an ejection seat (Courtesy of Douglas Aircraft Co.).

seconds, his seat ejects. In case the canopy fails to jettison, the seat will eject through the canopy.

Ejection initiation by the forward crewman will automatically jettison the canopy, power-retract both crewman's shoulder harnesses, and then eject both crewmen. The rear seat will eject 0.4 second after escape initiation, and the forward seat will eject 0.8 second after escape initiation. If the canopy fails to jettison, both seats will eject through the canopy at the set time intervals.

If desired, the canopy may be jettisoned independently by either crewman using the control handle provided in each cockpit. This action will not affect the status of the ejection seats, and subsequent escape will be accomplished in the sequence described.

These ejection seat systems are being further improved by incorporating a stabilization package, which controls the attitude of the seat once it is free of the aircraft and before the recovery parachute is deployed. The stabilization package utilizes a small gyroscope to sense the pitch of the ejected mass and to rotate a small vernier rocket to achieve trajectory control, which is necessary to eliminate random motion that could hinder the recovery sequence. Another new feature of these improved ejection seat systems is a seat-mounted sensor (pitot) that detects seat altitude and speed after ejection. It will switch a built-in electronic sequence programmer to the appropriate recovery mode for low speed, high speed and/or high altitude.

After ejection initiation, as the seat approaches the top of the rails, a sustainer rocket fires to accelerate the seat forward and up into the desired trajectory. The sequence programmer is activated and it, in turn, initiates the stabilization package and sets the correct parachute deployment mode.

If the ejection has taken place at low speed, i.e. below 250 knots, the programmer commands immediate deployment of the recovery parachute. With the seat rotated to a reclining position and the parachute aimed to the rear, a mortar is fired to extend the parachute to its full line length. The canopy, held partially closed for one second by reefing lines to obtain smooth filling, is then fully deployed. The seat is released and falls away from the crewman. A tether extracts a survival equipment container from the seat.

For high-speed ejection, i.e. 250 to 600 knots, the programmer delays deployment of the main parachute and instead activates a small drogue parachute for pitch and yaw stability during free flight. Less than one second later, the main parachute fires, and the drogue chute is automatically severed. The remainder of the sequence is the same as for the low-speed mode.

At high altitudes, regardless of speed, the drogue chute is deployed and is retained until the occupied seat descends to 15,000 feet before the programmer commands main parachute deployment and jettisoning of the drogue chute. A built-in oxygen supply supports the crewman at a high altitude.

Improved ejection systems of the type as described have the capability of ejecting the seat and saving the life of the crewman, even when the aircraft is in an inverted flight position.

Only one ejection system which utilizes a pulling rather than a pushing force to eject the crewman from the aircraft will be described briefly. The "Yankee" system, developed by Stanley Aviation Corporation, became known as representative of this second category of ejection systems.

The "Yankee" ejection system was developed with the objective to avoid the problems of inherent instability which can result from propulsion devices and systems that push rather than pull. This system consists of a rocket stored in a rocket launcher in the cockpit, a parachute system, and a light-weight foldable seat. After the canopy is removed, the rocket launcher is erected to the required attitude, and the rocket which is connected to the crewman's shoulders with the two pendant lines, one attached to each parachute riser, is launched at a speed of 115 feet per second. When reaching stretch of the pendant line in safe distance from the crewman, the rocket is ignited and pulls the man and the light-weight seat with the parachute up, while the main seat remains in the cockpit. The seat pan of the extracted seat folds down, still providing the required back and head rest. To protect the ejected crewman from the heat and noise of the rocket, he is suspended 12 feet below the rocket. Shortly before burn-out of the rocket, after a burning time of approximately 0.5 to 0.6 second, the rocket bridles are disconnected, and, propelled by the remaining energy, the rocket moves away from the area above the crewman. At this time, a pilot chute, fired from a drogue gun is deployed. The pilot chute, in turn, opens the main parachute which carries the crewman safely to the ground.

The rocket used in the "Yankee" ejection system is equipped with two nozzles on the leading end. The nozzles deflect backwards about 30 degrees from the rocket axis and skewed so that they will impart a rotation to the rocket during burning. This rotation creates a conical umbrella above the crewman. To avoid imparting any twist to the pendant lines, the rocket is attached by a ball bearing swivel so that it can spin freely.

Emergency Escape Capsules

For high speed and high altitudes exceeding the operational conditions of seat ejection systems, various types of emergency escape capsule systems have been developed in recent years.

One of the first capsule systems was designed and built for the General Dynamics B-58 Hustler. Each of the three crewmen was provided with a separate capsule which was open in normal flight. A conventional canopy covered the cockpit. In case of failure of the aircraft's oxygen or pressurization system, each crewman could close his capsule to provide oxygen and pressure. With the capsule closed, the pilot could continue to fly the aircraft, since the flight controls were arranged inside his capsule. All flight instruments were visible through a window. In case of emergency, each crewman would first close the capsule and then remove the canopy above him. He would then automatically be ejected up and away from the aircraft. The capsule would supply oxygen and pressurization at high altitude and protect the crewman from supersonic wind blast and extreme temperatures.

When the capsule had descended to 15,000 feet, an automatic baroswitch device would initiate deployment of the main parachute.

This parachute, having a diameter of 32 feet, would lower the capsule slowly to the ground. Another automatic device was provided for cushioning the impact of the capsule of the ground or ice. In case of a water landing, an automatic inflation system would inflate bladder-like flotation bags enabling the capsule to float. These capsules, which were developed by Stanley Aviation Corporation, were designed also to serve as shelters and to provide food and equipment necessary for survival on land, water and ice.

A sketch of an escape capsule system is shown in Fig. 11:8.

A similar emergency escape capsule system is being used in the F-111 aircraft with the main difference, as compared to the B-58 system, that the two-man crew sits side by side rather than one behind the other. The cockpit of the F-111 serves as an operating compartment, an escape capsule and a survival shelter and has been designed as a self-contained, independent vehicle within the aircraft. In case the crew is forced to abandon the aircraft, an explosive cutting cord shears the cockpit module from the fuselage, a rocket motor ejects it upward, and it descends by parachute to the ground or sea. If the capsule lands in a very remote area, it can serve as a survival shelter.

This escape capsule is designed to eject safely at any speed or altitude, even at zero speed and zero altitude. In case the airplane lands in the ocean and sinks, it should be possible to separate the capsule without firing the escape rocket or deploying the parachute. The capsule will rise to the water surface and remain afloat, and the crew will not get wet and will not suffer from effects of exposure in icy water.

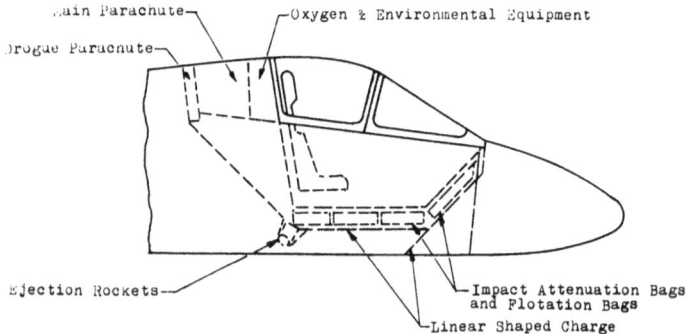

Fig. 11:8 Emergency escape capsule.

A rather unusual escape capsule system has been developed and tested by Soviet specialists. In case of emergency, on command from the pilot, the seat rotates into a horizontal position so that the back of the seat provides the bottom of a small capsule utilizing the aircraft canopy as the top of the capsule, which is sealed airtight. After separating the capsule from the airplane by detonation of an explosive charge, a small rocket located underneath the capsule bottom ejects the capsule upward with a thrust of several thousand pounds. Stabilizing fins and small parachutes are provided to ensure that the capsule is braked and remains in the correct position until the main parachute opens.

Capsules of this type have probably only been developed for small single-seater aircraft. Escape capsules similar to those described above, as used in the B-58 and F-111, are also being used in Soviet supersonic aircraft.

Ejectable escape capsules, especially nose capsules, may find increased application in the new generation of military and test aircraft which exceed Mach 2 speeds. These escape capsules do not only provide for safe recovery of aircraft crews, but they offer the added advantage of shirtsleeve environment for the crews, making it unnecessary to wear heavy pressure suits.

A new and very unconventional type of seat ejection system, the flying ejection seat, is at present being developed and tested for special military applications. The main objective of this development is to provide military pilots with a small jet-powered seat, which is suspended from a kite-like Rogallo wing, capable of flying at a speed of up to 90 mph to a safe area up to 50 miles away from the location of bail-out from their disabled aircraft. After reaching a safe area, the crewman would jettison his ejection seat and parachute to the ground. The flying ejection seat is shown in Fig. 11:9.

The Rogallo-wing version of the flying ejection seat, as shown, has a keel length of 6.7 feet, a span of approximately 13 feet, a wing area of 32 square feet, and its weight is less than 600 pounds. For flight control, a center of gravity system is used for pitch, and a wing bank system is used for turning.

Other types of flying ejection seats that are in the development and test stage of this AERCAB (Aircrew Escape and Rescue Capability) program include autogyro, Princeton sailwing with tail surfaces, and hot-air balloon concepts.

Fig. 11:9 Flying ejection seat (Courtesy of Brooks & Perkins, Inc.).

B. *Release Systems for Tanks, Stores and Equipment*

Military and research aircraft are often equipped with externally mounted fuel tanks, stores and similar devices to extend their performance capability. In case of emergency and after consumption of the fuel carried in these external tanks, the empty tanks and stores must be released from the aircraft to reduce the drag and weight and thus obtain the maximum aircraft performance. Especially in emergency situations, a rapid disposal of the tanks and stores and a safe recovery of this equipment for reuse is required. To fullfil this requirement, pyrotechnic release systems have been developed and are being used successfully.

The X-15-2 research aircraft was equipped with large external tanks which were suspended on both sides parallel to the fuselage underneath the wings, as shown in Fig. 11:10.

After the X-15-2 had been released from the B-52 mother ship, the fuel in the external tanks was used to give the aircraft the initial thrust on its climb trajectory. At an altitude of about 70,000 feet, at a

Fig. 11:10 External fuel tanks on x-15 research aircraft (Courtesy of Hi-Shear Corp.).

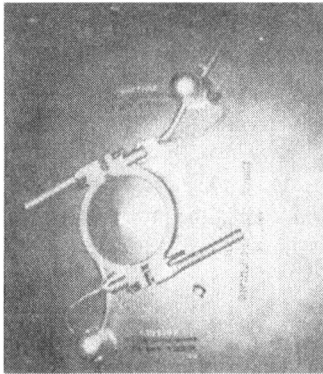

Fig. 11:11 Clamp separator for opening of valve at aft end of external tank. (Courtesy of Hi-Shear Corp.).

speed of Mach 2, the external tank release and recovery system was initiated, which could be accomplished either automatically or manually by the pilot. After initiation, once the tank release was begun, all subsequent functions of the system were automatic.

Each tank was equipped with a 5000-lb thrust rocket motor which, after release from the X-15, neutralized the airloads so that the release sequence could function as required. Immediately after tank dropoff, a pyrotechnic clamp separator, as shown in Fig. 11:11, opened a valve on the aft end of each tank to dump residual fuel and liquid oxygen. As the next sequence, a thruster on each tank ejected the drogue chute door on the lower portion of the tank's nose cone.

Then, a drogue gun, mounted at the nose cone, ejected the drogue chute. After deployment, the drogue chutes stabilized the tanks in vertical position, thus dumping the remaining fuel and slowing the descent of the tanks until they reached 15,000 feet. At this altitude, both drogue chutes were released by firing a separation nut, and then the nose cone was ejected. The next sequence was activation of a thruster in the nose cone of each tank to release the nose shrouds and deploy the main parachutes, which were used as main recovery devices, carrying the tanks at a reasonable descent velocity to the ground so that they landed with a mild impact to avoid serious damage. At ground impact, a pin puller in each tank fired to disconnect the parachutes to prevent the tanks from dragging over the ground, which could be caused by ground wind.

This tank release system utilizing a variety of pyrotechnic devices advantageously, functioned reliably and thus, contributed substantially to the success of the whole X-15-2 flight program.

For quick release of externally mounted tanks, stores and similar equipment from an aircraft, some non-electric stimulus transfer systems (NESTS) have been developed and applied as efficient and reliable release systems. The component arrangement of a pylon separation system, utilizing this non-electric stimulus transfer concept, by which four explosive studs of 3/4 inch diameter are separated, is shown in Fig. 11:12.

Only one initiator is required for this system. Each of the explosive studs is designed for a tensile load of 50,000 pounds. Confined

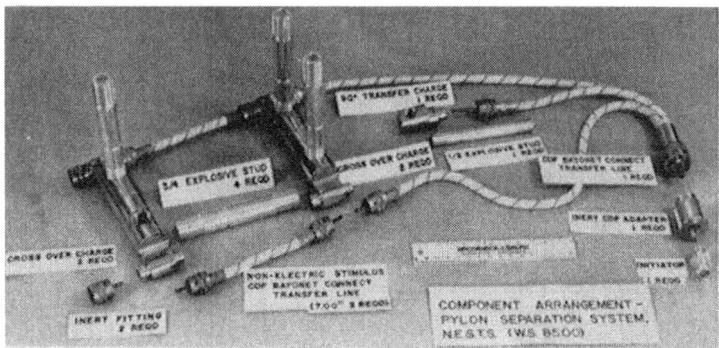

Fig. 11:12 Pylon separation system utilizing non-electric stimulus transfer system (Courtesy of McCormick-Selph).

detonating fuse (CDF) lines, containing only PETN secondary explosive, are used as transfer lines. To actuate this pylon separation system, a crewman in the cockpit pushes an initiation button, and within 10 milliseconds all explosive studs break simultaneously, thus releasing the pylons.

Pylon release systems of this type can easily be replaced after the aircraft returns to its base. The explosive studs and the confined detonating fuse (CDF) can be quickly installed in the aircraft at the same time the externally mounted equipment is replaced.

C. *Emergency and Rescue Systems*

With the development and introduction of large commercial jet aircraft, such as the jumbo jets with a capacity of 350 to 490 passengers, the urgent need for quick-operating and reliable egress systems arose. However, for several reasons, it cannot be expected that passenger or crew egress systems for emergency escape from flying commercial aircraft will be developed to be available within the next few years. A suitable egress and rescue system for aircraft emergencies on the ground has been developed and is being used as a standard safety equipment on these large airliners. Considerably improved systems for egress and rescue in case of ditching in the ocean and in case of airplane fires during ground operations must be provided in the near future to prevent the loss of numerous lives as a result of such aircraft accidents.

According to crash investigations which were conducted at the NASA Lewis Research Center, airline passengers have from one to four minutes to evacuate the aircraft in the event of a survivable crash involving fire. For the first 15 seconds after the airplane came to a stop, the action of the passengers was to sit quietly while any fuel mist surrounding the airplane was consumed. The arrangement and number of emergency exits and their dimensional clearances in current commercial airliners are provided according to the FAA safety regulations, which were issued as a result of the mentioned crash investigations and tests.

For safety in ocean ditchings, important factors are the behavioral characteristics of the airplane during impact and deceleration, the strength capacity of the bottom of the fuselage and its buoyancy. Immediate evacuation of the aircraft is not necessary if the listed capabilities are sufficient, and evacuation should only begin when the life rafts are launched and inflated.

For quick emergency evacuation of large commercial jet aircraft, inflatable escape slide systems have been developed as standard equipment. A typical escape slide system consists of a slide bag, a cool gas generator, aspirators, and connecting flexible hoses. The slide is usually made from a gas-tight elastomer-coated durable fabric and has a U-or V-shaped cross-section. The sides of the slide provide protection for the evacuating passengers against falling down over the edges.

Cool-gas generators, as described in Chapter 8 of this book, are ideally utilized for inflation of these escape slides, because they deliver the required amount of gas at the desired temperature within a few seconds, and because they are safe in operation, and they are of very small size and low weight.

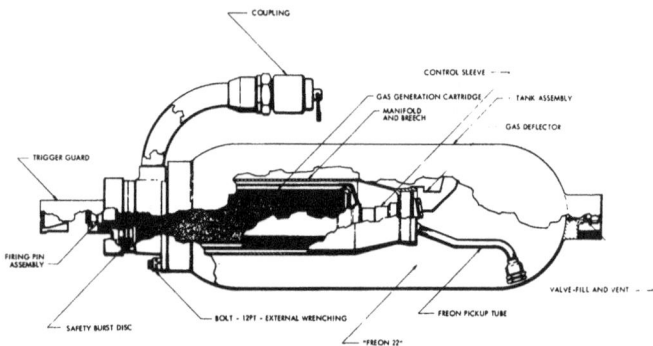

Fig. 11:13 Gas generator for escape slide (Courtesy of Rocket Research Corp.).

Fig. 11:13 presents a cross-section through a gas generator assembly, as used in an escape slide system for large jet airliners. The same gas generator with two aspirators is shown in Fig. 11:14.

When the gas generator is activated, the generated gas drives an aspirator which pumps air into the slide at high speed. Depending on the efficiency of the aspirator, the time span for inflation of the slide ranges from ·2 to 10 seconds. Escape slides for large jet airliners have a volume of 200 to 400 cubic feet and use a gas pressure from 1.5 to 3.0 psig. The gas generator-aspirator system remains attached to the slide after inflation, and thus, the gas generator can continue to

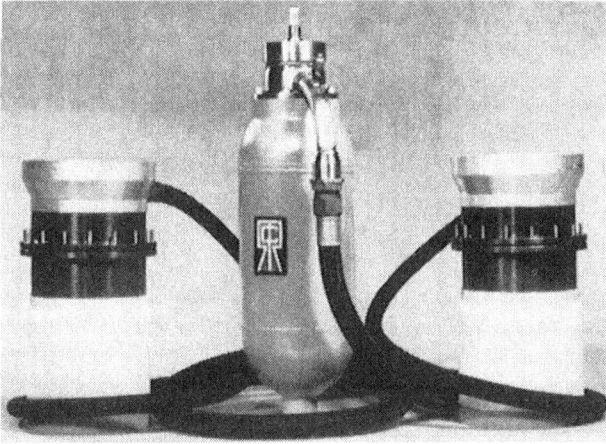

Fig. 11:14 Gas generator with aspirators (Courtesy of Rocket Research Corp.).

deliver a small amount of gas to the slide for several minutes to compensate for a possible pressure drop when the slide is exposed to extreme cold air or is immersed in cold water. These escape slide systems can be easily refurbished after they have been used. The total weight of an escape slide system, as described, is from 10 to 30 pounds.

A different type of escape system for evacuation of an aircraft is the "Inflatostair" system, developed by the Goodyear Aerospace Corporation. This system provides an inflated fabric stairway for the passengers to walk down rather than slide down to safety, in case of emergency. This system is equipped with chemical lighting which automatically illuminates the interior to aid in expediting evacuation. In stored condition, it requires a very small volume, and it inflates automatically at the turn of a switch. The near future will show which of both escape systems, the slide system or the inflated stair system, will provide the safest and quickest escape in emergency.

Similar gas generator and air aspirator systems are used to inflate life rafts and flotation bags for helicopters and and emergency escape capsules. A flotation system, as used on the Sikorsky S-58 helicopter, is shown in Fig. 11:15. A spherical flotation bag having a volume of 80 cubic foot is mounted in a pop-off hub cap on each wheel. A cool-gas generator for each side is attached to the wheel struts. When

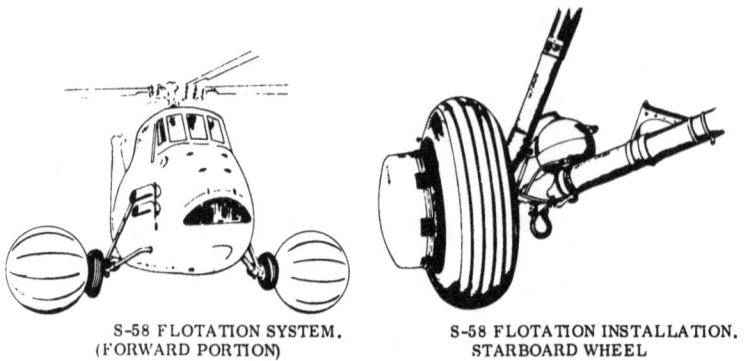

S-58 FLOTATION SYSTEM.
(FORWARD PORTION)

S-58 FLOTATION INSTALLATION.
STARBOARD WHEEL

Fig. 11:15 Helicopter flotation system (Courtesy of Walter Kidde & Co., Inc.).

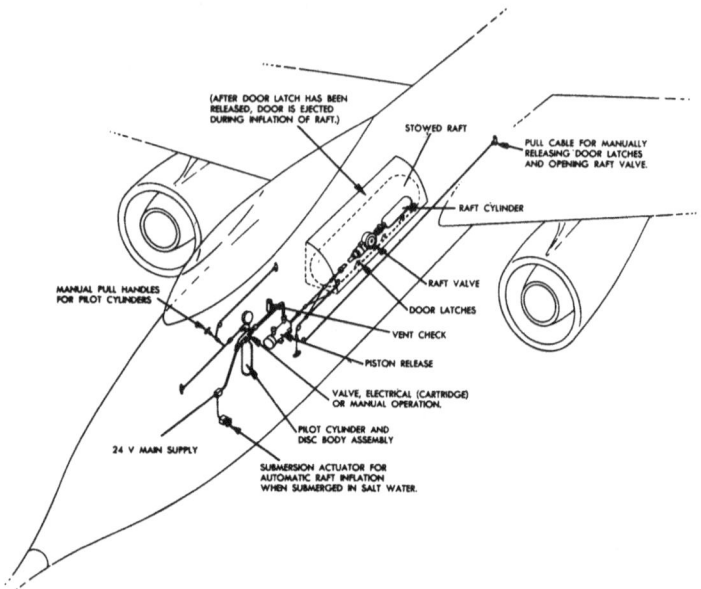

Fig. 11:16 Automatic life raft for aircraft-typical schematic installation (Courtesy of Walter Kidde & Co., Inc.).

actuated, the gas is delivered from the gas generator through a check valve mounted inside the hollow axle. When submerged, an immersion switch provided on each mount actuates both gas generators. The current for initiation is provided by the airplane's battery.

A similar, but manually-controlled flotation system is used on a Kaman helicopter. The flotation bags are attached to both sides of the fuselage. The system including the gas generators is stored in a compartment covered with a fiberglass door to reduce drag.

A schematic installation of a life raft inflation system, which can be activated either automatically or manually, is presented in Fig. 11:16. When the airplane ditches and the submersion actuator is in contact with salt water, the automatic system prepares and ejects the life raft. In any case, the system can also be initiated manually, for example, when the pilot must abandon the aircraft before it ditches.

Blow-out Doors and Panels

Quick and safe evacuation is only possible when the doors and emergency exits of the airplane can be opened quickly and without difficulty. Caused by landing gear failure or other accidents on the ground, the fuselage can get twisted, resulting in jamming the doors and making an egress impossible. To provide emergency exits which can be opened in milliseconds, blow-out doors and panels have been developed, and extensive tests with these emergency exit devices have been conducted on commercial airplanes.

A blow-out door consists of a skin panel equipped with a double-wall metal frame which serves as a charge holder. A flexible linear shaped charge (FLSC) is mounted in the hollow wall of the frame, as shown in Fig. 11:17, or it is bonded in place to the metal skins. Safe-arm devices are provided in the system to make accidental firing

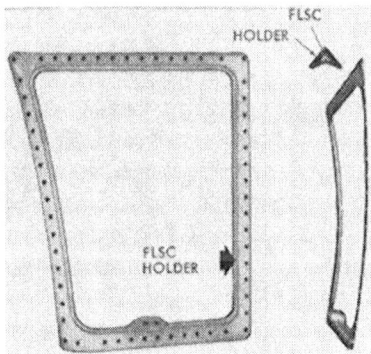

Fig. 11:17 Blow-out panel frame (Courtesy of McCormick-Selph).

impossible. When initiated, a high-energy shock wave concentrated in a thin line, caused by the explosive, results in a clean knife-like cut in the panel, and the panel is instantaneously ejected away from the aircraft. The safe-arm system is so designed that the charge cannot be initiated in flight, but only on the ground. The system is insensitive to environmental hazards, such as high or low temperature, lightning, impact, stray radiation or laser.

Blow-out panel systems can also be used to eject life rafts and valuable equipment, for example flight recorders and emergency beacons. To provide for quick and safe egress of the passengers, several blow-out doors and panels can be arranged at strategic locations in the aircraft. These panels can be connected by confined detonating fuse (CDF) transfer lines. In case of emergency, the blow-out panel system can be actuated simultaneously by an electrical signal to detonators to each blow-out panel assembly. A backup initiation system consisting of a pull-handle for firing a percussion primer to actuate the flexible linear shaped charge (FLSC) should also be provided to be able to use this emergency egress system in the event of electrical failure.

Airframe Severance Systems

Various emergency systems for safe landing of distressed aircraft and helicopters have been proposed and tested, utilizing similar methods as applied in the blow-out panel systems. As tests with military aircraft and helicopters proved, it is feasible to cut large parts and components off an airplane within a very short time by employing flexible linear shaped charge (FLSC). For example, the rotors of a helicopter in distress could be cut off, and large parachutes could then be deployed by a pyrotechnic sequencing system to land the main portion of the craft including crew and passengers safely. An artist's concept of a helicopter rescue, utilizing this principle, is shown in Fig. 11:18.

A similar method could be applied to release a failing engine or other equipment and structural parts from a large aircraft in flight, if necessary, or to eject very large fuselage panels from aircraft on the ground to facilitate mass evacuation of the passengers within the shortest possible time.

These rescue systems have not been approved for application in commercial aircraft yet, but their development is being continued

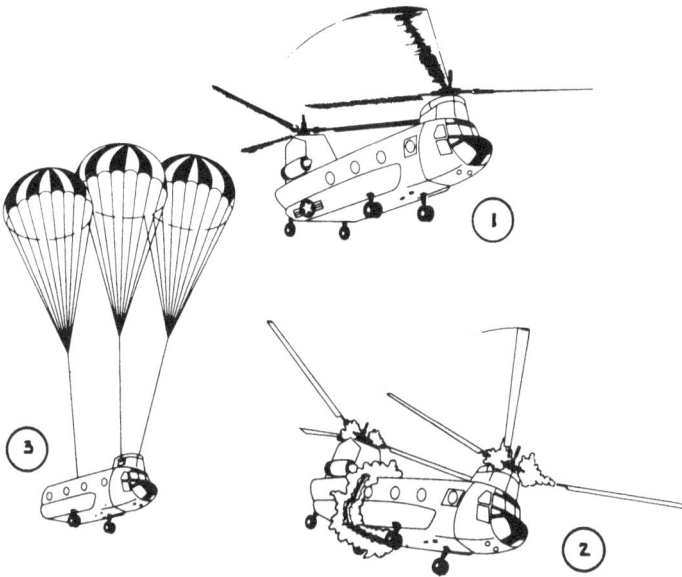

Fig. 11:18 Helicopter rescue system utilizing flsc for large scale cutting (Courtesy of McCormick-Selph).

with the objective to provide safe systems which will help to decrease the number of fatalities caused by aircraft accidents.

Emergency Hydraulic Systems

Hydraulic systems are used in aircraft mainly to provide backup power for landing gears, control systems and escape doors. Aircraft accidents as a result of failures of the aircraft hydraulic systems can be avoided by providing an emergency hydraulic system. A small, light-weight gas generator can be used in such an auxiliary system to pres-surize the hydraulic system. The gas generator can be installed in line with the conventional main hydraulic system. The gas delivered by the gas generator can be used either to drive a one-stroke piston pump which pushes the hydraulic fluid from a cylinder into the main system under the required pressure, or the generated gas can be delivered into a hydraulic accumulator which is connected to the main hydraulic system, thus eliminating the need for a piston device. To provide for easy refurbishing and cost reduction, replaceable propellant cartridges

209

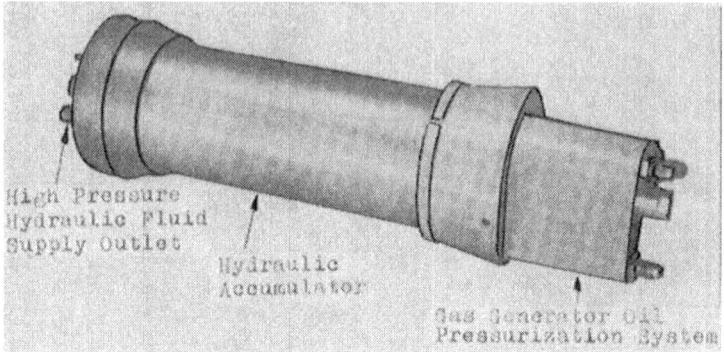

Fig. 11:19 Emergency hydraulic system (Courtesy of McCormick-Selph).

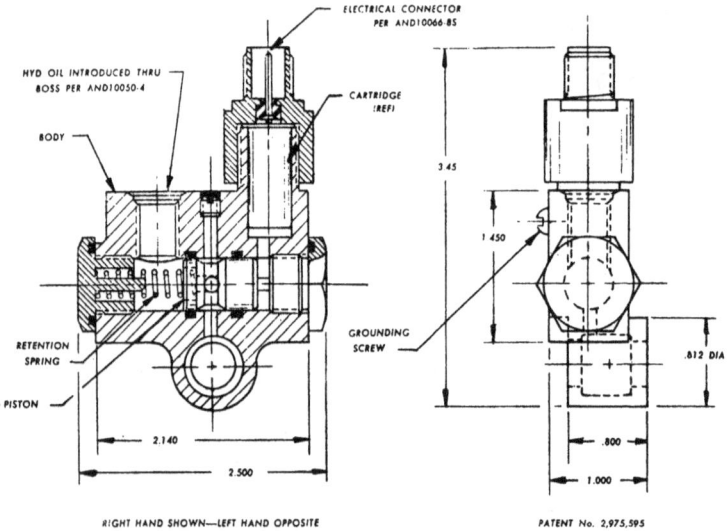

Fig. 11:20 Landing gear uplock actuator (Courtesy of Propellex, Div. of Chromalloy Corp.).

can be used in these gas generators. For activation of the gas generator, either an electrical initiator or a percussion primer, or for redundancy, both can be used. An emergency hydraulic system is shown in Fig. 11:19.

For lowering the landing gear in case of failure of the hydraulic or pneumatic system, a small and simple emergency uplock actuator has

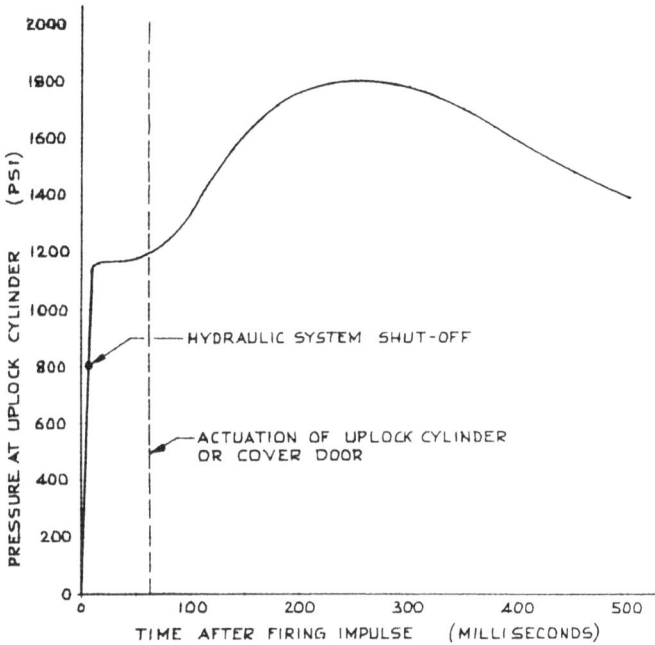

Fig. 11:21 Pressure vs. time diagram for landing gear uplock actuator (Courtesy of Propellex, Div. of Chromalloy Corp.).

been developed. This actuator which is designed on the piston principle can easily be adapted to existing hydraulic or pneumatic systems. In the event of failure of the hydraulic or pneumatic system which is used to actuate the landing gear uplock mechanism, the cartridge of the emergency uplock actuator is initiated and energizes the device, which seals off the hydraulic line and delivers a sufficient amount of gas to the uplock cylinder or cylinders to unlock the landing gear so that it can fall free into position. The same type of actuator can be used for emergency opening of the landing gear cover doors. Electrical or mechanical initiation can be used for these actuators. For electrical initiation, a firing voltage of 12 to 30 volts and a firing current of 10 amperes are recommended. The total firing energy required is 0.2 Watt-sec. The total weight of this small actuator is 0.78 pound. A cross-section through a typical landing gear uplock actuator, as described, is shown in Fig. 11:20, and a pressure-versus-time diagram of this device is presented in Fig. 11:21.

211

Drone Systems

Unmanned radio-controlled powered aircraft, known under the term "drones", are used by the armed forces as flying targets and as surveillance vehicles during day and night. When used as flying targets in military training, drones can simulate enemy aircraft. The small size, low weight and high reliability of pyrotechnic devices and systems are advantageously utilized in drones, especially since the load-carrying capability and volume of these small aircraft is limited.

For surveillance missions at night, drones are often equipped with magnesium flares mounted on the wing tips. These flares are ignited electrically from the drone's battery. High-speed target drones sometimes carry flash cartridges for spotting. (Flares and flash cartridges are described in Chapter 9 of this book).

Since drones must be capable of flying numerous missions without major refurbishing after each flight, they must be equipped with a highly reliable recovery system, which in most cases consists of a main parachute, a drogue chute and a pilot chute, or in simpler recovery systems, only a main parachute and a drogue chute are needed for a safe recovery operation. To initiate the recovery and landing of a drone, at a given radio signal, the recovery compartment door is opened by a solenoid, thruster or similar device. Then the pilot chute is ejected which, by a connecting line, extracts the drogue chute. Subsequently, after deployment of the drogue chute, the main parachute is extracted and deploys, carrying the drone safely to the ground. For ejection of small chutes, such as drogue chutes, small and light-weight ejector bags are often used. These bags are stowed in flat condition underneath the drogue chute in the recovery department. To eject the drogue chute, they are rapidly inflated through a connecting hose from a gas generator or a stored pressurized gas bottle.

In recovery systems for advanced, high-speed drones, drogue chute mortars or ejector guns are frequently used. A variety of drogue mortars has been developed. A simple mortar consists of a cylindrical housing which is closed at one end and which contains a piston. The aft end of the cylinder is equipped with a gas cartridge, and the small drogue chute, densely packed, is stowed in the cylinder in front of the piston. A light cover, attached to the cylindrical housing by shear pins, closes the forward end. When the cartridge is initiated, pressure builds up behind the piston. At a predetermined pressure level, the holding pins shear off, and the cover and the drogue chute are ejected.

Drones that are used near large bodies of water are also equipped with flotation bags and location aids to prevent loss of the valuable aircraft.

Drones that must be capable of landing at high ground wind conditions should be equipped with line cutters for the main parachute risers to release the parachute and thus, to prevent damage to the drone by getting dragged over the ground, caused by high wind.

12   Spacecraft Systems

Tremendous strides in space exploration and space travel have been made in recent years with unmanned and manned space vehicles. The great accomplishments in this field would have been impossible without utilizing of explosive-actuated devices and systems. Explosive power has found optimum use as the workhorse in numerous spacecraft systems not only because of its high power-to-weight ratio and high reliability, but because it delivers more energy in a shorter time than any other mechanical device. As electronics are utilized for the "brains" of spacecraft, explosive dynamics supply the power for the operation of a great variety of devices and systems in space vehicles.

Explosive-actuated devices and systems are advantageously used in almost all phases of a space mission, as for example for rocket motor ignition, launch pad separation, activation of electric batteries, initiation of reaction control systems, antenna deployment, fairing release, unbilical separation, ejection of equipment, stage separation, escape system initiation, ejection of recovery compartment doors, drogue chute ejection, parachute deployment, cutting of reefing lines, risers and cables, inflation of flotation bags, activation of location aids, and many other functions.

Explosive-actuated devices for space applications must be capable of withstanding extreme environmental conditions, such as high and low temperatures, high vacuum, humidity, shock and vibration. The success of a space mission depends to a high degree on the reliable functioning of the explosive-actuated devices under all conditions. To prevent outgassing and subsequent degradation of explosive compositions contained in devices to be used in space applications, a tight-closing seal on the charge-containing chamber is required. It can be expected that long-term exposure to space

215

environment will change critical properties of some explosives. Critical properties are sensitivity, explosive force, ignition temperature and thermodynamic properties. The estimated service life of some pyrotechnic devices for space applications is listed in Table 12:1.

Table 12.1

Estimated Service Life of Pyrotechnic Devices

Environment	Device			
	Sealed Squib	Parachute Disconnect (Shaped Charge)	Pressure Cartridge	Reefing Line Cutter
Vacuum Lunar Orbit Earth Orbit	45 days 45 days	14 days	45 days 45 days	45 days 45 days
Radiation Lunar Orbit Earth Orbit	2 years 2 years	2 years 2 years	2 years 2 years	2 years 2 years
Temperature Earth and Lunar Orbit	2 years	2 years	2 years	2 years
Zero G	2 years	2 years	2 years	2 years
Meteorites	2 years	2 years	2 years	2 years

The functions performed by pyrotechnic devices and systems in space-craft are classified as either crew critical or mission critical. The failure of a device used for a crew-critical function could result in the loss of the crew, whereas the failure of a device used for a mission-critical function could result in an alternate mission or in an aborted mission. Because of the high criticallity assigned to the various functions performed by pyrotechnic devices in manned spacecraft, maximum redundancy in devices and systems is required. When completely redundant systems cannot be considered because of space and weight limitations, redundant cartridges can be used. In cases, where redundant cartridges cannot be arranged, redundancy can be achieved by a single cartridge with dual initiators. To obtain the highest possible reliability in manned spacecraft pyrotechnic systems,

Fig. 12:1 Saturn V launch vehicle with Apollo (Courtesy of North American Rockwell Corp.).

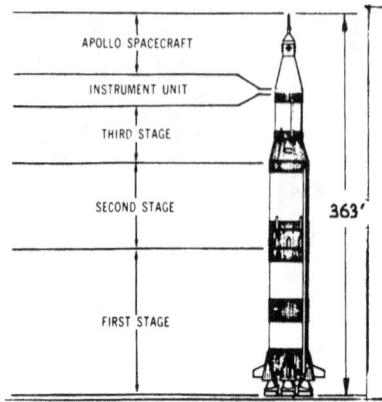

Fig. 12:2 Saturn-Apollo stage arrangement (Courtesy of McDonnel Douglas Corp.).

redundant electric circuitry and redundant control devices and batteries are provided.

The most complex pyrotechnic spacecraft system ever used is the Apollo-Saturn system. The Saturn V launch vehicle with Apollo during launch preparation is shown in Fig. 12:1, and the Saturn-Apollo stage arrangement is presented in Fig. 12:2.

The total height of this rocket-spacecraft combination is 363 feet. Its lift-off weight is 6,262,500 pounds, and its payload into a 115 statute-mile orbit is 280,000 pounds, or 100,000 pounds to the moon. The first stage of Saturn, whose engines develop 7.5 million pounds thrust, operates about 2.5 minutes to reach an altitude of about 200,000 feet (= 38 miles) at burnout. The second stage, which delivers more than one million pounds thrust, operates about 6 minutes from an altitude of about 200,000 feet (= 114.5 miles). Finally, the third stage, with 225,000 pounds thrust, operates about 2.75 minutes to an altitude of about 608,000 feet (= 115 miles) before second firing and 5.2 minutes to trans-lunar injection.

The reason for listing these data is to point out how substantial and important the tasks are which the small pyrotechnic devices must perform within very short time spans in these giant rockets and spacecraft.

The great variety and large number of pyrotechnic devices used in the Apollo spacecraft pyrotechnic system and their various functions are impressively shown in Fig. 12:3, and pyrotechnic devices and their applications in the Saturn-V pyrotechnic systems are presented in Fig. 12:4.

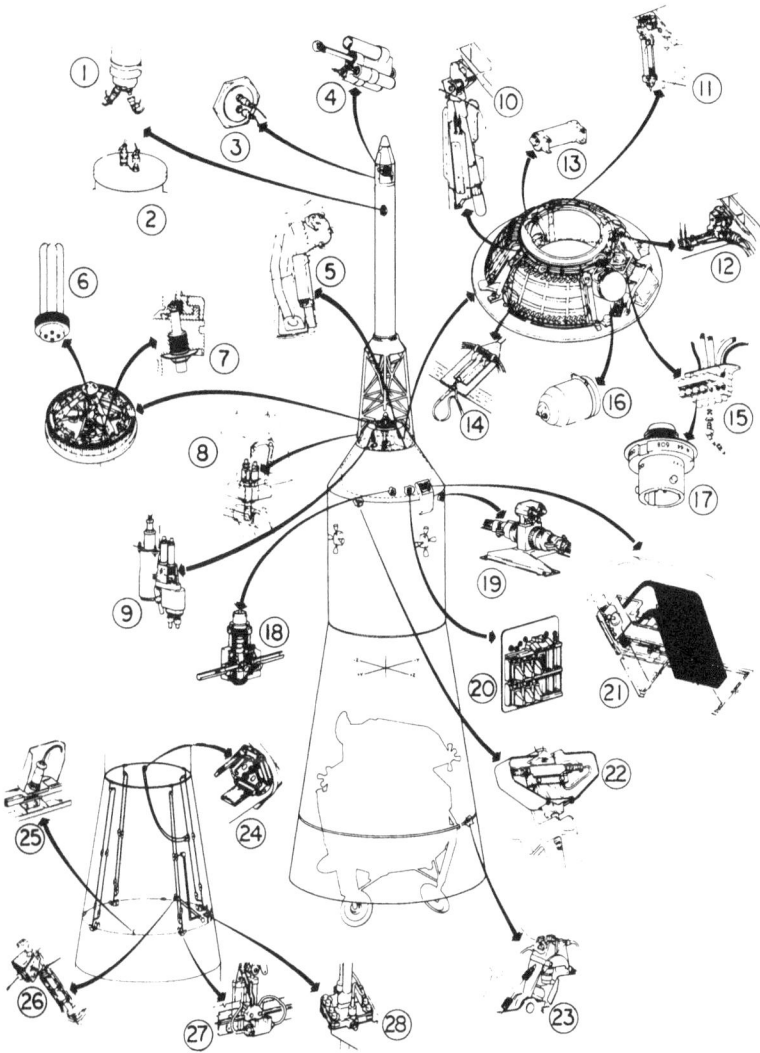

Fig. 12:3 Apollo spacecraft pyrotechnics

No.   Pyrotechnic Device

1   Tower Jettison Motor, 1 required
    initiated by Igniter Cartridge, 2 required

2   Launch Escape Motor, 1 required
    initiated by Igniter Cartridge, 2 required

3   Pitch Control Motor, 1 required
    initiated by Igniter Cartridge, 2 required

4   Canard Deployment Thruster, 1 required
    actuated by Cartridge, 2 required

5   Forward Heatshield Drag Parachute Deployment Mortar, 1 required
    actuated by Cartridge, 2 required

6   Docking Probe Retract Assembly 1 required
    actuated by Initiator, 4 required

7   CM Docking Ring Separation Assembly, 1 required
    initiated by Detonator, 2 required

8   Launch Escape Tower – CM Separation System Frangible Nut, 4 required
    initiated by Detonator, 2 required per assembly

9   Forward Heatshield Separation Thruster, 4 required
    actuated by cartridge, 4 required per vehicle

10 Recovery Beacon Light Deployment, 1 required
    released by Lanyard Line Cutter, 2 required

11 VHF Antenna No. 2 Deployment, 1 required
    released by Lanyard Line Cutter, 2 required

12 VHF Antenna No. 1 Deployment, 1 required
    released by Lanyard Line Cutter, 2 required

13 Pilot Parachute Deployment Mortar, 3 required
    actuated by Cartridge, 2 required per assembly

14 Drogue and Main Parachute Reefing Line Cutter, 26 required

15 Drogue and Main Parachute Disconnect Assembly, 1 required
    actuated by Cartridge, 5 required

16 Drogue Parachute Deployment Mortar, 2 required
    actuated by Cartridge, 2 required per assembly

17 Single Bridgewire Apollo Standard Initiator (Typ), 92 required

18 CM RCS (Helium/Propellant) Valves, 16 required,
    actuated by Cartridge, 1 required per assembly

19 SM Circuit Interrupter, 2 required
    actuated by Cartridge, 2 required per assembly

20 SM Circuit Interrupter, 2 required
    actuated by Cartridge, 2 required per assembly

21 CM–SM Umbilical Guillotine Assembly, 1 required
    initiated by Detonator, 2 required

22 CM–SM Tension Tie Cutter, 3 required
    initiated by Detonator, 2 required per assembly

23 LM Frangible Tie-Down Link, 4 required
    initiated by Detonator, 2 required per assembly

24 SM–SLA Umbilical Disconnect, 1 required
    initiated by Explosive Train, 2 required

25 SLA Panel Separation System (Cutting Charges), 1 required
    initiated by Detonator, 2 required

26 LM (GSE) Swing Arm Umbilical Guillotine, 1 required
    initiated by Explosive Train, 2 required

27 SLA Panel Deployment Thruster, 4 required
    actuated by Cartridge, 2 required per assembly

28 LM Lower Umbilical Guillotine, 1 required
    initiated by Detonator, 2 required

(Courtesy of North American Rockwell Corp.).

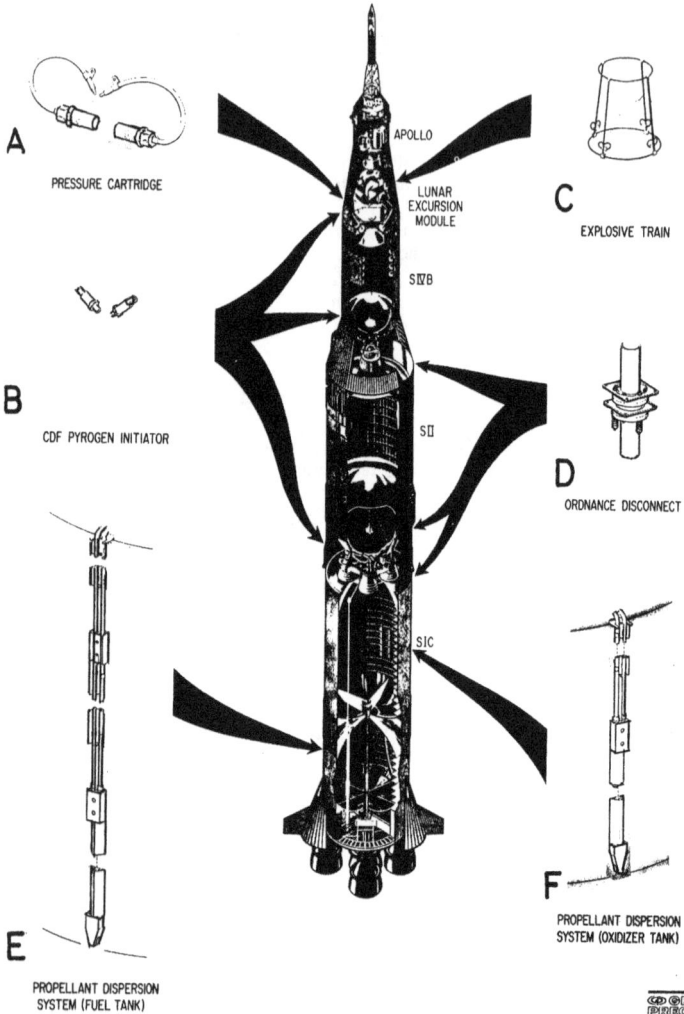

A PRESSURE CARTRIDGE

B CDF PYROGEN INITIATOR

C EXPLOSIVE TRAIN

D ORDNANCE DISCONNECT

E PROPELLANT DISPERSION SYSTEM (FUEL TANK)

F PROPELLANT DISPERSION SYSTEM (OXIDIZER TANK)

APOLLO

LUNAR EXCURSION MODULE

S IVB

S II

S IC

Fig. 12:4 Pyrotechnic system in Saturn V (Courtesy of General Precision, Inc., Link Ordnance Div.).

The complexity of the Apollo pyrotechnic system may be demonstrated by the fact that a total of 218 pyrotechnic devices, including 143 electrically initiated cartridged of 19 different types, are used in this system.

As the greatest innovation in the Apollo and Lunar Module, a modular concept of pyrotechnic devices and systems, using a standardized initiator, has been applied. The modular concept is based on the Apollo Standard Initiator ("ASI"), which is a fully qualified, standardized electro-explosive device. It is used as the basic energy conversion unit for all Apollo and Lunar Module pyrotechnic systems. While the term "ASI" is used for the concept, the standardized hardware is called "SBASI" (= Single-Bridgewire Apollo Standard Initiator). This standardization resulted in substantial cost reductions, shorter development times for higher assemblies, and higher reliability of the electro-explosive interface. It was found that dual bridgewires were not necessary, because the required redundancy could be achieved at higher levels in the system.

Three types of cartridges are used in the Apollo spacecraft: Igniter cartridges are used in rocket motors, pressure cartridges in mechanical devices, and detonator cartridges in high-explosive systems. Cartridge assemblies for special functions are obtained by adding booster modules to the standard SBASI.

The SBASI initiators incorporate great improvements, compared with conventional initiators. To obtain greater impact resistance at cryogenic temperatures, Inconel 718 is used as body material. The spark gap was moved to the interior of the initiator to increase the electrostatic survival capability from 9,000 volts to 25,000 volts, and to provide environmental and contamination protection. To increase the internal pressure capability to over 35,000 psi, a stepped header from Inconel 718 welded to the body is used, and the contact pins are glassed to the header.

To prevent improper installation, special-purpose cartridges with different outputs are equipped with different threads on the output ends. Cartridges which have the same output but which are located close to each other and are fired at different times are equipped with differently keyway-indexed connector ends.

The SBASI was developed in two body configurations:

a.   for use as a pressure cartridge, and

b.   for use as an ignition element of secondary charge in pyrotechnic assemblies. This configuration is obtained through the use of an integral washer welded to the initiator and to the next using device.

For redundancy, two firing circuits, A and B, which are completely independent and are electrically and physically isolated from each other and from other circuitry, are provided for the pyrotechnic system. The control components and the batteries that supply power for logic and firing are also redundant.

223

In a normal Apollo-Saturn lunar mission, pyrotechnic devices and systems are utilized for the following functions:

1. About 3 minutes after launch, the Launch Escape System (LES) is separated from the spacecraft and jettisoned immediately after ignition of the second stage booster by igniting the tower jettison motor and by firing the frangible nuts of the structural joint with Apollo (No. 1 and No. 8 in Fig. 12:3).

2. Upon completion of the first stage boost, the first stage is separated from the rest of the vehicle by actuating the structural disconnects (Illustration D in Fig. 12:4), and solid-propellant retro-rocket motors are fired to decelerate the spent stage as it breaks away. Ignition of these motors is accomplished by using confined detonating fuse (CDF) pyrogen initiators (Illustration A and B in Fig. 12:4).

3. After translunar injection by the third stage of the launch vehicle, the four panels of the Lunar Module Adapter (SLA) are separated from the spacecraft and jettisoned by firing redundant explosive trains and pyrotechnic thrusters, and the Command and Service Module (CSM) are separated from the launch vehicle. (Illustration C if Fig. 12:4 and No. 24 and 25 in Fig. 12:3).

4. After separation from the third stage, the Command and Service Module (CSM) returns and docks with the Lunar Module (LM). Through the docking interface, an umbilical is connected to the Lunar Module. To separate the LM, four frangible links connecting the Lunar Module to the fixed portion of the SLA are fired, and the umbilical is severed (No. 6 if Fig. 12:3).

5. In lunar orbit, prior to separation of the Lunar Module (LM) from the Command and Service Module (CSM), the LM landing gear is deployed by firing guillotines which sever tension straps that hold the gear in the retracted position. Springs are used to deploy the landing gear to the downlocked position. The LM landing gear uplock cutter is shown in Fig. 12:5.

6. To prepare for descent on the moon, the Lunar Module Reaction Control System (LM RSC) is functioned, and after system check, the Lunar Module (LM) separates from the Command and Service Module (CSM). Explosive valves are utilized in the Reaction Control System for pressurization of propellant tanks and numerous control operations.

7. During preparation for launch from the lunar surface, the explosive valves of the Ascent Propulsion System (APS) are fired.

Fig. 12:5 Landing gear uplock cutter for Lunar Module (Courtesy of Space Ordance Systems, Inc.).

8. Prior to launch of the LM ascent stage from the lunar surface, the LM stages are separated by actuating an explosive nut at each structural attachment point and by deadfacing the interstage electrical circuits and by severing the interstage umbilical by a guillotine (No. 26 and No. 28 in Fig. 12:3).

9. After lift-off of the LM ascent stage from the moon, and after rendevous, docking and transfer of the astronauts into the Command/ Service Module (CSM), the Lunar Module (LM) is separated from the CSM by actuating the LM Docking Ring Separation System and is jettisoned (No. 7 in Fig. 12:3, and Fig. 12:6).

10. Before entering the atmosphere of the earth at about 400,000 feet, the Command Module Reaction Control System (CM RCS) is activated by pressurizing the propellant tanks with helium which is released by opening explosive valves. The reaction Control System (RSC) is then used to orient the spacecraft (CSM) to separation attitude.

11. At separation of the Command Module (CM) from the Service Module (SM),the critical electrical circuits between both modules are deadfaced by circuit interrupters, the CM-SM umbilical is severed by a high-explosive operated guillotine, and the three structural tension ties connecting both modules are cut by actuating dual linear-shaped charges (No. 21 and No. 22 in Fig. 12:3).

Fig. 12:6 The Lunar Module separates for the Command/Service Module (Courtesy of the North American Rockwell Corp.).

Fig. 12:7 The Command Module separates from the Service Module (Courtesy of the North American Rockwell Corp.).

12. The Service Module (SM) moves away from the Command Module (CM), propelled by the +X thrusters of the Service Module Reaction Control System (SM RSC). (Fig. 12:7).

13. About 8 minutes after entry into the atmosphere, when the Command Module has reached an altitude of approximately 24,000 feet, the apex cover, which is also known as the forward heat shield, is jettisoned by a redundant thruster system. To prevent the loose cover from hitting the spacecraft or interfering with the drogue parachute deployment, a drag parachute, attached to the cover is fired from a lanyard switch-actuated mortar (No. 9 in Fig. 12:3).

14. Two seconds after jettison of the cover, two reefed drogue parachutes are deployed by mortars and are disreefed with a time delay of 10 seconds (No. 14 and No. 16 in Fig. 12:3).

15. About 40 seconds after drogue deployment, the drogue risers are severed by guillotines, and at the same time, the three pilot parachutes are deployed (No. 13 and No. 15 in Fig. 12:3).

16. The three main parachutes are deployed to a full-reefed condition by the pilot parachutes. Six reefing line cutters with a time delay of 8 seconds are actuated by the riser deployment of the main parachutes. These reefing line cutters also release a spring-loaded deployment mechanism of two VHF antennas and of a flashing beacon light provided to aid in recovery operations. Four reefing line cutters with 6

seconds time delay and two cutters with 10 seconds time delay on each of the three main parachutes are actuated at line stretch, thus disreefing the main parachutes in two stages to reduce the inflation shock loading on the parachutes (No. 10, 11, 12 and 14 in Fig. 12:3).

17. Immediately after splashdown, the main parachutes are disconnected by guillotines in the parachute disconnect assembly (No. 15 in Fig. 12:3).

These sixteen phases describe a successful Apollo-Saturn lunar mission. From launch to touchdown, pyrotechnic devices and systems of highest reliability contribute significantly to the success of such a mission.

A. *Launch and Control Systems*

Pyrotechnic devices and systems are ideally employed in spacecraft launch and control systems, mainly for rocket motor ignition at launch and for gas and fluid control in reaction control systems during flight.

In almost all solid-propellant rockets and in numerous liquid-propellant rockets, which are not ignited by hypergolic reaction or by an electric spark device, pyrotechnic igniters are used. Igniters are described in detail in part II, Chapter 3 of this book. A pyrotechnic igniter usually consists of an electric squib combined with a suitable solid-propellant igniter composition. After electrical ignition by a hot wire, the solid-propellant charge burns with a hot flame within the combustion chamber, heats the rocket propellant to its ignition temperature, and raises the pressure in the combustion chamber to the point where self-sustaining combustion takes place.

In applications where the rocket propellant is difficult to ignite, the chamber containing the igniter composition is covered with a nozzle closure which will only blow out when a specific minimum igniter chamber pressure has been attained.

The ignitibility of solid propellants depends to a high degree on the type of oxidizer used. Ammonium-perchlorate propellants are easy to ignite, potassium-perchlorate propellants and double-base propellants are of medium ignitibility, and ammonium-nitrate propellants are most difficult to ignite.

For applications where a very smooth and shockless ignition at low propellant temperatures must be achieved, igniter compositions containing magnesium, aluminum, or boron powder, and potassium nitrate or potassium perchlorate are used as an oxidizer to produce a hotter flame.

The igniter may be fitted onto the injector or the combustion chamber or may be mounted into the chamber through the nozzle throat of the thrust chamber. In a more sophisticated ignition system, a small precombustion chamber is arranged adjacent to the main combustion chamber and connected with the main chamber through an orifice. A small amount of fuel and oxidizer injected into the precombustion chamber is ignited, and the burning mixture, torch-like entering the main chamber, ignites the propellant which is injected into the main chamber. This method has been used successfully with liquid oxygen-gasoline and oxygen-alcohol thrust chambers. It offers the advantage that it permits repeated starting of variable thrust engines.

In the first stage of the Saturn V, the five F-1 rocket engines, in which RP-1 (kerosene) is used as fuel and liquid oxygen as oxidizer, are equipped with a gas generator system to provide hot gases for driving the velocity-compounded turbine which drives the fuel and oxidizer pump. Propellants entering the gas generator through a valve and injector are ignited in the combustion chamber by dual pyrotechnic igniters, which also serve the additional purpose of re-igniting the fuel-rich turbine exhaust gases as they exit from the nozzle extension.

The second stage boost of the Saturn consists of two steps. When all F-1 engines of the first stage have cut off, the first stage separates. To provide positive acceleration to the second stage prior to ignition of the five J-2 engines, eight ullage rocket motors are arranged around the periphery of the interstage structure between the first and second stage fire for approximately 4 seconds. Each of these ullage motors, having a diameter of 12:5 inches and a length of 89 inches, provides 22,500 pounds of thrust. The nozzles of these ullage motors are canted 10 degrees outward to reduce exhaust impingement against the structure. To provide the desired high performance and optimum mechanical properties under conditions encountered in space, a special formulation of Flexadyne solid propellant is utilized in the ullage motors. An illustration of an ullage motor assembly is presented in Fig. 12:8. Through-bulkhead initiators with confined detonating cord, as shown in Fig. 12:9, are used for simultaneous ignition of the ullage motors.

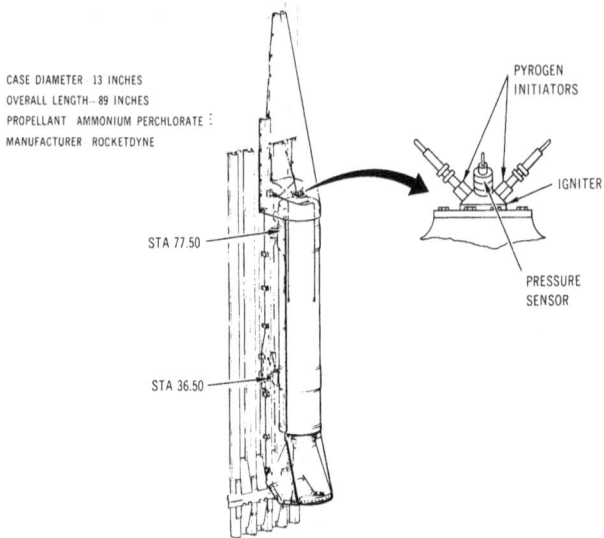

CASE DIAMETER 13 INCHES
OVERALL LENGTH 89 INCHES
PROPELLANT AMMONIUM PERCHLORATE
MANUFACTURER ROCKETDYNE

PYROGEN INITIATORS

IGNITER

STA 77.50

PRESSURE SENSOR

STA 36.50

Fig. 12:8 Ullage rocket motor assembly (Courtesy of North American Rockwell Corp.).

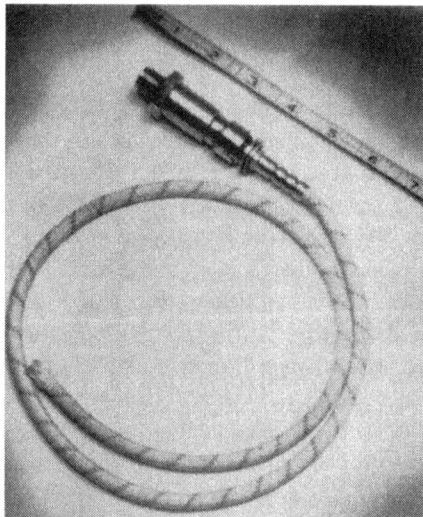

Fig. 12:9 Initiator with a confined detonating cord for small rocket motor ignition on Saturn (Courtesy of Link Ordnance Div., General Precision Inc.).

The successful operation of the Lunar Module (LM) also depends to a great extent on the proper functioning of pyrotechnic devices and systems. Prior to earth launch, all propellant tanks in the Lunar Module are only partly pressurized at less than 230 psia, so that the tanks are maintained within a safe pressure level under temperature changes experienced during launch and earth orbit. Before initial engine start, additional pressurization of the ullage space in each propellant tank is required. A relatively small amount of helium stored at ambient temperature and at intermediate pressure is used for this initial pressurization. The path from the ambient helium tank to the propellant tanks is opened by firing three normally-closed explosive valves: an ambient helium isolation valve and two propellant compatibility valves which prevent backflow of propellant vapors 'from degrading upstream components. The ambient helium passes through a filter and then enters a pressure regulator which reduces the helium pressure to approximately 245 psi.

After initial pressurization, supercritical helium is used to pressurize the propellants by opening a normally-closed explosive valve which is fired 1.3 seconds after the LM descent engine is started. This time delay prevents the supercritical helium from entering the fuel/helium heat exchanger until propellant flow is established so that the fuel cannot freeze in the heat exchanger. After the explosive valve opens, the supercritical helium enters the two-pass fuel/helium heat exchanger where it is slightly warmed by the fuel. The helium then flows back into a heat exchanger in the supercritical helium tank where it increases the temperature of the supercritical helium in the tank, causing a pressure rise and thus ensuring continuous expulsion of helium throughout the entire period of operation. Finally, the helium flows through the second loop of the fuel/helium heat exchanger where it is heated to operational temperature before it is regulated and routed to the propellant tanks. The helium pressure reduction system consists of two parallel, redundant regulators. In case that one pressure regulator fails, the astronauts close the malfunctioning line and open the redundant line to restore normal propellant tank pressurization.

After completion of their mission on the lunar surface, the astronauts take off in the ascent stage of the LM. The propulsion system of the ascent stage consists of a pressurization section, a propellant feed section, and an engine assembly, similar to the propulsion system of the LM descent stage. The engine, which develops 3500 pounds of

Fig. 12:10 Lunar Module (LM) on lunar surface (Courtesy of Grumman Aircraft Engineering Corp.).

thrust in a vacuum, can be shut down and restarted, as the mission requires.

Prior to initial ascent engine start, the propellant tanks must be fully pressurized with gaseous helium, which is stored in two tanks at a pressure of 3050 psia at a temperature of $+21°C$. The outlet of each helium tank is equipped with a normally closed explosive valve. To accomplish pressurization of the propellant tanks, the astronauts fire six explosive valves simultaneously: the two helium isolation valves and four propellant compatibility valves, of which two are connected in parallel for redundancy in each pressurization path.

If a pressure check before firing of the valves shows an unusual

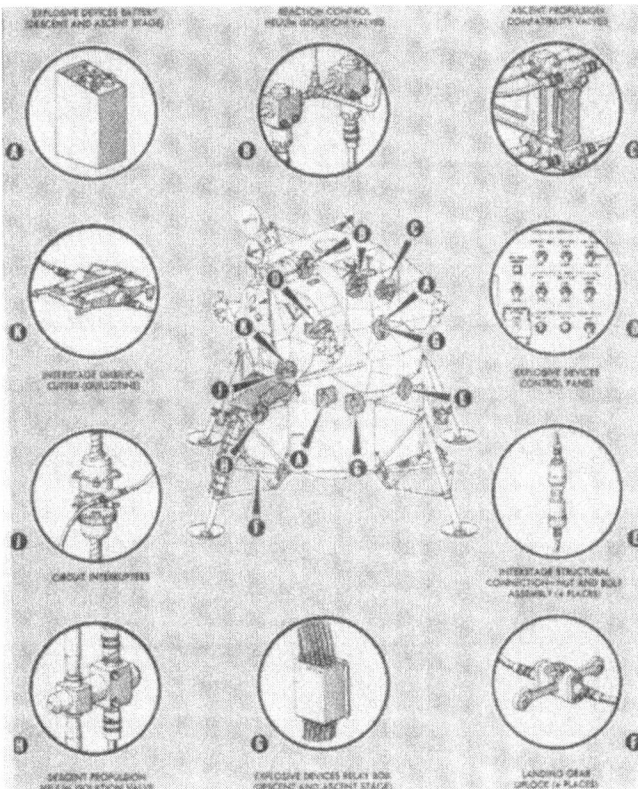

Fig. 12:11 Major explosive devices in the lunar module (Courtesy of Grumman Aircraft Engineering Corp.).

low reading of one tank, indicating leakage, the appropriate helium isolation valve can be excluded from the fire command, thus isolating the faulty tank from the pressurization system and preventing helium loss.

An illustration of the Lunar Module (LM) is presented in Fig. 12:10, and the location of major explosive devices in the Lunar Module is shown in Fig. 12:11.

The LM explosive devices subsystem is presented in form of a systems diagram in Fig. 12:12, and an LM ascent propulsion control diagram is shown in Fig. 12:13.

Explosive valves are also utilized in the LM reaction control subsystem and in the Command Module (CM) reaction control system of Apollo for helium tank isolation. As shown in Fig. 12:14, the LM ascent stage is equipped with reaction control quads mounted on extensions opposite to one another.

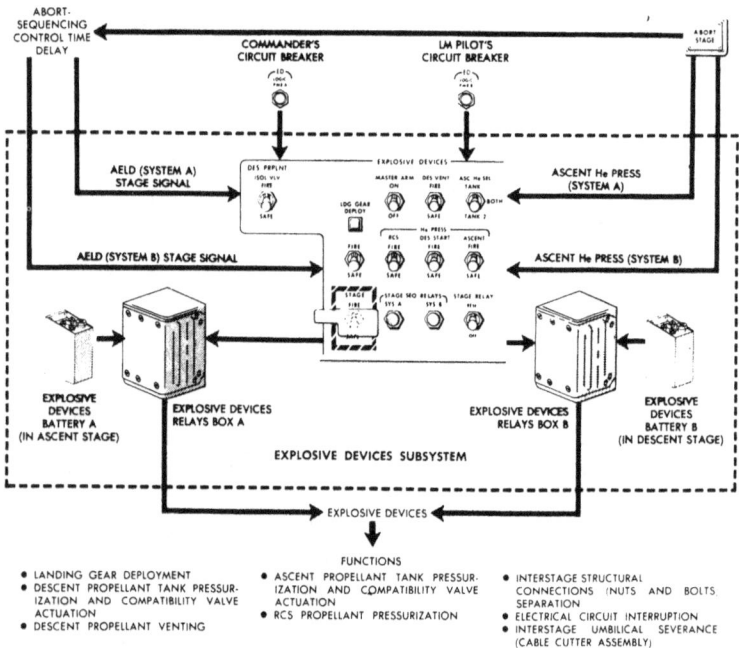

Fig. 12:12 Diagram of LM explosive devices subsystem (Courtesy of Grumman Aircraft Engineering Corp.).

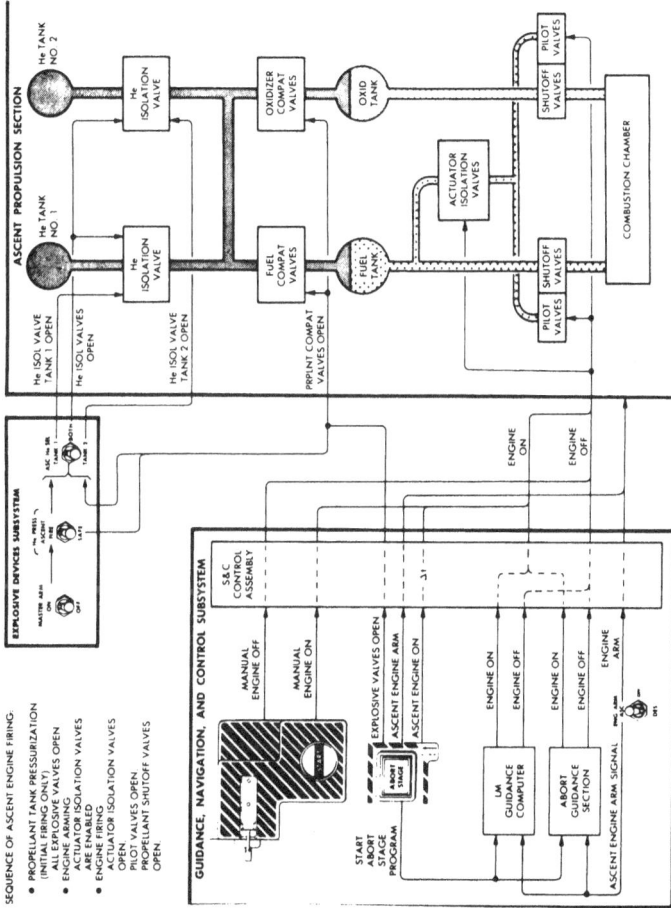

Fig. 12:13 Lunar Module ascent propulsion control diagram (Courtesy of Grumman\Aircraft Engineering Corp.).

235

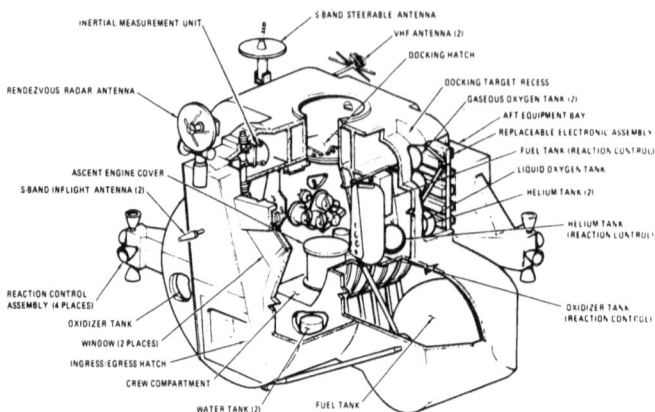

Fig. 12:14 Ascent stage of Lunar Module (Courtesy of North American Rockwell Corp.).

The Apollo Command Module Reaction Control System (CM RCS) provides the thrust for normal and emergency attitude maneuvers, and it operates in response to automatic control signals from the stabilization and control subsystem in conjunction with the guidance and navigation subsystem. It can also be controlled manually by the crew.

The CM reaction control system is used after separation of the Command Module (CM) from the Service Module (SM) and for certain

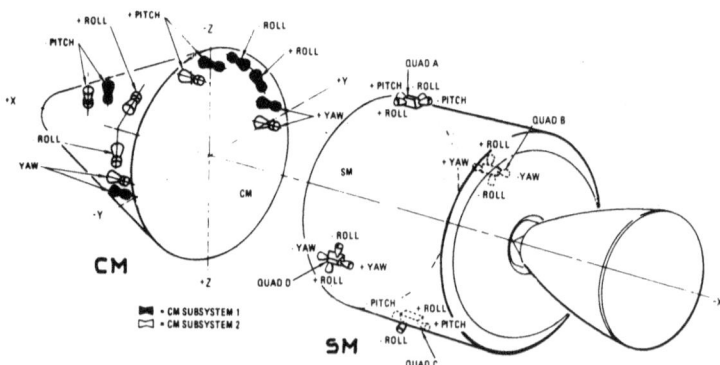

Fig. 12:15 Reaction control engines on Command Module and Service Module (Courtesy of North American Rockwell Corp.).

abort modes. It provides three-axis rotational and attitude control to orient and maintain the Command Module in the proper entry attitude, and during entry into the atmosphere, it provides the torque necessary to control roll attitude.

The reaction control system consists of two independent redundant systems. Each system contains six engines, helium and propellant tanks, and a dump and purge system. Each system is capable of pro-

Fig. 12:16 Apollo Command Module joined with Service Module (Courtesy of North American Rockwell Corp.).

viding all the impulse needed for the entry maneuvers, and both systems can also operate in tandem. Normally, only one system is used.

Ten of the twelve engines of this system are located in the aft
department and two in the forward compartment, as shown in Fig.
12:15. Each engine produces approximately 93 pounds of thrust. The
nozzle of each engine is ported through the CM heat shield, matching
the mold line.

Monomethyl hydrazine and nitrogen tetroxide are used as propellant, and helium is used for propellant tank pressurization. The total
helium for each system is contained in a spherical storage tank of
about 9 inches diameter which contains 0.57 pounds of helium at
4150 psia pressure. Two explosive helium isolation valves are provided
in the line from each helium tank to confine the helium into as small
an area as possible to reduce helium leakage. The high-pressure helium
flows through pressure regulators and check valves to the propellant
tanks, where it maintains pressure around the expulsion bladders. To
prevent inadvertent firing of the engines, the propellants are forced
into the engine feed lines through a burst diaphragm, and to the
engines.

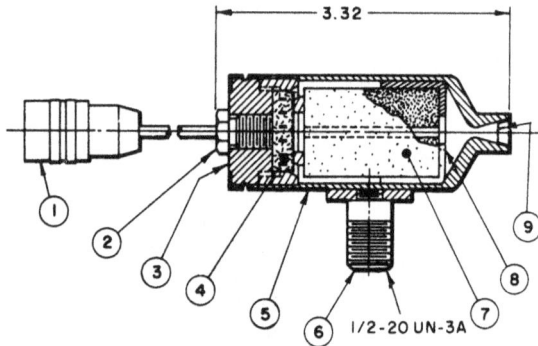

COMPONENTS:

1. Connector PT06P-8-3P
2. Primer
3. Igniter Assembly
4. Igniter Charge
5. Motor Case
6. Mounting Stud
7. Propellant Grain: Ammonium Perchlorate
8. Spacer
9. Nozzle Closure

Fig. 12:17 Small rocket motor for spin stabilization of satellite (Courtesy of Bermite, Div. of Whittaker Corp.).

The Apollo Service Module (SM) is equipped with a similar reaction control system with the engines in quad arrangement, as shown in Fig. 12:15 and 12:16, however, explosive valves are not utilized in this system.

Pyrotechnics are also ideally used for stabilization of satellites. A variety of types and sizes of small rocket motors for spin stabilization of satellites has been developed. A typical small rocket motor for such application is presented in Fig. 12:17.

B. *Emergency Systems*

An entirely new spectrum of safety requirements has been generated by space flight operations. Since the reliability of large liquid rockets and of some new unique and relatively untried spacecraft subsystems cannot approach the reliability of manned aircraft systems, it is essential that the reliability of spacecraft emergency and escape systems be of a higher order. The problems resulting from space environmental conditions, extreme velocities and altitudes add to the requirement for the development of highly reliable emergency and escape systems. The hazardous nature of some fuels used for spacecraft propulsion and reaction control is an important factor contributing to failure probability. Critical emergency situations can also be caused by failures in the equipment pressurization and cooling systems. Since malfunction during boost can occur with a rate of onset exceeding man's response time, the use of an automatic escape system is required during the boost phase, and a separation with safe distance within a prescribed time interval is necessary so that the crew escapes the fireball and the propagating pressure wave following the explosion.

Other danger points in space flight operations are malfunction during take-off and staging, in-flight emergencies, malfunctions during re-entry and crash at landing impact. The best chance for survival in each of these events may be given by immediate escape.

Pad Abort Systems

The Mercury spacecraft was equipped with a pad abort system, utilizing an escape tower as shown in Fig. 12:18. The escape tower, attached with bolts and explosive nuts to the forward end of the spacecraft, contains an escape rocket and a pitch control rocket motor

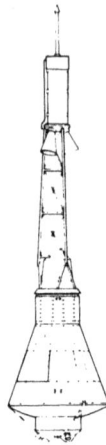

Fig. 12:18 Mercury spacecraft with escape tower for pad abort (Courtesy of North American Rockwell Corp.).

in the top. In case of an emergency with the spacecraft on the launch pad, action to fire the escape rocket could be initiated by either the astronaut or the launch crew. The escape rocket separated the spacecraft from the launch vehicle and carried it to an altitude of about 2500 feet. Near apogee, the drogue parachute was automatically deployed, and after an interval of two seconds and after the spacecraft had been oriented to a normal re-entry attitude, the main parachute was deployed. This system was designed for launch aborts occurring at altitudes below 21,000 feet. A normal reentry sequence for drogue parachute deployment can be accomplished at this altitude range with this pad abort system.

A diagram of a launch pad abort of spacecraft Mercury is presented in Fig. 12:19.

The launch pad abort system of the spacecraft Gemini consisted of two ejection seats, one for each crew member. Since in Gemini's booster, the Titan II, the fuel was of hypergolic type, which does not explode, a launch escape tower as used with Mercury as not required. The Gemini escape system was designed to eject both astronauts up and away from an on-the-pad booster malfunction in less than two seconds. The system functioned completely automatically after pilot initiation. Actuation could be accomplished by either crew member by pulling a "D" ring built into the front portion of each ejection

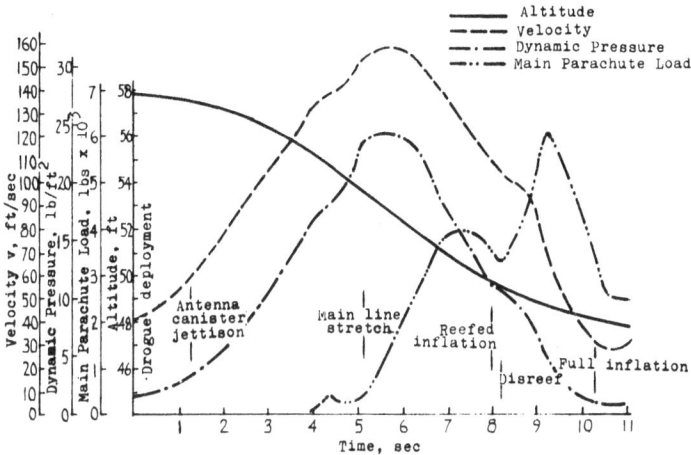

Fig. 12:19 Mercury launch pad abort.

seat. The system was designed for an altitude range from 0-70,000 feet and for a maximum launch force of 24 G. Gemini ejection seat trajectories for 5th and 95th percentile astronauts are shown in Fig. 12:20.

The Apollo launch escape system (LES) is designed to carry the command module (CM) containing the astronauts away from the

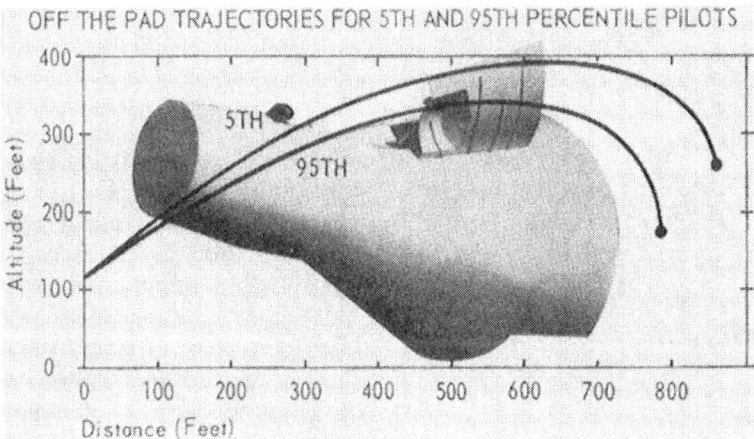

Fig. 12:20 Gemini ejection seat trajectories (Courtesy of Weber Aircraft).

launch vehicle to a sufficient height and to the side in case of emergency on the launch pad or shortly after launch. The system consists of a 33 foot long escape tower structure which is attached to the command module by means of a bolted connection utilizing explosive frangible nuts. The escape tower structure houses three solid-propellant rocket motors and a pitch control motor, deployable canards, instruments and ballast in the forward portion, as shown in Fig. 12:21.

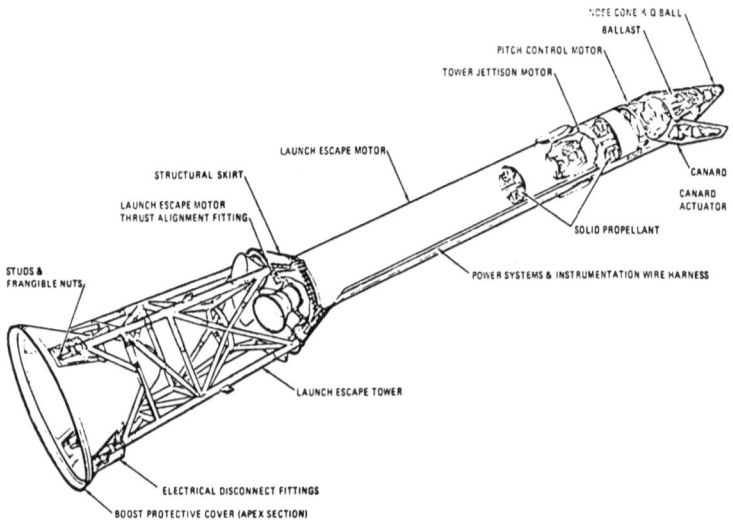

Fig. 12:21 Apollo launch escape subsystem (Courtesy of North American Rockwell Corp.).

At the lower end, a boost protective cover, which completely covers the command module, is attached to the escape tower to protect the command module from the rocket exhaust and from the heating caused by launch vehicle boost through the atmosphere.

An emergency detection system activates the launch escape system automatically during the first 100 seconds. The system can also be activated manually by the astronauts at any time from the pad to jettison altitude, which is at about 295,000 feet, or about 30 seconds after ignition of the second stage, when Saturn V is used as booster. With Saturn IB as booster, the jettison altitude is at about 275,000 feet, about 20 seconds after second stage ignition.

After an abort signal has been received, and after 40 seconds of flight, the booster is cut off, the command module-service module (CM-SM) separation charges are fired, and the launch escape motor is ignited. The launch escape motor lifts the command module (CM), and the pitch control motor, which is used only at low altitudes, directs the flight path off to the side to a safe distance from the launch pad. Eleven seconds after the abort is initiated, two canards, which are wing-like surfaces at the top of the launch escape tower are deployed. Caused by the aerodynamic forces acting on the canard surfaces, the command module is turned so that its blunt end faces forward. Three seconds after this operation on very low-altitude aborts, or at about 24,000 feet on high-altitude aborts, the escape tower separation devices are fired, and the tower jettison motor is started to carry the launch escape subsystem assembly away from the landing trajectory of the command module. The earth landing sub-system of the command module is activated four-tenths of a second

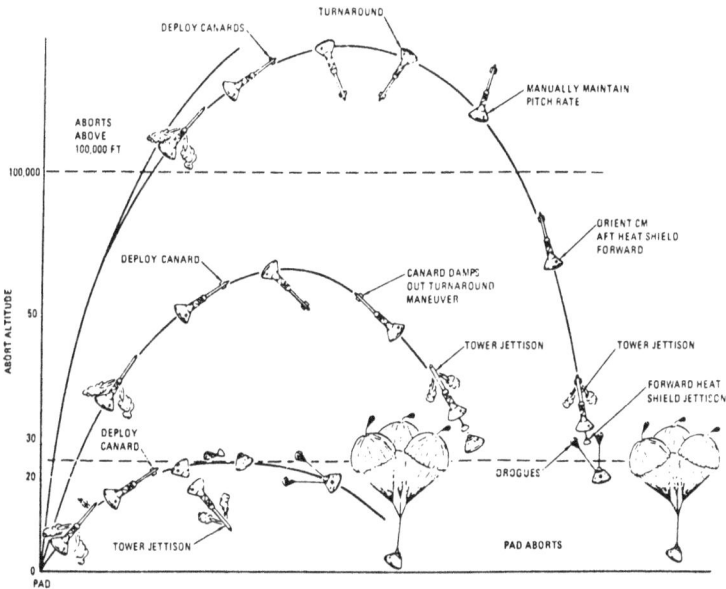

Fig. 12:22 How launch escape subsystem operates at different altitudes (Courtesy of North American Rockwell Corp.).

after escape tower jettisoning, and the sequence of operations begins to land the command module safely.

The operation of the Apollo launch escape subsystem at different altitudes is shown in Fig. 12:22, and a test of the launch escape subsystem's ability to carry the command module to safety during pad abort is presented in Fig. 12:23.

The launch escape subsystem can be used for aborts at three altitude regions:

Fig. 12:23 Pad abort test of Apollo launch escape subsystem (Courtesy of North American Rockwell Corp.).

244

1. Low altitude (pad abort to 30,000 feet)
2. Intermediate altitude (30,000 to 100,000 feet)
3. High altitude (100,000 feet to jettison of the launch escape subsystem).

For the first few seconds of abort flight, the sequence of events is common to aborts in any of the three regions and consists of:

a. Abort initiation by an astronaut or by the emergency detection subsystem (EDS)
b. Booster engine cutoff (only for aborts after 30 seconds of launch vehicle flight time)
c. Command module/service module interface separation
d. Ignition of launch escape and pitch control motor and reaction control system propellant dump (only during aborts initiated up to 42 seconds after lift-off)
e. Canard deployment 11 seconds after abort initiation.

The earth landing system can be initiated either 16 seconds after abort initiation or during descent at approximately 24,000 feet if the abort altitude is above 30,000 feet. A special procedure has been established for aborts initiated above 100,000 feet because of the problem caused by the lower aerodynamic stability at that altitude. Immediately after motor burn-out, the astronauts must establish a specific pitch rate using the reaction control system to avoid acquiring an undesirable trim condition and adverse acceleration during descent.

Special pyrotechnic functions are utilized only in aborts between manning of the spacecraft (crew insertion) on the launch pad and orbital insertion of the spacecraft. Highly complex sequences of pyrotechnic events are required for aborts from the launch pad and at low altitudes. The sequence and combination of these pyrotechnic events are functions of altitude. On-board automatic control, on-board manual control, and ground control of abort initiation is provided to minimize risk to the crew.

An abort from the launch pad to 30,000 feet altitude begins with the following pyrotechnic functions: Time T = 0.

1. Deadfacing of the command module/service module (CM/SM) umbilical by actuating four circuit interrupters.
2. Pressurization of the propellant of the command module reaction control subsystem (CM RCS) by opening four explosive valves.
3. Interconnection of helium, fuel, and oxidizer of the com-

mand module reaction control subsystem (CM RCS) by opening four explosive valves.

4. Dumping of the oxidizer of the command module reaction control subsystem by opening two explosive valves.

5. Structural separation of the command module (CM) from the service module (SM) by firing three dual linear-shaped charges (LSC).

6. Separation of the command module/service module (CM/SM) umbilical by firing a guillotine.

7. Ignition of the launch escape motor by firing two initiator cartridges.

8. Ignition of the pitch control motor by firing two initiator cartridges.

Time T + 5 seconds: Initiation of the fuel dump of the command module reaction control subsystem (CM RCS) by opening two explosive valves.

Time T + 11 seconds: Deployment of the canards in the launch escape subsystem (LES) by activating a thruster to reverse the attitude of the command module (CM) for jettisoning of the LES and parachute deployment.

T + 14 seconds: Separation and jettisoning of the docking ring with the launch escape tower to which it is attached.

T + 14.5 seconds: Jettisoning of the apex cover, as in a normal landing.

T + 16 seconds: Deployment of the drogue parachutes.

T + 18 seconds: Purging of the reaction control subsystem (RCS) fuel and oxidizer lines, and dumping of the residual helium pressurant by opening four explosive valves.

T + 27 seconds: Disreefing of the drogue parachutes.

T + 28 seconds: Deployment of the main parachutes and disconnect of the drogue parachutes.

The deployment of the recovery aid devices, the descent and landing of the spacecraft are the same as in a normal mission.

The pyrotechnic functions in an abort from 30,000 feet to jettison of the launch escape subsystem (LES) are similar to those in an abort as described above. Different in this abort is mainly the jettison of the reaction control subsystem propellant, which is more rapid than in the lower-altitude abort, and in general, the time interval between some events are slightly different.

In an abort above 100,000 feet, the astronauts may elect to jettison the launch escape subsystem (LES) and prepare for a landing by following normal procedures.

In a normal Apollo mission, the launch escape subsystem (LES) is jettisoned immediately after ignition of the second stage, which occurs about three minutes after launch. Both of the escape tower jettison switches will normally be used to initiate this function, however, either switch will initiate the jettison circuits. At activation of the jettison switches, two detonators in the frangible nut assemblies, by which the escape tower legs are attached to the command module (CM), are fired. Simultaneously, the jettison circuit also ignites the escape tower jettison motor.

When a mission must be aborted after tower jettison and prior to separation of the command and service module from the launch vehicle, the service propulsion system of the service module (SM) is used. Separation of the command and service module from the launch vehicle is then accomplished as in a normal mission.

The canards mounted on two hinges near the nose cone of the escape tower are deployed by a gas-operated actuator. When a time delay of eleven seconds after the abort signal is received by the master events sequence controller, the cartridges of the actuator are fired by an electric current, and the generated gas from the cartirdges causes a piston in the actuator to retract, thus operating the canard opening mechanism. When fully opened, the canards mechanically lock in place.

At the lower end of the launch escape subsystem, the apex section of the boost cover is attached to the launch escape tower legs. The apex of the boost cover is connected with the docking probe of the command module (CM). During a normal launch, the cover is pulled away by the thrust of the tower jettison motor, and the tension tie breaks, leaving the docking probe with the command module. However, during an abort, the pyrotechnic devices that separate the docking ring from the command module are fired by signals from the master events sequence controller. During jettison of the launch escape assembly, the tension tie pulls the docking probe and docking ring away with the boost protective cover.

The four legs of the escape tower fit in wells in the forward structure of the command module and are connected by frangible nuts which contain small explosive charges. Simultaneously with jettison of the launch escape subsystem, the charges in these frangible nuts are fired, breaking the nuts and thus separating the escape tower from the command module.

The leg attachment of the launch escape tower to the command module is shown in Fig. 12:24.

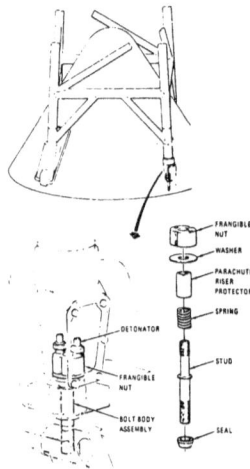

Fig. 12:24 Leg attachment of launch escape tower (Courtesy of North American Rockwell Corp.).

Another unique spacecraft emergency system is the propellant dispersion system, as used on the Saturn V vehicle. When the mission is aborted, this propellant dispersion system splits the oxidizer and fuel tanks open and safely disperses the unused propellant. The system is attached to the exterior of each of Saturn's six fuel and oxidizer tanks. The tanks are severed by linear-shaped charge which is initiated through a cross-over block by a sheathed detonating cord in carrier assemblies attached to the Saturn's skin. The support structure is designed to withstand the environments of the Saturn flight.

The linear-shaped charge (LSC) is 175 grain per foot loading, having a capability of severing the 0.500 inch thick walls of the aluminum tanks. All Saturn/Apollo vehicles have been equipped with this dispersion system, and they are qualified for man-rated flights with a 0.9997 reliability at 95% confidence level.

A safety and arming device prevents inadvertent initiation of the explosive train by providing a positive isolation of the EBW detonator and explosive train until arming is commanded. At the firing center, visual and remote indications of SAFE and ARMED conditions are displayed at all times. Upon proper command, EBW firing units activate the EBW detonators which, in turn, explode the linear-shaped charge. This propellant dispersion system is shown in Fig. 12:4.

As this chapter shows, the presently available spacecraft emergency systems are only suitable for on-pad or early-booster phase aborts. Malfunctions requiring escape can be expected to occur throughout the mission. Therefore, reliable emergency systems that provide safe escape for the astronauts during all phases of a space mission should be developed.

C. *Stage Separation Systems*

Most currently used boosters for spacecraft and satellites are multistage vehicles. Each stage of these boosters propels the spacecraft and the upper booster stages to a specific altitude at which the burnt-out stage must be separated from the vehicle and the next higher stage must be ignited.

Major requirements for stage separation, to accomplish the mission goals, are the proper orientation of the vehicle prior to stage separation, stabilization of the vehicle and separated stage, and provision for a propulsion system to thrust the separated stage away from the main vehicle to avoid damage to the vehicle by interference with the spent stage and by flying fragments.

Booster stages used with manned spacecraft are usually separated when the thrust of the rocket motor has decreased to a level at which the separation system can operate effectively. The separation of booster stages used with unmanned spacecraft, however, can in some cases only be accomplished with a delay, after solar cell panels, booms and antennas have been deployed. In other cases, when the separation is monitored from the earth, a spent stage cannot be separated from the spacecraft at any time, but only when the spacecraft is in the range of a ground station.

In selecting and designing the best suitable stage separation system for a given spacecraft application, it must be taken into consideration that a major constraint is usually the amount of energy required to achieve the desired relative separation velocity between the spent stage and the spacecraft.

The separation system energy E is, based on the conservation of kinetic energy and linear momentum, determined by:

$$E = \frac{V_r m_1 m_2}{2\, m_1 + m_2}$$

where V_r = the relative separation velocity
m_1 = the mass of the spacecraft
m_2 = the mass of the spent stage.

This equation shows that the same amount of energy is required to separate large masses at a low relative velocity as to separate small masses at a high relative velocity.

In cases where a relatively small amount of energy for separation is required, spring or cold-gas devices are generally used, whereas in cases where a large amount of energy is required, explosive devices are utilized. In conducting trade-off studies to determine the best suitable devices and systems for stage separation, the weight, complexity, size, and reliability of operation of these devices and systems must be considered.

For separating two stages, not only a reliable system for cutting the vehicle's skin, but also for simultaneously separating all electrical cable and fluid and gas line connections between both stages must be provided. The most commonly used pyrotechnic devices for cutting the skin are the flexible linear-shaped charge (FLSC) and the mild-detonating fuse (MDF). Both devices are described in detail in Part I, Chapter 1 of this book. Flexible linear-shaped charge usually consists of an inverted V-shaped seamless metal sheath containing a column of explosive material. When detonated, the shock wave is directed by the shape of the explosive charge through the open face of the Vee of the shaped charge and thus provides the cutting action. The flexible linear-shaped charge is installed in a molded retainer around the periphery of the skin. The retainer serves the purpose of absorbing shock from the detonation. A back-blast shield is mounted inside the upper stage to prevent damage of equipment in the vehicle by flying fragments. A cross-section through a typical flexible linear-shaped charge installation is shown in Fig. 12:25.

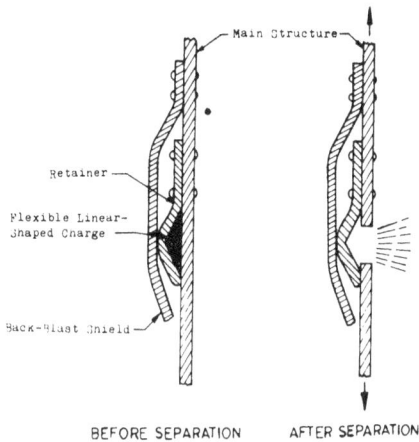

BEFORE SEPARATION   AFTER SEPARATION

Fig. 12:25 Flexible linear shaped charge for stage separation.

For applications where the flexible linear-shaped charge (FLSC) cannot be considered because of the high shock it produces, a mild-detonating fuse (MDF), in the form of a metal-clad detonating cord, can be used. Since the explosive energy of MDF is spread over a larger area, the shock levels produced are significantly lower than those produced by flexible linear-shaped charge (FLSC). However, greater fragmentation results from the wider spreading of the explosive energy in MDF. For this reason, the explosion must be confined internally to prevent damage to the spacecraft and equipment.

Explosive compositions commonly used in the flexible linear-shaped charge (FLSC) and in the mild-detonating fuse (MDF) are RDX (cyclotrimethylenetrinitramine), PETN (aliphatic nitrate) and HMX (homocyclonite).

For applications where two stages or structural assemblies must be mechanically connected and during a certain phase of the mission be separated, a splice plate or splice ring joint, as shown in Fig. 12:26, can be used. In this separation system, the retainer for the mild-detonating fuse (MDF) or flexible linear-shaped charge (FLSC) consists of two parts, both having a Z-shaped edge, one of which is a part of the splice plate. Before separation, the noses of both Z-shaped edges cover one another, thus providing a tight enclosure for the charge. At the height of the charge, a recess in the wall of the upper stage is provided. When the charge is detonated, the splice plate de-

251

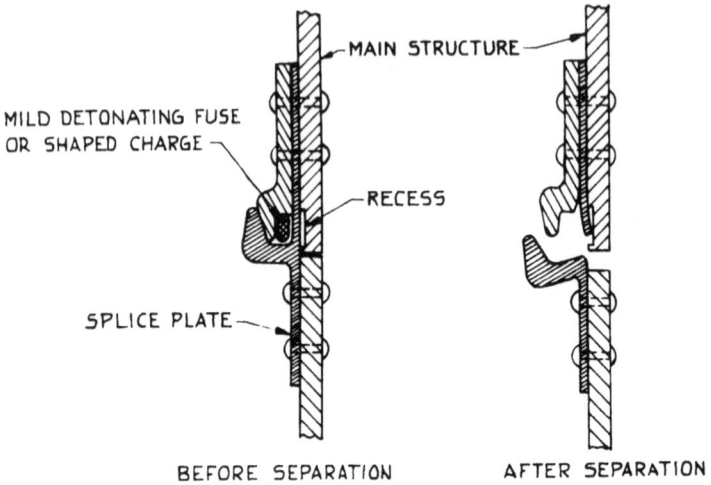

MAIN STRUCTURE

MILD DETONATING FUSE
OR SHAPED CHARGE

RECESS

SPLICE PLATE

BEFORE SEPARATION   AFTER SEPARATION

Fig. 12:26 Splice plate joint for stage separation (Ref: NASA-owned U.S. Patent; Tech Brief. No. 65-10166).

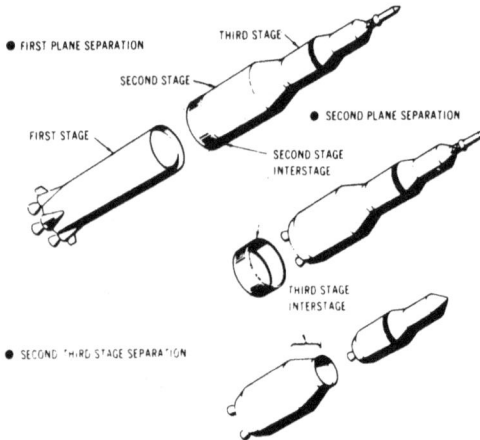

● FIRST PLANE SEPARATION

THIRD STAGE

SECOND STAGE

FIRST STAGE

● SECOND PLANE SEPARATION

SECOND STAGE
INTERSTAGE

THIRD STAGE
INTERSTAGE

● SECOND THIRD STAGE SEPARATION

Fig. 12:27 Saturn stage separation (Courtesy of North American Rockwell Corp.).

forms into the recess and fractures, thus separating both stages or structural assemblies.

A simple separation system, as used in the Ranger satellite, utilizes three explosively-actuated pin pullers which release a centrally

Fig. 12:28 First stage separation during an Apollo/Saturn V shot (Courtesy of the Boeing Co.).

located separation spring. Satisfactory results were achieved with this system, however, placement of satellite components is very critical when such a separation system is employed, because whock levels vary by over an order of magnitude, depending on location in the vehicle.

The Saturn V rocket used in the Apollo program consists of three stages plus a second stage interstage and a third stage interstage. The functions of the three main stages are described in the introduction to Chapter 12 of this book. The arrangement of the Saturn V stage separation is presented in Fig. 12:27, and separation of the first stage is shown in the illustration in Fig. 12:28.

The first stage is parted from the second stage by a dual-plane separation, involving an interstage. When the propellants in the first stage have been consumed, an engine cutoff signal is initiated, and the two stages are severed by a linear-shaped charge. Exploding bridgewire initiators are utilized for initiation of the linear-shaped charge. In order to decelerate the first stage and complete the stage separation, the retrorockets on the first stage are simultaneously ignited. To provide good propellant flow to the five J-1 engines of the second stage after separation of the first stage, eight solid-propellant ullage motors mounted on the aft interstage of the second stage are ignited. Thus, positive vehicle acceleration and proper propellant settling is achieved. Interstage separation is initiated on receiving a signal when the out-

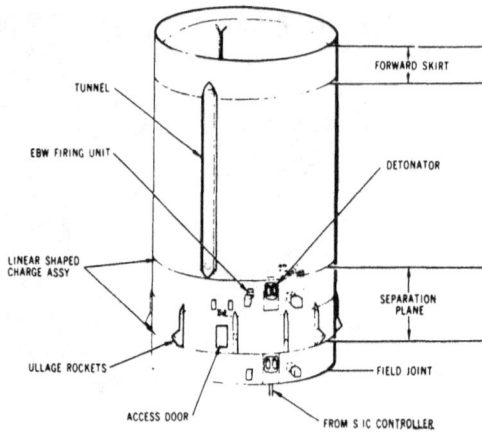

Fig. 12:29 Saturn stage separation ordnance system (Courtesy of North American Rockwell Corp.).

board engines on the second stage reaches 90% of maximum thrust.

The arrangement of the Saturn second stage separation ordnance system is shown in Fig. 12:29.

The stage separation system for separation of the second stage from the third stage consists of a severable tension strap, mild-detonating fuse (MDF), explosing bridgewire (EBW) detonators and EBW firing units. The severable tension strap contains two redundant MDF cords in a "V" groove which circumvents the stage at the separation plane between the aft skirt and aft interstage. About three seconds after second stage engine cutoff, a signal from the second stage sequencer triggers the ignition of the MDF cord through the EBW and EBW firing units. Each firing unit provides on command a 2,300 vdc pulse to fire a specific EBW detonator which initiates the MDF explosive train. The explosive force of the MDF generates at a rate of 23,000 feet per second. The high voltage pulse requirement for ignition was established to provide the necessary safety for the system from random ground power or vehicle electrical power.

A systems diagram for separation of the second stage from the third stage is presented in Fig. 12:30.

To assure clean separation of the third stage from the second stage by decelerating or braking the spent booster, four solid-propellant retrorockets mounted at equal spacing around the aft inter-stage assembly are fired. Two EBW firing units on the aft interstage

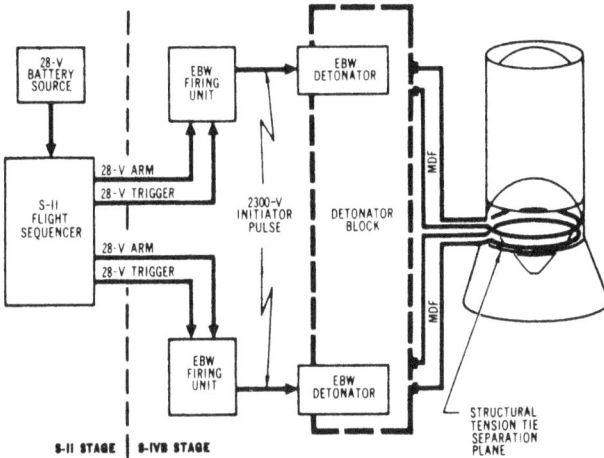

Fig. 12:30 Saturn stage separation systems diagram (Courtesy of McDonnel Douglas Astronautics Co.).

which are initiated by a signal from the second stage ignite two detonator manifolds which in turn ignite the retrorockets through redundant pairs of confined detonating fuse and pyrogen initiators. Each retrorocket having a weight of 384 pounds and a burn time of 1.5 seconds has a nominal thrust rating of 35,000 pounds.

A diagram of this retrorocket system is shown in Fig. 12:31. The pyrogen initiators used in this system are also shown in the illustration Fig. 12:4B.

Similar to the second stage, the third stage is also equipped with an ullage rocket system. Two solid-propellant ullage rockets are mounted on the aft skirt of the third stage forward of the separation plane. On a signal from the stage sequencer the ullage rockets are ignited by EBW initiators. After firing and after the propellant flow to the J-2 engine of the third stage has been provided, the burned-out ullage rocket casings and fairings are jettisoned to reduce the stage weight by detonating two forward and aft frangible nuts by which each ullage rocket motor and its fairing is mounted on the stage. A confined detonating fuse which is initiated by a command from the stage sequencer is used to detonate the frangible nuts and thus to free the ullage rocket assembly from the stage.

A systems diagram of this ullage rocket system is presented in Fig. 12:32.

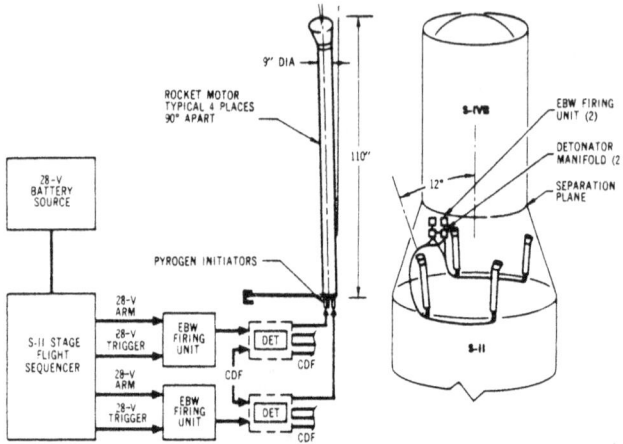

Fig. 12:31 Saturn retrorocket system (Courtesy of McDonnel Douglas Astronautics Co.).

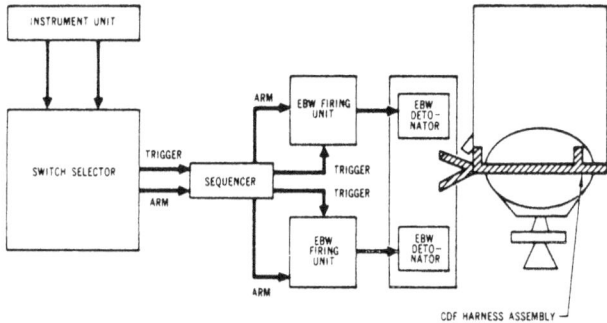

Fig. 12:32 Saturn ullage rocket system (Courtesy of McDonnel Douglas Astronautics Co.).

Major staging operations of the Apollo spacecraft on lunar missions are the separation of the Command and Service Module (CSM) from the launch vehicle Saturn, the separation of the Lunar Module (LM) from the Command and Service Module (CSM), and the separation of the Command Module from the Service Module. After translunar injection by the third stage of Saturn, the Command and Service Module (CSM) is separated from the launch vehicle. To accomplish this task, the spacecraft-LM adapter, which protects the Lunar Module

(LM) during launch and provides the structural attachment of the spacecraft to the launch vehicle, must be separated from the Service Module (SM). The spacecraft-LM adapter (SLA) is connected with the Service Module (SM) by bolts through a flange which extends circumferentially around both structures. For triggering the separation devices, an umbilical cable from the Service Module to the SLA is provided. The spacecraft-LM adapter (SLA), consisting of eight aluminum-honeycomb panels arranged in two sets of four panels of equal size, is separated from the Service Module (SM) by firing an explosive train which cuts through the metal connecting both stages. The explosive train consisting of 28 charge holders, each of which contains two strands of detonating cord, is mounted on the flange, which connects the Service Module (SM) with the spacecraft-LM adapter (SLA), and on the splice plates which connect the forward panels. To assure that the entire explosive train fires, a booster charge is mounted at the end of each charge holder and at each cross-over point.

During the panel splice-plate separation, the umbilical between the Lunar Module (LM) and one SLA panel is severed by firing a high-explosive guillotine, and the umbilical arm is retracted to the panel by a spring reel for jettison with the panel. The SM-SLA umbilical is separated by an explosive frangible link disconnect aft of the Service Module. To assure separation, redundancy is provided in the separation initiation signals, in the detonators and in the explosive cord trains. The charges are "sympathetic", which means that detonation on one charge sets off another.

After separation of the Command and Service Module (CSM) from the third Saturn stage (S-IVB), the CSM docks with the Lunar Module (LM). Through the docking interface, an electrical umbilical is connected with the LM firing circuits. To separate the Lunar Module (LM) from the spacecraft-LM adapter (SLA), the mild charges in the four frangible links which connect the LM with the fixed portion of the SLA are fired. The electrical umbilical is severed by a high-explosive guillotine with a time delay of 30 milliseconds after firing of the frangible links. The separation sequence is initiated by receiving signals from two redundant lunar module separation sequence controllers. Immediately after cutting the umbilical, both stages are being pushed apart by four spring thrusters which are mounted on the lower SLA panels.

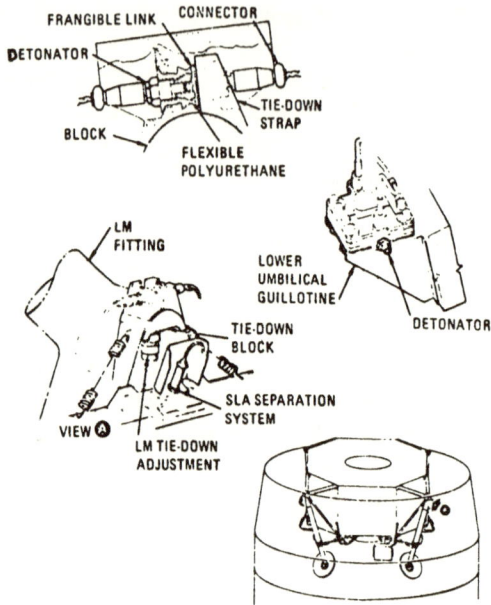

Fig. 12:33 Apollo lunar module adapter-lunar module (SLA–LM) separation system (Courtesy of North American Rockwell Corp.).

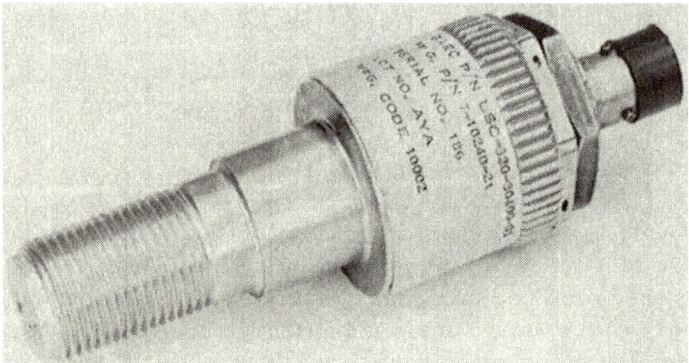

Fig. 12:34 Explosive bolt for stage joint of lunar module (Courtesy of Space Ordnance Systems, Inc.).

Details of the Apollo SLA-LM separation system are shown in Fig. 12:33.

After completion of the mission on the lunar surface, prior to launch, the ascent stage of the Lunar Module (LM) must be separated from the descent stage. This is accomplished by firing an explosive nut and bolt at each structural joint between both stages. The interstage umbilical which combines the electrical and fluid lines is severed by a guillotine cutter, and the interstage electrical circuits are deadfaced by two circuit interrupters.

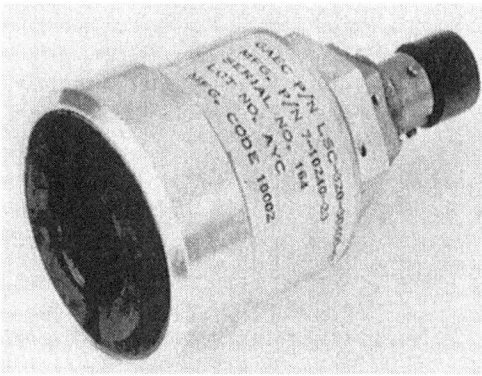

Fig. 12:35 Explosive nut for stage joint of lunar module (Courtesy of Space Ordnance System, Inc.).

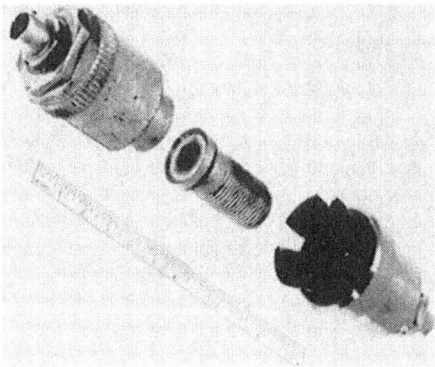

Fig. 12:36 Explosives bolt and nut of LM stage joint after separation (Courtesy of Space Ordnance Systems, Inc.).

The explosive bolt and nut used for this structural stage joint are presented in Fig. 12:34 and Fig. 12:35, respectively, and the same explosive bolt and nut after separation are shown in Fig. 12:36.

After separation, the LM ascent stage lifts off from the lunar surface for rendezvous and docking with the Command and Service Module (CM-SM) during its lunar orbit to transfer the Lunar Module crew to the Command Module. After completion of this maneuver, the LM ascent stage is separated from the Command and Service Module and jettisoned. This separation is accomplished by firing a redundant explosive train of mild-detonating fuse (MDF), which is located around the periphery of the docking ring. This explosive train charge consists of dual strands of 6 grains per foot high-temperature explosive core in a silver sheath, terminating in transfer boosters which are soldered to each end. The charge separates the ring structure between the forward heat shield and the probe mounting and leaves the docking hardware with the LM ascent stage. Normally, the charge is initiated by a signal from the Command Module for LM ascent separation just before injection of the Command and Service Module (CM-SM) on the trans-earth flight. The Lunar Module (LM) ascent stage then remains in orbit around the moon.

Modes of docking probe separation, including abort modes (see Chapter 12 B of this book) are shown in Fig. 12:37.

A short time before entry into the atmosphere, the Service

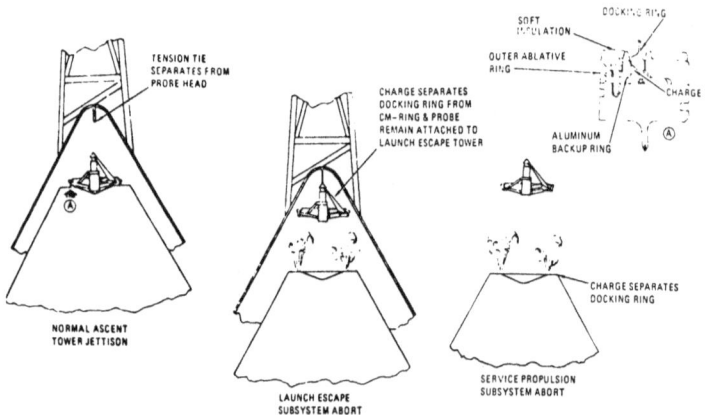

Fig. 12:37 Modes of separating docking probe of Apollo (Courtesy of North American Rockwell Corp.).

Module (SM) must be separated from the Command Module (CM). The events required for a proper separation must occur in a rapid sequence and they are controlled automatically by two redundant Service Module jettison controllers. Major events of this separation sequence are: Physical separation of all connections between the Command Module and the Service Module, transfer of electrical control, and firing of the reaction control motors of the Service Module to move the modules apart.

Prior to separation, the crew in the Command Module (CM) transfers the electrical control to the Command Module reaction control subsystem in order to pressurize it and check it out in preparation for the entry maneuvers. After the electrical control has been transferred back to the Service Module (SM) subsystem, separation is initiated manually by activation of either one of two redundant switches which send signals to the Service Module jettison controllers. First, a signal to fire ordnance devices which activate the CM-SM electrical curcuit interrupters is sent by the controllers. After an interval of a tenth of a second after the electrical wires are deadfaced, the controllers send signals to fire ordnance devices which sever the physical connections between the Command Module and the Service Module. The modules are connected by three tension ties which extend from the aft heat shield of the Command Module to compression pads on the Service Module. These tension ties consist of straps made from 4340 carbon steel of high tensile strength, 0.170 inch thick, 2.5 inches wide and 4 inches long, and they are bolted at one end to the Command Module (CM) and at the other end to the Service Module (SM). Both modules are also connected through the CM-SM umbilical containing the electrical power cables and tubing for oxygen, water and water-glycol. At the time of separation, the tubing lines in the umbilical are closed by valves.

Tension tie cutters utilizing 100 grains per foot linear-shaped charge (LSC), set off by detonators, provided in each tension tie assembly sever the connecting steel straps to separate the modules. The charges, the detonators, and the signals that set off the detonators are all redundant. At the same time when the tension ties are severed, a guillotine device is fired to cut the wires and tubing of the umbilical. The guillotine device contains two stainless steel blades either one of which will cut the umbilical connections. Redundant detonating cord, set off by detonators, is used to drive the guillotine.

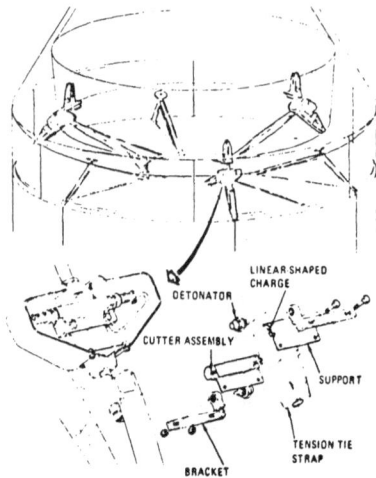

Fig. 12:38 Apollo CM–SM separation system (Courtesy of North American Rockwell Corp.).

Simultaneously, at the instant of severing the tension ties and the umbilical connections, the controllers send signals which fire the reaction control motors of the Service Module. To alter the course of the Service Module from that of the Command Module, roll control motors are fired for 5 seconds, whereas the thrust motors are fired continuously until the propellant is consumed or the fuel cell power is expended. Through these maneuvers, the Service Module is moved far away from the entry path of the Command Module. The Service Module enters the earth's atmosphere after separation and burns up, while the Command Module with the crew returns to earth. The Command Module-Service Module (CM-SM) separation system is shown in Fig. 12:38.

D. *Fairing Release Systems*

Fairings or shrouds are normally used on spacecraft to protect delicate payload and equipment, such as satellites, instruments and antennas from ascent heat and from other possible damaging environmental effects, and to reduce the drag during the launch and boost phase to a minimum by an aerodynamically shaped shell. After the spacecraft has reached a specified altitude, the satellite carried in the nose of the vehicle must be separated from the launch vehicle to start its space mission, or the instruments or antennas must be made operable by exposing them, which requires that the fairing or shroud must be released and moved away from the spacecraft.

Major requirements for fairing release systems are light weight, low volume, simplicity and high reliability. Explosive systems used for fairing release must function without causing fragmentation to avoid damage to the spacecraft and its payload.

Important factors for optimum design and construction of separable fairings or shrouds are the size, shape, and complexity of the payload to be protected, the mission profile of spacecraft and payload, and the spacecraft's shape and structural characteristics.

On spacecraft of medium size, clam-shell type fairings consisting of two half-shells which separate along the vertical axis have been used successfully. The Surveyor spacecraft, the unmanned exploration vehicles which landed on the moon, were protected by such a fairing mounted on the nose of the Atlas-Centaur launch vehicles.

A clam-shell type fairing consisting of two halves which are separated by a linear-shaped charge or by a linear explosive charge is shown in Fig. 12:39.

On large spacecraft, the fairing or shroud is usually divided into four or more panels, which results in parts of reasonable size which can be separated and moved away from the vehicle easier than clam-shell halves. To thrust the released fairing parts away from the spacecraft, gas systems and spring devices are commonly employed.

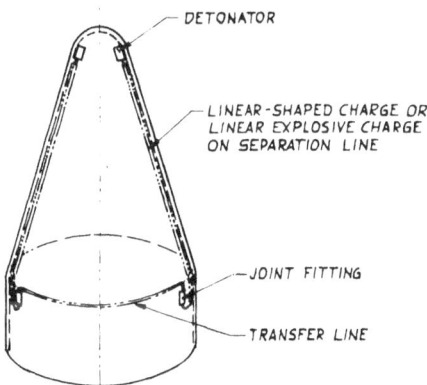

Fig. 12:39 Two-part clam-shell fairing.

263

An unusual fairing concept was used on the unmanned Soviet Luna 9 spacecraft which landed on the moon in February 1966. This spherical-shaped capsule which contained transmitters, receivers, temperature control system, programming equipment and batteries, was equipped with several antennas folded inside a four-segment fairing. The four segments which had trapezoidal shape were hinged at the equator of the spherical capsule. They were opened just before impacting on the lunar surface, thus unfolding the antennas. These four hinged fairing segments served the additional purpose of dampening the impact on the lunar surface.

A very large separable spacecraft fairing is used on the Saturn-Apollo vehicle in the form of the Spacecraft-Lunar Module adapter (SLA) which is located between the third stage of Saturn (S-IVB) and the Service Module of the Apollo spacecraft. The main purpose of this SLA adapter is to protect the stowed Lunar Module (LM) and to hold it firmly in place on its long voyage to the moon. To separate the Command and Service Module (CM-SM) from the launch vehicle, after translunar injection, and to prepare the Lunar Module (LM) for its mission, the SLA adapter, which is divided into four panels, must be released. This is accomplished by firing redundant explosive trains which are attached on the forward, aft, and inner and outer longitudinal splice plates of these 0.040 inch thick aluminum panels. The arrangement of this explosive train is shown in Fig. 12:4 C (Chapter 12).

The explosive train consists of 28 charge holders. Redundancy is achieved by providing in each charger holder two strands of detonating cord, either one of which will sever the joint. The charge holders, which consist of aluminum strips to which the detonating cord is bonded, are mounted on the splice plates which join the panels longitudinally and on the horizontal flange provided for connecting the SLA adapter with the Service Module (SM). At the ends of each charge holder and at each crossover point, boosters are arranged to assure that the entire explosive train fires and all four panels are separated.

For deploying and jettisoning the panels, two sets of thrusters, one pyrotechnic and one spring thruster, are employed. During the separation of the fairing, each of the four panels is rotated outwardly around a center hinge by a pyrotechnic thruster which is powered by dual cartridges. Each of these pyrotechnic thrusters is equipped with two pistons, one acting on each panel. Thus, each panel has two pistons thrusting against it, one at each end.

The pressure cartridges in the pyrotechnic thrusters are initiated by the explosive train which is used for separating the panels and which is also routed to these pressure cartridges. By igniting the pressure cartridges, the pistons in the thrusters are driven against the panels, and thus start deployment of the panels. The impulse applied by the pyrotechnic thrusters to the panels results in only 2 degrees of rotation, which is sufficient to assure deployment. The speed imparted by the thrust of the pistons, which is 33 to 60 degrees per second of angular velocity, remains constant. After rotating about 45 degrees, the hinges which connect the upper panels to the lower panels disengage, and as a result, the panels are freed from the aft portion of the SLA adapter.

On the outside of the upper panels, spring thrusters are provided. When the hinges separate, the springs in the spring thrusters push against the lower panels and move them away from the spacecraft at

Fig. 12:40 Preparation of SLA adapter panel (Courtesy of North American Rockwell Corp.).

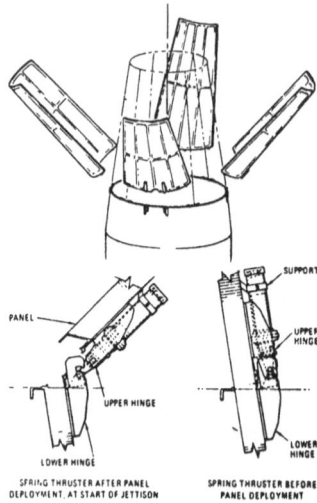

Fig. 12:41 Release of SLA adapter panels (Courtesy of North American Rockwell Corp.).

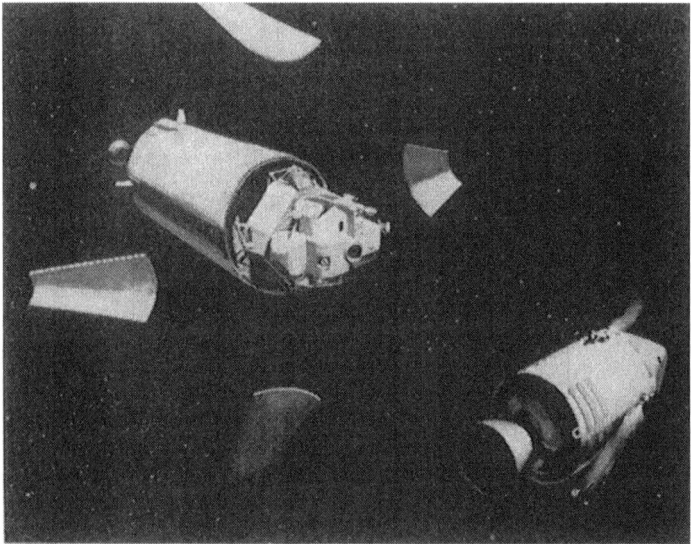

Fig. 12:42 Jettison of SLA adapter panels (Courtesy of North American Rockwell Corp.).

an angle of 110 degrees to the vehicle centerline and at a velocity of about 5.5 miles per hour.

Simultaneously with firing the explosive train to separate the panel splices, a high-explosive driven guillotine device is fired to sever the umbilical which connects one SLA panel to the Lunar Module (LM). The umbilical arm is then retracted to the panel by a spring reel for jettison with the panel. A high-explosive charge in a frangible-link disconnect is used to separate the SM-SLA umbilical just aft of the Service Module (SM).

The panel separation system is explosively interconnected and redundancy is achieved by providing dual detonators which initiate explosive trains and confined detonating cords used to connect the explosive trains to the SM-SLA umbilical disconnect and to the LM-SLA guillotine device. The structure and the large size of one of these SLA panels is shown in Fig. 12:40.

A schematic of the release of these Lunar Module Adapter panels is presented in Fig. 12:41, and jettison of the panels after lunar injection is shown in Fig. 12:42.

E. *Recovery and Landing Systems*

One of the most critical phases of a space mission is the recovery and earth landing of the space vehicle. Various recovery and landing systems for unmanned and manned spacecraft have been developed with the objectives of the highest possible reliability, low weight and small volume. The safe return of manned spacecraft from earth orbits and lunar missions and the retrieval of numerous instrumented space capsules, providing a wealth of scientific data, are accomplished with such recovery and landing systems.

Recovery and landing of spacecraft usually consist of five phases:
* Re-entry into the atmosphere
* Deceleration and stabilization
* Descent at low velocity
* Impact on water or ground
* Locating and retrieving

Pyrotechnic devices are ideally utilized in recovery and landing systems in all these phases. Typical pyrotechnic devices used in recovery and landing systems are: thrusters for cover ejection, drogue mortars, reefing line cutters, shaped charges and igniters for heat

shield separation and for similar applications, gas generators for inflation of flotation bags, parachute release devices, and pin pullers or line
cutters for deployment of antennas and location aids.

Most earth landings of spacecraft were made on water, using vertically descending parachutes as decelerators. Concurrent with the
development of these systems, beginning with the Gemini program,
steerable decelerators, such as flexible wings, Rogallo wings and steerable parachutes for land landing of manned spacecraft, rather than
water landing, were developed and tested. These steerable decelerators
offer the advantage over conventional parachutes that the spacecraft
can land at a predetermined spot.

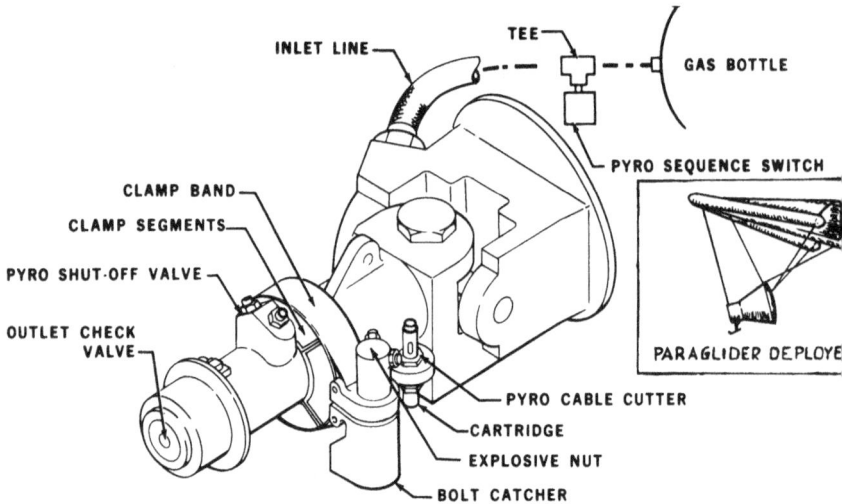

Fig. 12:43 Pyrotechnic system for paraglider inflation and deployment.

A typical pyrotechnic system application for the inflation and
deployment of a Rogallo wing is shown in Fig. 12:43. The wing
disconnect fitting, combining a structural assembly with a pneumatic
system, is shown in inflation condition. At a signal to inflate the wing,
a sequencer switch opens a pyro outlet valve at a nitrogen gas bottle in
the vehicle. After inflation is completed, which takes 20 seconds, the
pyro shutoff valve in the fitting is actuated, and the outlet valve is also
closed to shut off the gas flow into the wing. Immediately thereafter,
the explosive nuts which hold the disconnect segment clamp closed by

means of a steel clamp band, are actuated, and with a very short time delay, the cable cutter fires to sever the electric cables leading from the vehicle to the shutoff valve and to the clamp nuts. The clamp opens and disconnects the inflated wing from the main fitting so that only the short separated portion of the disconnect fitting containing the pyro shutoff valve and the outlet check valve remains with the wing. It was an absolute requirement for this pyro assembly that no fragmentation occurs, since wing and spacecraft are very susceptible to damage.

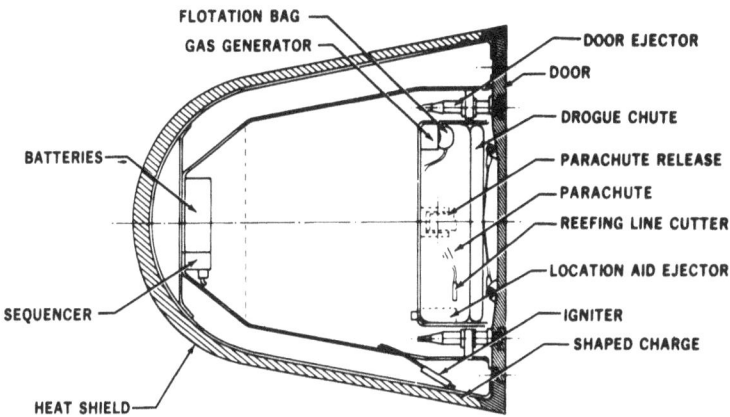

Fig. 12:44 Pyrotechnic devices for recovery system of data capsule.

An ideal application and arrangement of various pyrotechnic devices for a recovery system of an unmanned spacecraft, a data capsule, is shown in Fig. 12:44. During reentry into the atmosphere, the sequencer switch, sensed by aneroids, actuated two pyro thrusters utilized as door ejectors. The door is released, and by means of the attached drag line, extracts and deploys the drogue chute which, after full deployment, extracts the main parachute. To avoid too high opening shock loads, the main parachute is deployed to a reefed condition (i.e., a smaller than fully open diameter). After an interval of several seconds achieved by a time-delay train built into the reefing line cutter, the main parachute is disreefed. In order to reduce the landing impact, the heavy heat shield is released prior to landing by igniting the separation shaped charge. At the next phase, the gas generator is actuated to inflate the flotation bag which, connected to the capsule, is ejected onto the water surface. Finally, the main parachute is separated from the capsule by activating the parachute release, and the antennas and location aid devices are deployed by activating a pin puller (or a line cutter). Since this space capsule is of extremely small size, having a diameter of about 19 inches and a length of 22 inches, the advantages of utilizing pyrotechnic devices for such a variety of different functions and of arranging them in such a very small available volume are obvious.

In spacecraft recovery systems where ejection of the recovery compartment door or canister cannot be utilized to extract the pilot chute, a drogue gun is commonly used. By firing an electrically initiated squib, a gun projectile, which is connected with the pilot chute by a bridle, is forced out of the drogue gun and thus, pulls the pilot chute out of the spacecraft. Subsequently, the pilot chute extracts and deploys a small first stage deceleration parachute at an altitude below 100,000 feet at supersonic velocity. At an altitude between 5,000 and 15,000 feet and a descent rate of about 135 feet per second, the first stage deceleration parachute extracts and deploys a large final stage parachute in reefed condition to keep the opening shock at an acceptable level. With a predetermined time delay of one second or a few seconds, the parachute is disreefed by built-in reefing cutters, reducing the descent rate to about 15 to 20 feet per second.

In a typical recovery system for a spacecraft of about 500 lbs., a ribbon parachute of 9 feet diameter is used as first stage deceleration parachute, and an extended skirt parachute of 52 feet diameter is used as final stage parachute. Prior to water impact, a short distance above the water surface, a flotation bag is deployed from the descending spacecraft, and a SOFAR bomb is released. After water impact, the final stage parachute is separated from the spacecraft by a disconnect device. Simultaneously, the location aid devices, consisting of an antenna, a radio beacon and a flashing light, are deployed and activated, and shark repellent and dye marker are ejected into the water.

It is a desirable feature of drogue guns that their reaction force is low. A typical drogue gun contains a one-pound projectile which is ejected at a velocity of 200 feet per second for a minimum of 3 feet, and it has a reaction force of less than 4000 lbs. The pressure output of the cartridge used in this drogue gun is 5000 PSIG in a 20 cc closed chamber in 35 milliseconds. This drogue gun has a length of 6.75 inches and a diameter of 1.85 inches.

Performance diagrams of drogue guns with various projectile weights are presented in Fig. 12:45.

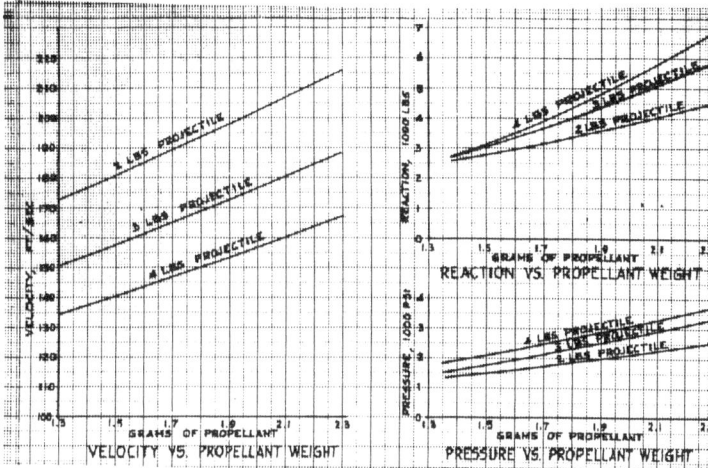

Fig. 12:45 Performance diagrams of drogue parachute guns (Courtesy of Gould Laboratories).

At supersonic velocities, mortars are utilized advantageously for deployment of drogue chutes and parachutes. A mortar usually consists of a short cylindrical tube of a large diameter. One closed end of the mortar tube is equipped with a breech assembly in which two pressure cartridges are mounted. The mortar tube contains a piston-like sabot and the stowed parachute in a cylindrical bag. After firing of the cartridges, the generated gas pressure pushes the sabot and the parachute pack toward the end of the mortar tube. Caused by the inner pressure, the end cover shears off, and the parachute is ejected. The bridle connecting the parachute with the spacecraft is pulled taut by the ejection force, and the parachute deploys. Mortars are frequently used to eject drogue chutes which, subsequently, extract the next stage larger parachutes.

The weight of a mortar can be estimated with fair accuracy from the weight of the drogue chute as:

$$W_M = 11\ \frac{W_D{}^{2/3}}{25}$$

where W_M = Weight of mortar, and
W_D = Weight of drogue chute.

Typical parachute mortar assemblies and their components are shown in Fig. 12:46.

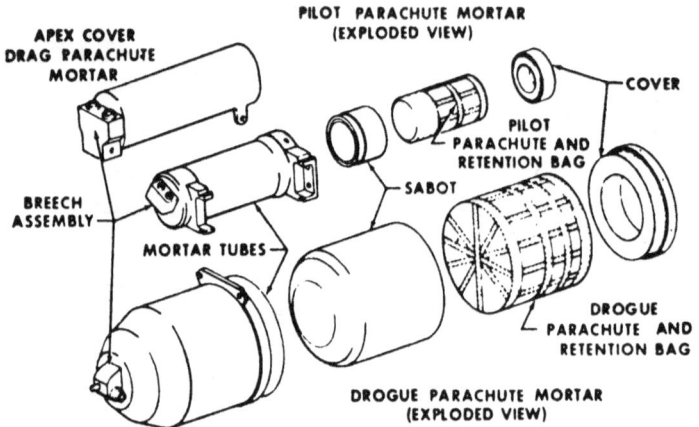

Fig. 12:46 Parachute mortar assemblies (Courtesy of NASA).

Deployment of a recovery system can be initiated by ground command, by flight conditions, such as engine failure, loss of control signal, or loss of braking signal, or by an altitude velocity sensor which is armed by a timer at a predetermined interval, usually 0.5 seconds, after separation of the capsule from the booster stage. By providing two independent electrical circuits complete with timers, control devices and batteries, system redundancy is achieved, so that in case of failure of one circuit, the recovery system will still function.

For landing on solid ground, rather than on water, the spacecraft is equipped with a suitable landing shock absorption system, such as a crushable platform, air bags, or penetration spikes.

Recovery and Earth Landing System for the Apollo Spacecraft

The Apollo recovery and earth landing system is probably the most advanced and the most thoroughly developed and tested recovery system. The main design requirements were to provide a system of the highest possible reliability, capable of functioning after a completed lunar mission at normal re-entry, maximum dynamic pressure escape, and also in case of pad abort. The system must stabilize the Command Module, which houses the crew, during post-entry descent, and it must decelerate the spacecraft to a vertical landing velocity of 30 feet per second at an altitude of 5,000 feet. The system must also reduce the impact acceleration so that neither the Command Module structure nor flotation is impaired.

To provide the required redundancy, two drogue parachutes and three main parachutes, which are deployed by pilot parachutes, are used. Two of the main parachutes will be sufficient for providing the required low descent rate for water landing. The drogue parachutes, each having a diameter of 16.5 feet, and the 7.2-foot diameter pilot parachutes are ejected by mortars to assure that these parachutes are ejected beyond the turbulent air surrounding and following the Command Module.

The main components of the recovery and earth landing system are located in the forward compartment of the spacecraft and are

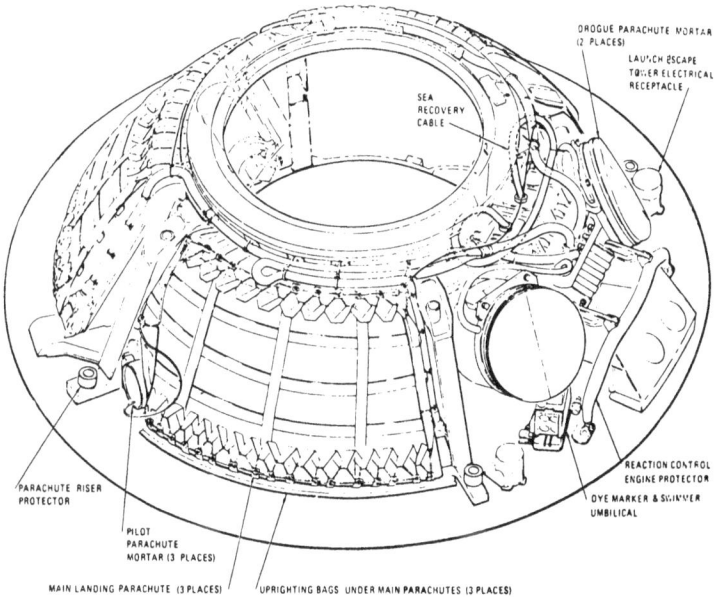

Fig. 12:47 Location of main components of the Apollo Recovery and earth landing system (Courtesy of North American Rockwell Corp.).

Fig. 12:48 Installation of a main parachute in the recovery systems bay adjacent to two drogue parachute mortars (Courtesy of Northrop Corp., Ventura Div.).

arranged circumferentially around the docking tunnel on the upper deck above the crew compartment, as shown in Fig. 12:47 and Fig. 12:48.

The forward heat shield, shown in Fig. 12:49 and Fig. 12:50 covers the apex of the vehicle and provides the necessary protection from environmental effects during space flight.

The sequence of operation of the Apollo recovery and earth landing system is presented in Fig. 12:51. At an altitude of approximately 24,000 feet, about eight minutes after re-entry into the atmosphere, the forward heat shield is jettisoned by a redundant thruster system, as shown in Fig. 12:52. An especially developed pressure cartridge, as used in this thruster system, is shown in Fig. 12:53. After initiation, the generated pressure in each thruster assembly forces two pistons apart, thus breaking a tension tie used to connect the forward heat shield with the forward compartment structure. The lower piston in each of the thrusters is pushed against a stop, and the upper piston is forced out of the thruster cylinder. The upper piston rod ends are attached to fittings on the forward heat shield, which is ejected away from the Command Module. Only two of the thruster assemblies are equipped with breeches and pressure cartridges. The breeches are connected with thrusters which are arranged as diametrically opposite

274

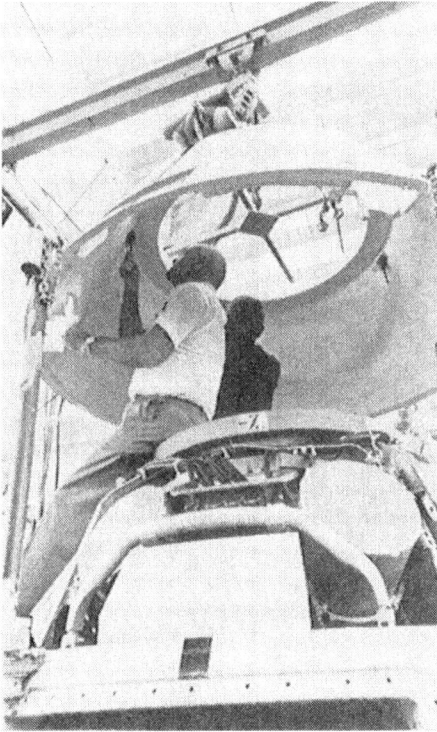

Fig. 12:49 Preparation for attachment of forward heat shield (Courtesy of Northrop Corp., Ventura Div.).

structural members of the Command Module. To prevent the forward heat shield from recontacting the Command Module and from interfering with the deployment of the drogue parachute, a ring-slot drag parachute of 7.2 feet diameter is ejected from a mortar in the heat shield by a lanyard-operated switch. This parachute drags the forward heat shield out of the area of negative air pressure.

Two seconds after jettison of the forward heat shield, the two reefed drogue parachutes are simultaneously ejected by their mortars and are deployed. The built-in reefing line cutters are actuated at line stretch and disreef the drogue parachutes with a time delay of ten seconds. Each drogue parachute is equipped with two reefing lines and with two cutters per line, whereby disreefing is prevented in case one reefing line cutter fires prematurely. The reefing lines are being held in

Fig. 12:50 Two main parachutes installed in their bays. In lower center, a pilot parachute mortar (Courtesy of Northrop Corps., Ventura Div.).

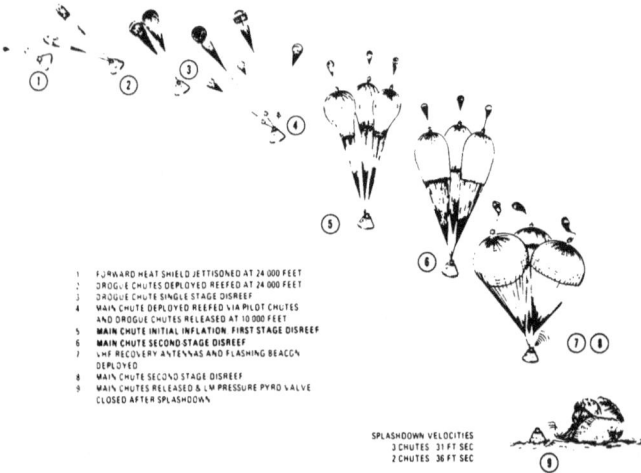

1   FORWARD HEAT SHIELD JETTISONED AT 24 000 FEET
2   DROGUE CHUTES DEPLOYED REEFED AT 24 000 FEET
3   DROGUE CHUTE SINGLE STAGE DISREEF
4   MAIN CHUTE DEPLOYED REEFED VIA PILOT CHUTES
    AND DROGUE CHUTES RELEASED AT 10 000 FEET
5   **MAIN CHUTE INITIAL INFLATION FIRST STAGE DISREEF**
6   **MAIN CHUTE SECOND STAGE DISREEF**
7   VHF RECOVERY ANTENNAS AND FLASHING BEACON
    DEPLOYED
8   MAIN CHUTE SECOND STAGE DISREEF
9   MAIN CHUTES RELEASED & LM PRESSURE PYRO VALVE
    CLOSED AFTER SPLASHDOWN

SPLASHDOWN VELOCITIES
3 CHUTES 31 FT SEC
2 CHUTES 36 FT SEC

Fig. 12:51 Sequence of operation of the Apollo recovery and earth landing system (Courtesy of North American Rockwell Corp.).

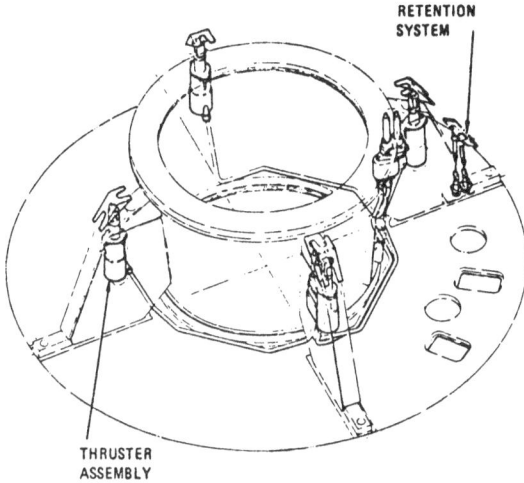

Fig. 12:52 Thruster system for ejection of forward heat shield (Courtesy of North American Rockwell Corp.).

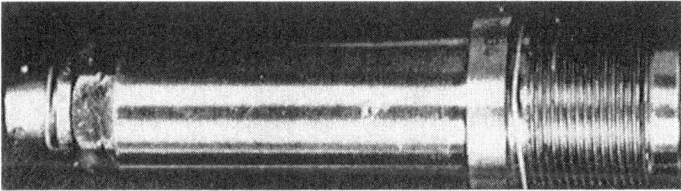

Fig. 12:53 Pressure cartridge for forward heat shield thruster system (Courtesy of General Precision Systems Inc., Link Group).

place in rings, which are sewn to the inside of the parachute skirts, and they run through the reefing line cutters. When the parachute suspension lines pull taut, an attached lanyard pulls the sear release from the reefing line cutter, and hereby, the pyrotechnic time delay train in the cutter is initiated. At the end of the burning delay train, the propellant in the cutter is ignited, driving the cutter blade through the reefing line. The drogue parachutes orient the Command Module properly and provide initial deceleration, and thus, they set up dynamic conditions safe for the subsequent deployment of the main parachute cluster. This phase of the recovery and landing sequence is shown in Fig. 12:54.

Fig. 12:54 Drogue parachutes open at an altitude of 24,000 feet (Courtesy of North American Rockwell Corp.).

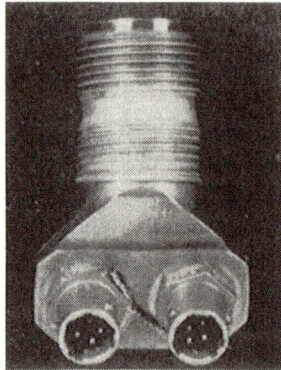

Fig. 12:55 Dual cartridge for pilot parachute mortar (Courtesy of General Precision Systems Inc., Link Group).

SPACECRAFT SYSTEMS

At an altitude of about 10,000 feet, approximately 40 seconds after deployment, the drogue parachutes are released by severing their risers with propellant-gas operated guillotines, and simultaneously, the three pilot parachutes are deployed by their mortars at a muzzle velocity of 100 feet per second. A pilot parachute mortar is shown in the lower centre of Fig. 12:50. A redundant initiator cartridge assembly, as used in the breeches of these pilot parachute mortars, is presented in Fig. 12:55. After mortar ejection, the pilot parachutes inflate instantly and pull the main parachutes from the forward compartment of the Command Module. This phase of the recovery is shown in Fig. 12:56.

Fig. 12:56 Pilot parachutes deploying the main parachutes at an altitude of 10,000 feet (Courtesy of North American Rockwell Corp.).

The main parachutes are of the ring-slot type and have a diameter of 83.5 feet. To reduce the opening shock, the main parachutes are inflated in two stages. Each main parachute is equipped with three reefing lines, two of which are severed with a time delay of six seconds to allow for partial inflation of the canopy, and the second

279

reefing line is severed with a time delay of ten seconds and allows the canopy to inflate fully. The required redundancy is achieved by providing three reefing line cutters for each line, six cutters for each parachute, a total of eighteen reefing line cutters for the main parachute system.

The reefing line cutters that have been developed especially for the Apollo recovery and earth landing system are presented in Fig. 12:57 and Fig. 12:58. The mounting arrangement of one of these cutters on the skirt band of a main parachute is shown in Fig. 12:59.

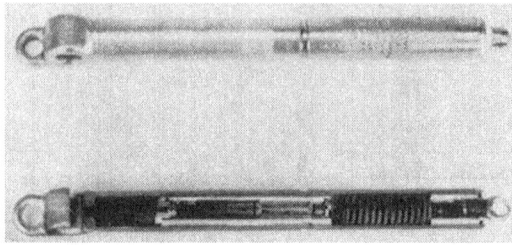

Fig. 12:57 Reefing line cutter for Apollo recovery system (Courtesy of Bermite, North Hollywood Branch, Div. of the Whittaker Corp.).

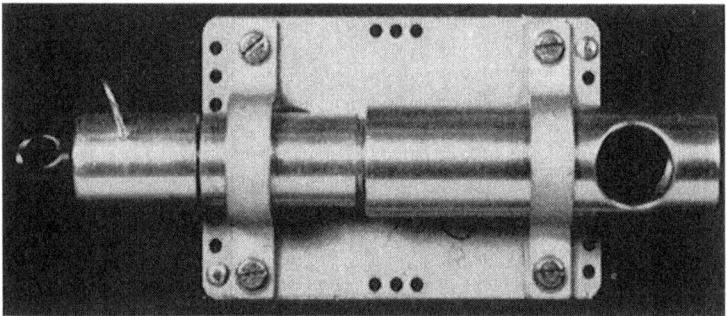

Fig. 12:58 Apollo reefing line cutter with mounting plate (Courtesy of General Precision Systems Inc., Link Group).

The requirement that the nine six-second delay reefing line cutters and the nine ten-second delay cutters must operate as one, at the same instant, has necessitated a high degree of sophistication previously unknown in the design of reefing line cutters. The time delay composition in these cutters must react much more uniformly and consistently that in conventional cutters, as used in single parachute

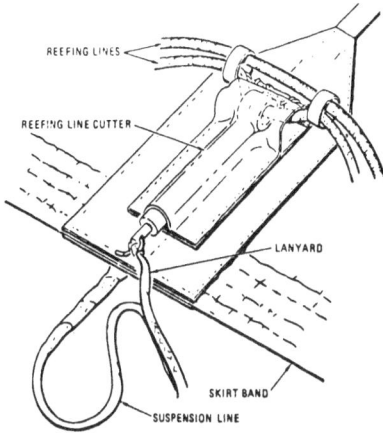

Fig. 12:59 Reefing line cutter attached to the skirt band of main parachute (Courtesy of North American Rockwell Corp.).

landing systems. A substantially higher mechanical precision of the components and of the cutter assembly than that of convential units is an additional requirement to achieve the high accuracy in operation as necessary in the Apollo recovery and landing system.

The final descent of the spacecraft on the main parachutes lasts about five minutes. The Command Module is suspended from the

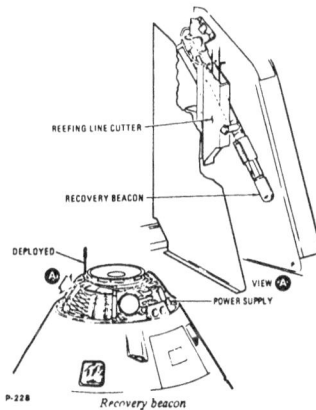

Fig. 12:60 Location aid devices and their deployment mechanism (Courtesy of North American Rockwell Corp.).

Fig. 12:61 Splashdown of Apollo and release of main parachutes (Courtesy of North American Rockwell Corp.).

main parachutes at an angle of 27.5 degrees, so that the spacecraft will hit the water at the last impact load.

By the deployment of the main parachute riser, six reefing line cutters with an eight-second time delay are actuated to release a spring-loaded deployment mechanism on two VHF antennas and on a flashing beacon light used as location aids to assist in recovery operations. Parachute rigging cord, which passes through the reefing line cutters, is used for securing these recovery devices. The arrangement of these location aid devices and their deployment mechanism are shown in Fig. 12:60.

Immediately after splashdown, the main parachutes are disconnected by firing five guillotines in the parachute disconnect assembly,

Fig. 12:62 Apollo spacecraft after splashdown (Courtesy of North American Rockwell Corp.).

also called "flower pot". This phase of the recovery operation is presented in Fig. 12:61.

The main parachute disconnect and the recovery aid subsystem is set in operation by the crew. The recovery aid subsystem consists of an uprighting system utilizing three spherical flotation bags, a swimmer's umbilical for communication with the crew in the spacecraft, a powder fluorescine sea dye marker, and a shark repellent. Additionally, a sea recovery sling of steel cable, which will spring into position after deployment of the parachutes, is provided for lifting the Command Module aboard a recovery ship. Fig. 12:62 shows the Apollo spacecraft after splashdown, floating on the ocean, with the location aids deployed, prepared to be lifted aboard a recovery ship after a successful space mission and a safe recovery and landing.

13   Missile Systems

Pyrotechnic devices are utilized in missiles for numerous functions in a similar manner as in spacecraft. Major differences in the requirements and utilization of pyrotechnic devices in missiles as compared with spacecraft result mainly from the purpose and use of the vehicle. While spacecraft, in general, are used for scientific exploration and observation and are sent on extensive outer space missions and earth orbits of long duration to accomplish their tasks, it is the objective of missiles to hit a certain target at a known distance, which they reach after a flight of a very short duration. As a result of these major differences, the required capability of pyrotechnic devices to withstand extreme environmental conditions for various time spans and the required degree of their reliability can vary widely.

Two basically different types of missiles are used: unguided missiles or rocket projectiles, and guided missiles. Both types of missiles, consisting of a shell containing a military payload, such as an explosive or smoke charge, are propelled by a rocket motor during the initial part of their flight, and they continue their path in free flight without propulsion to reach their target. The military payload is normally contained in a sealed compartment in or near the nose cone, the propellant near the center, and the rocket motor and nozzle at the aft end of the missile. In multi-stage missiles, each stage contains the propellant in the main portion of the stage shell, and the rocket motor nozzle at the aft end of the stage. In guided missiles, the guidance system consisting of an inertial navigation stabilized platform with gyroscopes and a solenoid system is usually located in a sealed compartment between the military payload and the propulsion compartment.

In rocket projectiles, solid propellant is generally used, while both solid and liquid propellants are used in guided missiles. Rocket projectiles are normally of smaller size and less complexity and have a considerably lower target accuracy and shorter range than guided missiles.

Pyrotechnic devices in missiles are mainly used for prelaunch holddown, rocket motor ignition, umbilical disconnect, spin motor ignition, shroud separation, stage separation, data package ejection, operation of auxiliary power units (APU) and gyroscopes, thrust termination, thrust reversal, safe and arm, and destruct.

Application of pyrotechnic devices and systems for the most important functions in missiles, and typical pyrotechnic devices developed for missile applications are described in the following sections of this chapter.

A. *Safety and Arming Systems*

The safety of missile operations depends largely on the use of highly reliable safety and arming devices and systems which prevent premature initiation of pyrotechnic devices utilized for such functions as rocket motor ignition, shroud separation, stage separation, self-destruct, and thrust termination or thrust reversal. While these safety and arming devices, which have the two stable positions "safe" and "armed", when set in "safe" position, make an inadvertent or premature firing impossible, they must initiate the pyrotechnic devices, when set in "armed" position, after receiving a firing signal. For easy status checking, safety and arming devices are usually equipped with visual, mechanical or electrical "safe" and "armed" indicators.

The required safety and arming feature in a typical solid-propellant rocket motor igniter is achieved mechanically by a rotatable venting device located at the external end of the unit, on the side of which the initiator is mounted. When set in "safe" position, and in case the initiator should be fired inadvertently, this venting device prevents the products of combustion of the initiator from reaching the firing train of the igniter and from initiating it. Only when rotated into the "armed" position, the products of combustion will reach the firing train through a connecting duct to ignite the solid propellant. An additional safety feature is usually provided by the electrically shorted initiator.

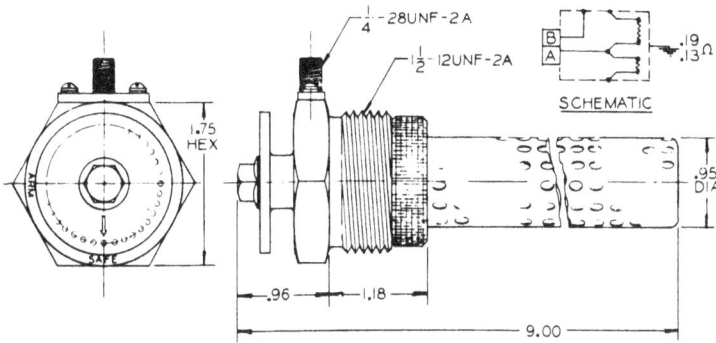

Fig. 13:1 Safe/arm igniter for solid-propellant rocket motor (Courtesy of Bermite, North Hollywood Branch, Div. of the Whittaker Corp.).

A safe/arm igniter for a solid propellant rocket motor of a missile, as described, is shown in Fig. 13:1.

Safety and arming devices for in-flight operations, such as shroud or stage separation, thrust termination, thrust reversal, and self-destruct are usually also designed on the closed and open-port principle. Since they must be remotely actuated, they are driven by solenoid actuators.

A typical safe/arm device for such application is equipped with a rotary switch which is attached to a motor-driven locking cam. The cam rotates about 90 degrees from the "safe" to the "armed" position, and it is held in either position by a flat surface or detent at each end. When the device is actuated, the rotary actuator opens the outlet port, and the initiator is fired, and thus the required operation is safely accomplished.

To prevent failures in missile operations, it is required that remotely-actuated safe/arm devices are equipped with highly reliable remote "safe" and "arm" indicators for ground control. It is a further requirement that the electric actuator circuits are shielded, because electro-mechanical actuators generate radio-frequency noise, caused by closing and interrupting electrical circuits. To attain the required high reliability, too great complexity in the design of safe/arm devices must be avoided, because most of these devices have to function flawlessly under extreme environmental conditions, such as shock, vibration, acceleration, high and low temperatures, and vacuum conditions.

B. *Ignition Systems*

Ignition systems utilizing pyrotechnic igniters are used in most solid-propellant rocket motors and in many liquid propellant motors. Pyrotechnic igniters for missile applications are usually equipped with a safe and arm switch. They contain an initiator, ignition powder, and a main ignition charge and are initiated by the energy from an electrical impulse. The main charge in pyrotechnic igniters used for ignition of solid-propellant rocket motors in missiles usually consists of flame- and gas-producing charge material pellets which are contained in a cylindrical basket. This arrangement offers the advantage that a large surface area of the main charge is provided close to the solid propellant, resulting in rapid ignition of a relatively large surface area. Two typical basket-type pyrotechnic igniters for solid-propellant rocket motors are shown in Fig. 13:1 and Fig. 13:2.

Under the heading "Igniter and Initiator Cartridges", a detailed description of pyrotechnic igniters is presented in Part II, Chapter 3, Section A-2, of this book.

Ignition systems for solid-propellant rocket motors must be selected and designed to meet specific function time and energy

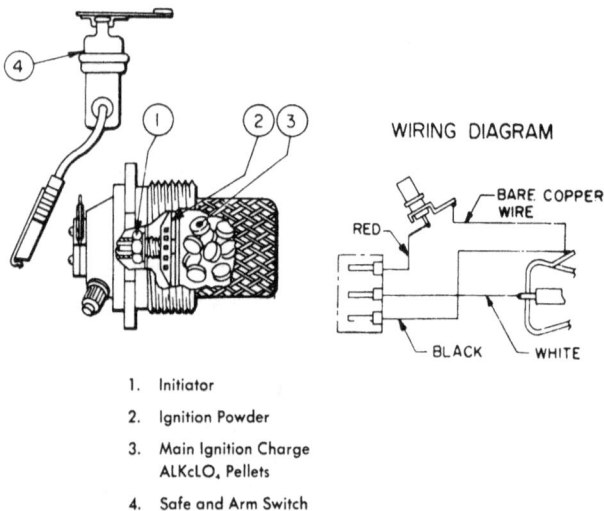

WIRING DIAGRAM

1. Initiator
2. Ignition Powder
3. Main Ignition Charge ALKclO₄ Pellets
4. Safe and Arm Switch

Fig. 13:2 Basket-type igniter with safe/arm switch (Courtesy of Bermite, North Hollywood Branch, Div. of the Whittaker Corp.).

requirements according to operation requirements of the missile, and also to the propellant characteristics, mainly the ignitability of the propellant, which affect the ignition cycle and the ignition time. The ignition cycle is the sequence consisting of spreading the igniter flame over a sufficient surface area of the propellant and of developing rocket chamber pressure until the required operating pressure has been attained.

Pyrotechnic igniters containing conventional solid propellant grain, rather than pyrotechnic pellets, known as "pyrogen igniters", are mainly used in ignition systems for large solid-propellant rocket motors. A typical pyrogen igniter is shown in Fig. 13:3. For solid-propellant rocket motor applications that require spreading the flames from the whole surface of the main charge container of the igniter, a pyrogen igniter with a perforated tube can be used.

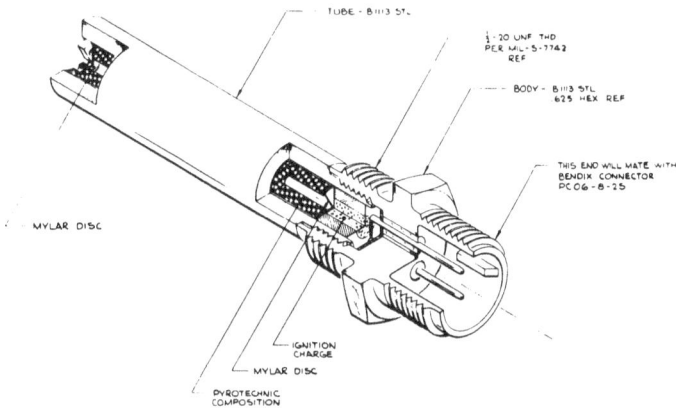

Fig. 13:3 Pyrogen igniter (Courtesy of SDI, Special Devices Inc.).

Igniters in solid-propellant rocket motors are usually mounted in the head of the motor case, which requires a mounting surface in the form of a boss, and also a seal to hold the combustion chamber pressure during the operating cycle of the motor. After ignition has been accomplished, the igniter with its installation components and the structural mounting boss represent dead weight.

Studies and tests showed that these disadvantages inherent in front-end-ignited rocket motors can be avoided and that the mass fraction can be improved by using an aft-end ignition system. To ignite a single-stage missile or the first stage of a multi-stage missile, the igniter can be arranged to be a component of the launch pad. In multi-stage missiles, to ignite an upper stage, the igniter can be mounted at the forward end of the lower, preceding stage. In both arrangements, the igniter does not stay with the rocket motor after ignition has been accomplished.

It was found that the aft-end arrangement of the igniter results in reasonably short ignition delays and short ignition pressure gradients, and that the shock wave forms farther upstream in the port, thus improving gas penetration, when the igniter is fitted with a nozzle exit cone. In some test applications, the igniter gases formed a stagnation plane in the port and did not penetrate to the head of the rocket motor, the flame propagation was lower than for head-end igniters, and chamber pressure built up slowly during ignition, which are disadvantageous results when using aft-end igniters. However, by increasing the zone penetrated by the igniter gases, the rate of pressure buildup can be improved.

To prevent ignition overpressure, the clearance area between the igniter and the exit cone of the rocket motor has to be larger than the throat area of the motor. Aft-end ignition is well suited for clustered rocket motors. When manifolded aft-end pyrogen igniters are used in a clustered rocket motor, all motors will be ignited, even if one igniter does not function.

Tests have also been conducted with a hypergolic aft-end ignition system. Ignition was accomplished by squirting oxidizer from the end of a long tube deep into the port cavity of a solid-propellant rocket motor. The tests showed that this ignition method, for which very simple and reliable safe-and-arm devices can be used, is probably a safer method than the pyrotechnic or pyrogen-type ignition.

A simple, but unconventional method for the ignition of solid-propellant rocket motors is the use of an exothermic bimetallic ignition material, such as "Pyrofuze", which is an aluminum-core, palladium-clad matrix. This bimetallic material is available in form of sheets, foil, tubes, solid and braided wire, granules, and in some other forms. A detailed description of this exothermic material is presented in Part I, Chapter 1, Section J, of this book under the heading "Bi-Metallic Exothermically Alloying Compositions". The presently avail-

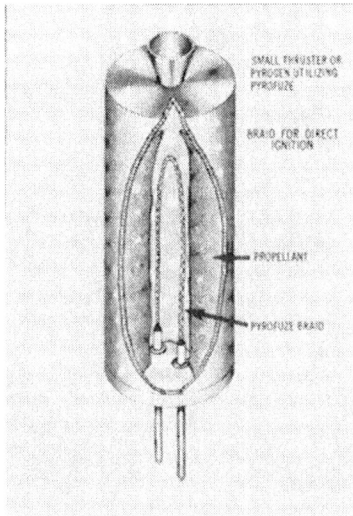

Fig. 13:4 "Pyrofuze" braid igniter (Courtesy of the Pyrofuze Corp.).

able bimetallic material Pyrofuze can be used in braided form as an igniter in solid-propellant rocket motors, as shown in Fig. 13:4.

The exothermic action of the alloying function of Pyrofuze releases 2890 calories per cubic centimeter, or 325 calories per gram. When Pyrofuze solid or braided wire is used as an igniter for a solid propellant, the extended wire material should be attached in tension to the propellant, because inconsiderable brisance occurs during ignition.

In braided or coiled form, surrounded by a roll of perforated foil, this exothermic bimetallic material can also be utilized as igniters, as shown in Fig. 13:5.

Another possible application of exothermic bimetallic material is to embed exothermic wire in straight or coiled form, or in straight and coiled form combined, in the solid propellant to enhance and control burning, as shown in three examples in Fig. 13:6.

The exothermic bimetallic igniter systems, as described, eliminate an igniter-cause ignition spike, and they also minimize the rate of pressure rise. They further eliminate the need for igniter augmentation to the mass discharge of the rocket motor. It can be expected that

291

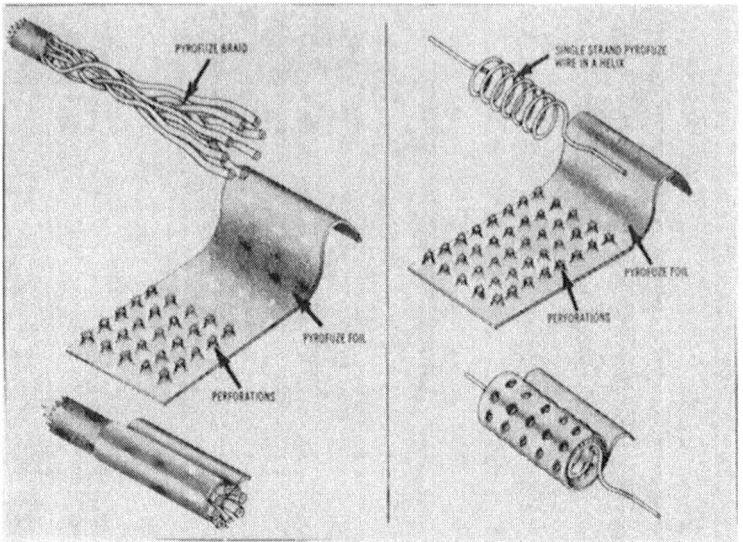

Fig. 13:5 "Pyrofuze" foil igniters (Courtesy of the Pyrofuze Corp.).

further tests and developmental work of exothermic bimetallic igniters will result in a very simple and efficient ignition technique for solid-propellant rocket motors.

Solid-propellant rocket motors offer the advantage that they are simple in design, which is an important feature for achieving the desired reliability and low cost. Their disadvantage is, however, the fact that they cannot be restarted, once they have been stopped. Since many missiles do not need restart capability, solid-propellant motors are sufficient for their operation.

For missiles that must have restart capability, liquid rocket motors are used. Ignition of liquid-propellant rocket motors presents considerably greater difficulties than ignition of solid-propellant motors because of possible variations in flow rate, pressure, temperature, and ratio of mixture of the liquid propellant and the oxidizer. At low temperatures, for example, the ignitability of most liquid propellants decreases considerably. Liquid-propellant rocket motors which do not use hypergolic fluids, are ignited either by pyrotechnic igniters or by spark igniters. The ignition system in a liquid-propellant rocket motor includes one or several igniters, valves, ducts, switches, and an

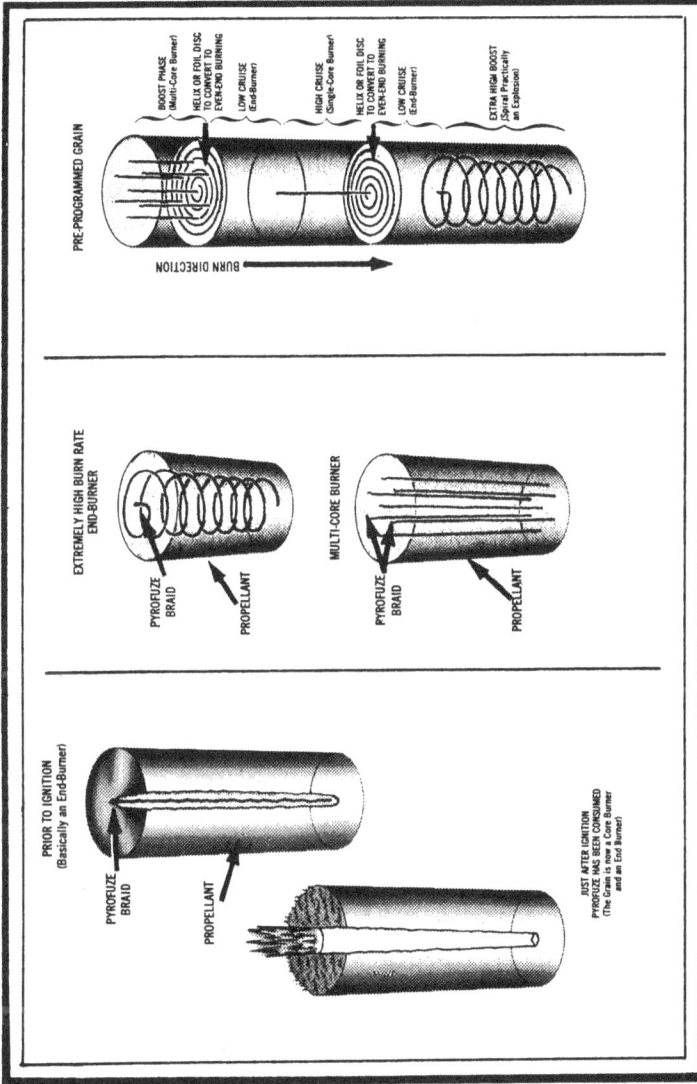

Fig. 13:6 Application of "Pyrofuze" wire in solid propellant to enhance and control burning (Courtesy of the Pyrofuze Corp.).

ignition indicator and detection device, or a system of ignition indicator and detection devices.

When restart capability is required in a liquid-propellant motor, and pyrotechnic ignition is to be used, a group of pyrotechnic igniters is provided and arranged in such a manner that only one, or when redundancy is required, two igniters are initiated for one start. The remaining igniters are then to be used for later starts.

It may appear much simpler to utilize a spark igniter when restartability is a requirement. However, spark igniters are susceptible to carbon deposits when hydro-carbons are used as propellant. These carbon deposits can easily cause ignition failures.

When hypergolic fluids are used in a rocket motor, the required restartability can be provided by a solenoid-actuated valve system, which must function reliably under all expected conditions.

C. Control Systems

Guided missiles are usually launched in a vertical direction and must instantly be tilted into their programmed ballistic flight trajectory to reach the target. For accomplishing the control operations necessary to meet the target objective, guided missiles are equipped with a suitable control system which is connected to a sophisticated inertial guidance system. Considering the fact that guided missiles have a range of up to 6000 miles and an average velocity of 15,000 miles per hour, the guidance and control systems must be designed to react and operate at an extremely high speed and with the greatest accuracy.

Various methods and systems for missile control have been developed. Early missiles, which had to operate within the atmosphere, were controlled by aerodynamic control devices, such as folding fins, which opened by inertia. Missiles or payload stages can be stabilized in flight by a spinning motion, which can be induced by a series of small nozzles that are canted at an angle and thus, impart torque and thrust to the missile, or by small spin motors, which are small rocket motors.

A suitable method for missile control is by means of deflection vanes mounted in the exhaust of the rocket nozzle. To achieve pitch, yaw, and roll control, three pairs of vanes are used for control movements about all three axes. Deviations from the programmed flight trajectory are sensed in flight by gyroscopes within an automatic control system, and corrective signals are sent to servo-motors which operate the vanes. Carbon and molybdenum are mainly used as heat-resistant materials for the vanes. Disadvantageous features of jet vane control systems are a decrease of the thrust caused by drag of the vanes, and erosion of the vane material caused by the hot exhaust gas.

A result of further development of the control system, as described, is the jetevator, a ring-shaped deflector which is arranged at the nozzle periphery and which can be rotated about a few degrees into the edge of the rocket jet, as required for control operation. This design concept requires that the exposed components of the jetevator actuation mechanism be well protected from the hot exhaust of the nozzle. Pitch and yaw control can be achieved with one gimballed jetevator. For roll control movements, a minimum of two nozzles with gimballed jetavators is required.

Another efficient method of control is gimbal-mounting of the rocket motor thrust chamber or of the nozzle. This method was used successfully in several intermediate-range ballistic missiles (IRBM) and intercontinental ballistic missiles (ICBM). In gimballed control systems, the thrust vector can usually be deflected about 10 degrees to either side of the center line of the missile.

Roll control of missiles, which are equipped with only one jetevator nozzle or gimballed nozzle, can be accomplished by small jet nozzles which are tangentially arranged on the side of the missile and which can be fed from a pressurized gas bottle.

In spite of the fact that pyrotechnic devices, at present, find only limited application in missile control systems, descriptions of these basic systems and their functions are presented, presuming that in future control systems, pyrotechnics may play a more important role than presently. It is felt that, when the designer fully understands the functions to be performed, he will be best prepared to develop suitable actuation devices and systems.

Pyrotechnic devices are utilized in missile control systems mainly for the following operations: roll control, vernier control, attitude control, spin stabilization, spin cancellation, thrust reversal, thrust termination, and provision of power for turbo-pumps in fuel systems. In many missile control systems, roll control, vernier control, attitude control, spin stabilization, and spin cancellation are achieved by the thrust from a small rocket motor. A typical motor used for spin stabilization of a medium-size missile has a diameter of 1.5 inches, a length of 6.5 inches, and a loaded weight of 0.8 pounds. The igniter is located in the head end of the combustion chamber, which contains solid propellant. This small rocket motor, which has a burning time of approximately one second, provides an average thrust of approximately 45 lbf and a total impulse of about 50 lbf-sec. Similar small rocket motors, sized for specific operations, are ideally used to accomplish other one-shot control functions in missiles.

At a specific point of its flight path, the thrust of a missile must be terminated, so that the vehicle continues on its trajectory in free flight to reach the target. Thrust must be terminated or reversed, for example, in preparation for stage separation. Various methods are used to accomplish thrust termination. A simple and commonly used method is to open the forward end of the combustion chamber, which can easily and reliably be achieved by a thrust termination sector assembly, as shown in Fig. 13:7. This thrust termination device consists of a thrust termination retainer ring, an initiator assembly containing two initiator cartridges, and mild-detonating fuse (MDF) containing 10 grains per foot. When fired by electric current, the mild-detonating fuse (MDF) and the sector segment disintegrate and thus allow the thrust termination retainer ring to collapse. Subsequently, the thrust termination closure port is expelled by chamber pressure,

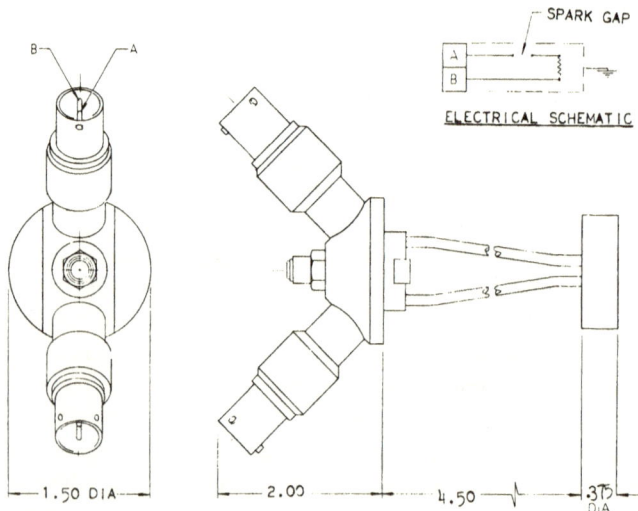

Fig. 13:7 Thrust termination sector assembly (Courtesy of McCormick-Selph).

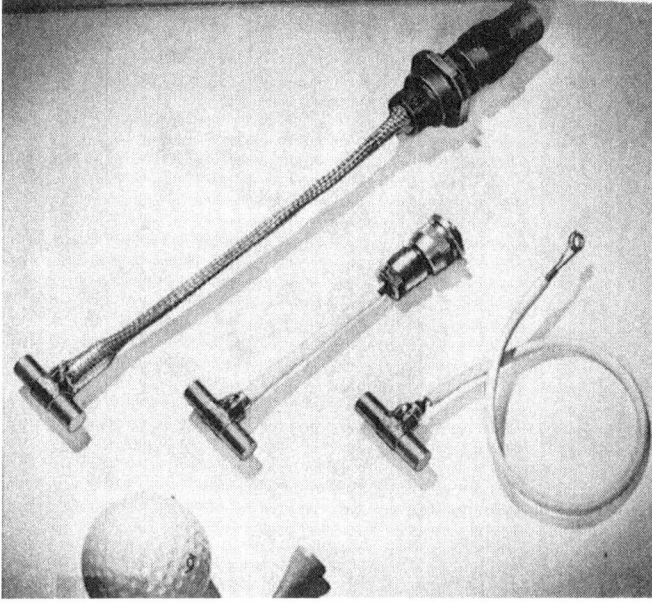

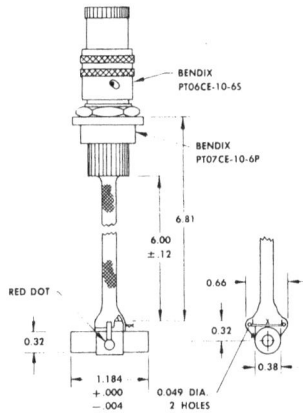

Fig. 13:8 EBW thrust termination sector (Courtesy of Singer General Precision, Link Group).

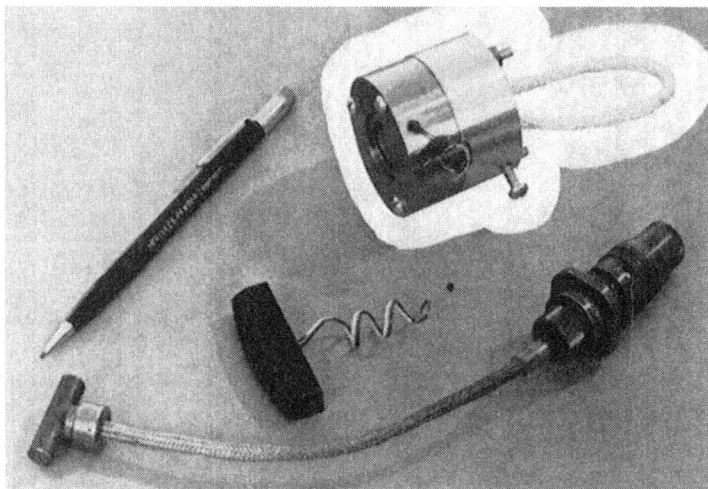

Fig. 13:9 Frangible snap-ring sectors and safe/arm mechanism (Courtesy of Hercules Inc.).

which results in venting of the rocket motor and in terminating of its forward thrust.

Another type of thrust termination sector, which also consists of a frangible sector, a retainer ring, and an adapter, is shown in Fig. 13:8 and in Fig. 13:9. The sector, made from Meehanite, holds a closure port in the rocket motor combustion chamber in place by means of a retaining ring. When initiated, the frangible sector disintegrates, and the retainer ring collapses, thus releasing the closure port, venting the motor, and terminating the thrust.

Besides two thrust termination sectors, a Primacord safe and arm mechanism for missile and spacecraft systems is shown in Fig. 13:9.

In some thrust termination systems for rocket motors, the forward end of the combustion chamber is closed by a burst diaphragm. A flexible linear-shaped charge (FLSC) arranged circumferentially around the sealed-off port, held in an external mounting flange is utilized to burst the diaphragm, thus venting the combustion chamber and terminating the thrust. A mild-detonating fuse (MDF) can ideally be used as a transfer line to initiate the charge from a remotely located initiator.

Other applications of pyrotechnic devices in missile control systems are, for example, the utilization of cartridge-actuated valves for

opening and closing of fluid and gas lines or containers, and the utilization of gas generators as a power source for turbo-pumps in fuel
systems for rocket motors. Cartridge-actuated valves are described in
Part II, Chapter 4, Section F, and gas generators in Part II, Chapter 8,
of this book.

D. *Stage Separation Systems*

Large, long-range guided missiles usually consist of several stages. The
last, topmost stage contains the detonating charge or warhead, while
the lower stages serve as boosters to propel the top stage into a programmed flight path, so that it will hit the target with the highest
possible accuracy. The multi-stage arrangement results in greatly improved performance and higher velocity, as compared with single-stage
missiles, mainly because the expended booster stages are not being
carried along on the flight path, after they have served their purpose;
or, in other words, the mass ratio of the missile is substantially improved when using a multi-stage system.

The various stages must be designed to provide, connected with
one another, a sound structural shell for the contained missile systems, and they must be capable of withstanding high loads caused by
acceleration, shock and vibration during launch and flight.

Critical phases during the flight of a multi-stage missile are the
stage separations, which must be accomplished flawlessly at specific
time intervals, depending on the predetermined thrust and burning
time of each stage. A successful stage separation can only be expected
after the kinematics and dynamics of the separation process have been
analyzed. Major steps of a stage separation sequence are:
1. Thrust termination of the lower booster motor,
2. Separation of the two stages, and
3. Ignition of the upper-stage booster motor.

It is important that the missile prior and during stage separation,
has adequate flight stability and control and that during the critical
phase of stage separation sufficient clearance exists between the two
stages.

The area between the head end of a lower stage and the aft end of
an upper stage is usually closed by an adapter connecting both stages.
When stage separation has to be accomplished at high dynamic pressure, the upper stage rocket motor should be ignited immediately,

which necessitates the provision of blowout panels or vent doors in the adapter to prevent harmful pressure buildup and possible damage. Blowout panels or vent doors can be released or opened by explosive bolts which are fired simultaneously with ignition of the upper motor. Prior to ignition of the upper motor, the explosive bolts which are fired simultaneously with ignition of the upper motor. Prior to ignition of the upper motor, the explosive bolts hold the blowout panels in closed position. In adapters where explosive bolts cannot be used advantageously, linear-shaped charges may be utilized for this door release operation. Outside the atmosphere or at low dynamic pressure, the upper stage rocket motor can be ignited with a short delay after stage separation.

When the upper stage is ignited with a time delay after stage separation has been accomplished, the upper stage must be accelerated to attain the necessary separation velocity. Small rocket motors mounted on the head end of the lower stage and on the aft end of the upper stage are used in several missiles to move the upper stage away from the separation area and to decelerate the lower stage, and thus, an interference of the separated lower stage with the flight path of the upper stage is avoided. In some missile stage separation systems, only the lower stage is equipped with small retrorockets to decelerate the stage, and the upper stage continues its flight by its own momentum until its rocket motor is ignited.

To correct disturbances which might occur during the separation of two stages, such as unsymmetrical booster cutoff, thrust-vector misalignment of the upper stage rocket motors, unsymmetrical thrust, or similar disturbances, the upper stage in missiles is often equipped with small control jet nozzles for pitch, yaw, and roll control.

Stage separation within the atmosphere can usually be accomplished without utilizing retrorockets on the lower stage and auxiliary rockets on the upper stage. The drag of the expended, empty lower stage is normally higher than that of the heavier upper stage, resulting in a smooth and safe parting of both stages.

In many stage separation systems, compression springs are arranged in the separation plane attached to the lower stage. Prior to separation, the springs are held compressed by the joint of the stages. When stage separation is initiated, the upper stage is propelled forward by the energy of the compression springs.

Pyrotechnic devices for stage separation, similar to those as described in Part III, Chapter 12, Section C, of this book are utilized for missile stage separation.

Explosive bolts have been used in numerous missile stage separation systems mainly for small or medium-size missiles. With the increase of the size of a missile, the required number of explosive bolts for a separable stage joint would increase proportionally. A greater number of pyrotechnic devices to be initiated to accomplish stage separation, however, would result in a lower reliability than by using only one or two devices.

An ideal solution is the use of a linear-shaped charge for instantly cutting the skin of the missile's shell, resulting in a clean cut without dangerous fragmentation. One initiator is only required for this operation. Simultaneously with cutting the skin by initiating the linear-shaped charge (LSC), the electrical cable bundles and fluid lines that connect both stages during flight prior to stage separation are cut by firing pyrotechnic cable and pipe cutters, immediately after necessary switching for circuit breaking and valve closing of the fluid and gas system have been accomplished by pyrotechnic switches and valves, or by solenoid devices.

E. Destruct Systems

In case of malfunction, the flight of a missile must be terminated, and the missile must be destroyed in a safe area, because extensive damage on the ground could be caused by a straying or otherwise malfunctioning missile, if it would be allowed to continue its flight. To accomplish destruction of a missile in flight, linear-shaped charges are commonly used. In some earlier missile destruct systems, one explosive charge was placed in a sealed cavity in the liquid oxygen tank and one charge was placed in a similar cavity in the fuel tank, and both charges were ignited simultaneously on receiving a destruct signal. On the German V-2 missiles that were launched at White Sands after World War II, primacord, arranged circumferentially around the midsection of the missile was used as a destruct device.

In missile destruct systems, linear-shaped charges are usually attached lengthwise on the outside of the fuel and oxidizer tanks, and in some systems also on the head ends of the tanks. For initiation of multiple explosive charges, a mild-detonating fuse is used advantageously in missile destruct systems.

A typical example of a destruct system designed to rupture the fuel and oxidizer tanks by an externally attached linear-shaped charge is the propellant dispersion system used on the Saturn V booster, as described in Part III, Chapter 12, Section B, of this book. In destruct systems for missiles using hypergolic propellants, linear-shaped charges can be mounted on the booster adapter and on the ends of the tanks. Destruct is then achieved by the rupturing of the tanks, which causes mixing of the hypergolic propellants, and explosion.

A typical missile destruct system consists of two identical and redundant destruct subsystems, each of which is capable of destroying the missile on receipt of a ground command destruct signal. The most important components in each destruct subsystem are a safe/arm mechanism with two detonator cartridges, two batteries, two antennas mounted opposite to one another on the sides of the missile, a command receiver, two or more destruct charges, and connecting wire assemblies.

The destruct system can be initiated either by a signal from the command receivers during the flight of the missile prior to separation of the final stage from the booster stage, or automatically, in case a premature final stage separation occurs before the booster engine has been cut off. Under normal conditions in a successful flight, the automatic portion of the destruct initiation system is made inactive at booster engine cutoff to achieve a normal separation of the booster stage from the final stage.

At lift-off, the safe/arm mechanisms in the destruct subsystems are armed by lanyards, or by similar means. During flight, destruction of the missile can be commanded by the range safety officer, if required, by transmitting a coded signal to the command receivers in the missile. When receiving a destruct signal, each command receiver sends an electric signal to the two detonators provided in the missile safe/arm mechanisms and to the initiators of the destruct charges. The system is so designed that either detonator in a safe/arm mechanism will initiate the destruct charges. To ensure destruction of both stages, a time-delay is provided in the safe/arm mechanisms of the booster stage, so that destruct of the booster stage is initiated with a delay of 0.1 second after initiation of the final stage.

A missile destruct system mainly consisting of two ring-shaped

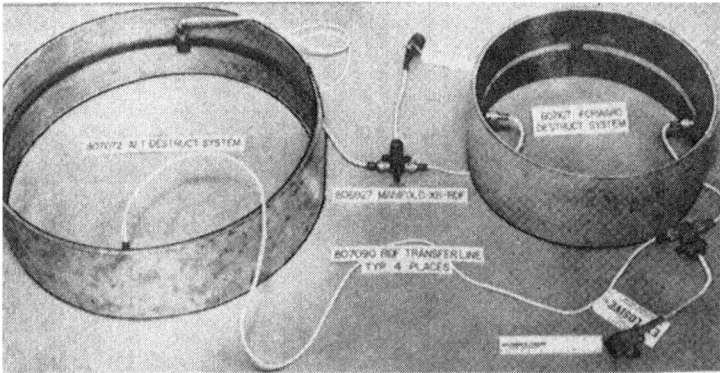

Fig. 13:10 Missile Destruct system (Courtesy of McCormick-Selph).

subsystems, a forward and an aft destruct ring, is presented in Fig. 13:10. The forward destruct ring consists of 50 grain per foot RDX lead-sheathed flexible linear-shaped charge (FLSC), which is bonded into a phenolic back-up ring, and the aft destruct ring contains 50 grain per foot flexible linear-shaped charge (FLSC). This missile destruct system is initiated by one initiator through a manifolded mild-detonating fuse (MDF). This arrangement offers the advantage that the initiator can be mounted in a safe area on the missile and that good accessibility to the initiator is provided.

Destruction systems and destructors are not only used to destroy missiles in case of malfunction, but also aircraft, drones and especially important electronic equipment, such as circuit boards, and electrical, optical and mechanical systems and equipment of military importance, in case of a landing in enemy territory.

A typical incendiary destructor, as shown in Fig. 13:11, developed for destruction of electronic circuit boards and similar equipment, contains besides a squib mixture four pyrotechnic compositions: an initiator mixture, a booster charge, a transition charge, and an incendiary charge, in a flat case having a size of only about 6 x 3 inches, and 0.38 inches thick. The destructor is fastened to the equipment which is to be destroyed in case of emergency. An exploding bridgewire of 1.0 ± 0.1 ohm resistance is used for initiation. The all-fire current for this destructor is 3.5 amperes for 10 milliseconds, and the no-fire current level is 1 ampere for 5 minutes. When initiated, destruction occurs in the direction normal to either face of the de-

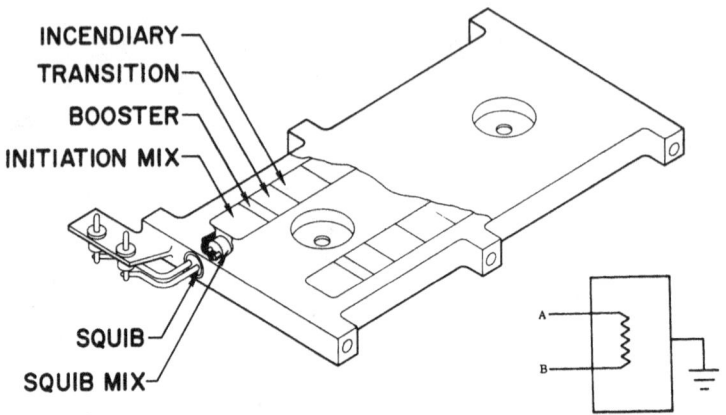

Fig. 13:11 Incendiary destructor (Courtesy of Hi-Shear Corp.).

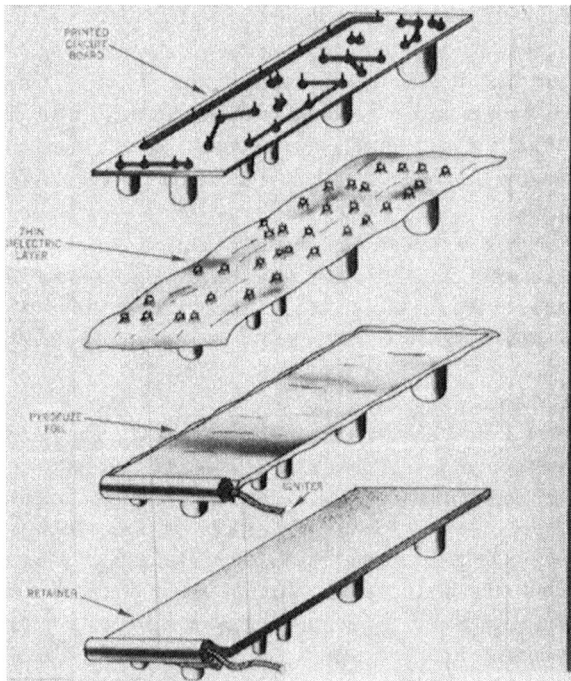

Fig. 13:12 Destructor utilizing "Pyrofuze" foil (Courtesy of Pyrofuze Corp.).

structor plate. This destructor is designed for a heat output of 90,000 calories at a rate of approximately 1500 calories per second, and for a burn duration of 60 to 90 seconds at a temperature range from −54°C to +121°C. The auto-ignition temperature of the incendiary powder used in this destructor is in excess of 260°C.

In another type of destructor for electronic devices, as shown in Fig. 13:12, bimetallic exothermically alloying Pyrofuze foil is utilized as deflagrating material. This type of destructor is made in the size of the electronic device to be destroyed, and is directly mounted to it.

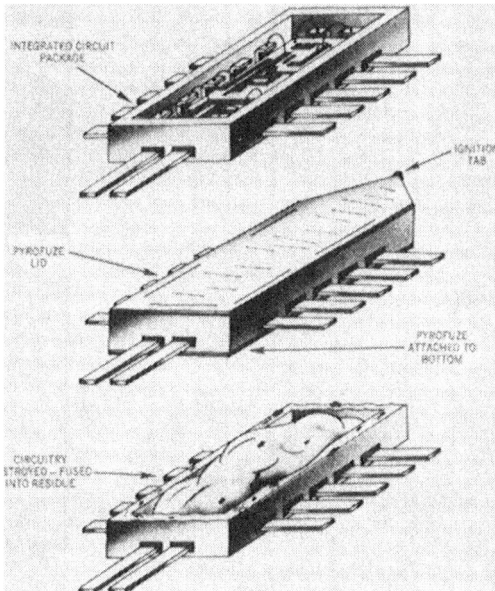

Fig. 13:13 "Pyrofuze" destructor for integrated circuit package (Courtesy of Pyrofuze Corp.).

In a different version of a destructor for an integrated electronic circuit package or similar device, as shown in Fig. 13:13, a Pyrofuze sheet is used as a deflagrating material. When initiated, these destructors instantly and completely destroy the devices which they are attached to, leaving only a fused residue.

These examples show that for a specific application the best suitable pyrotechnic destructor or destruction system can be selected from a number of available types.

Part IV

Reliability and Testing

Pyrotechnic devices and systems must be so designed that they will function reliably under specific environmental conditions, and the materials, including joints and seals used for pyrotechnic devices must be capable of withstanding the expected environmental effects during their lifetime.

Pyrotechnic devices, which are not available as shelf items and which are not qualified, and pyrotechnic systems, have to undergo extensive testing under simulated environmental conditions to attain the required high reliability and thus, to avoid failures in their operation which, in many cases, could be catastrophic.

The possible effects of environmental conditions on pyrotechnic devices, a method for determining the reliability of pyrotechnic devices, and various test methods used to achieve the high reliability level, as required, are described in the following chapters.

14 Effects of Environmental Conditions

As described in the previous chapters of this book, pyrotechnic devices and pyrotechnic systems are primarily used to perform important functions in aircraft, drones, spacecraft, missiles, and underwater systems. The environments in which these vehicles and systems are used, can be categorized as

a. Atmosphere,

b. Outer space, and

c. Deep sea.

The utilization of pyrotechnic devices and systems in aircraft and drones presents the least severe problems with regard to environmental effects, since the temperatures and pressures in the altitude range in the atmosphere, in which these vehicles operate, are at a reasonable level and are only slightly different from the conditions on earth.

Pyrotechnic devices that are to be utilized in aircraft and drones must be capable of withstanding shock and vibration induced by the operation of the vehicle, and temperature ranges somewhat different from those on the ground. These requirements are met by most pyrotechnic devices designed for operation on the ground.

Space environmental conditions that affect spacecraft and missiles during their flight are primarily vacuum, high and low temperatures, and radiation. Pyrotechnic devices and systems used in these vehicles must be capable of withstanding the effects of these environments. To be able to design and build the devices and systems according to these requirements and to select the best suitable materials, a thorough understanding of the effects of space environmental conditions is necessary.

Major effects of a vacuum environment are vaporization and sublimation which will change the physical properties of chemical compositions, elastomers, lubricants, and also pyrotechnic compositions. It was found that reliable functioning of pyrotechnic devices after exposure to an ultra-high vacuum environment depends mainly on the quality of the seal used to confine the explosive composition within the devices. If an explosive charge is not provided with a proper tight seal, sublimation effects and chemical degradation effects in a high vacuum environment would occur over an extended time period, which would alter the explosive material's critical properties, such as sensitivity, explosive force, thermodynamic properties, and ignition temperature. Tests with an unsealed explosive device, containing nitro-cellulose, nitro-glycerin, and lead styphnate, after an exposure to a pressure of 10^{-6} torr for 20 hours at a temperature of $25°C$, showed that the styphnate became more sensitive to thermal initiation.

According to the specification for pyrotechnic devices used in manned spacecraft systems, these components shall test to a maximum leak rate of 10^{-6} cc per second by the Radiflo method. Actually, many pyrotechnic devices that were received and tested to be used in the manned Apollo spacecraft had leak rates averaging 10^{-8} cc per second.

Effects of ultra-high vacuum environments on metals are loss by vaporization and/or sublimation of a volatile component of an alloy, and cold welding of metal surfaces in intimate contact. Cadmium, zinc, and magnesium have a high volatility rate under space vacuum conditions, and they are, therefore, not used as materials for spacecraft components or structures. Aluminum, iron, nickel, chromium, molybdenum, titanium and manganese do not present problems with regard to sublimation or vaporization and are extensively used in spacecraft structures and systems, and also in pyrotechnic devices.

Loss of material through vaporization or sublimation in a perfect vacuum can be approximated by the Langmuir equation:

$$W = \frac{p}{17.14}\sqrt{\frac{M}{T}}$$

where W = the weight loss in $gm/cm^2/sec$
                p = the vapor pressure in mm Hg

Table 14.1

Sublimation Rates versus Temperature for Various Elements
Used in Manned Spacecraft Landing System

Element	10^{-5} cm/yr	10^{-3} cm/yr	10^{-1} cm/yr	Melting Point	Use in Space-craft System
	°C	°C	°C	°C	
Cd	40	80	120	320	Not used
Se	50	80	120	220	Not used
Zn	70	130	180	420	Not used
Mg	110	170	240	650	Not used
Te	130	180	220	450	
Li	150	210	280	180	
Sb	210	270	300	630	Solder-potted
Bi	240	320	400	270	Solder-potted
Pb	270	330	430	330	Solder-potted
In	400	500	610	160	
Mn	450	540	650	1240	Alloy
Ag	480	590	700	960	Elect.-potted
Sn	550	600	800	230	Solder-potted
Al	550	680	810	660	Alloy
Be	620	700	840	1280	
Cu	630	760	900	1080	Alloy-el.-ptd.
Au	660	800	950	1060	Elect.-potted
Ge	660	800	950	940	
Cr	750	870	1000	1880	Alloy-Cres
Fe	770	900	1050	1540	Alloy
Si	790	920	1080	1410	Alloy-el.-ptd.
Ni	800	940	1090	1450	Alloy-coating
Pd	810	940	1100	1550	
Co	820	960	1100	1500	
Ti	920	1070	1250	1670	Alloy
V	1020	1180	1350	1900	Alloy
Rh	1140	1330	1540	1970	
Pt	1160	1340	1560	1770	
B	1230	1420	1640	2030	
Zr	1280	1500	1740	1850	
Ir	1300	1500	1740	2450	
Mo	1380	1630	1900	2610	Alloy
C	1530	1680	1880	3700	Steel ($\simeq$ 1%)
Ta	1780	2050	2300	3000	Alloy ($<$ 1%)
Re	1820	2050	2300	3200	

(Courtesy of Northrop Corporation, Ventura Division)

M = the gram molecular weight or gram atomic weight
of the gas phase

T = the absolute temperature

The rate of evaporation is highly dependent on temperature. The main effect of a temperature increase is to increase the rate of vaporization, since the vapor pressure increases logarithmically as the temperature increases.

Before using alloys, it is advisable to investigate whether volatilization of one or more volatile components of the alloy could cause a problem. Sublimation rates at a wide range of temperatures for various elements are listed in Table 14:1. The sublimation rate for suitable spacecraft materials is less than 10^{-5} cm per year at temperatures up to 450°C, according to Table 14:1.

The sublimation rates of metals and alloys to be used in spacecraft, obviously, do not present a problem at present for space missions of limited duration. However, before selecting metals and alloys for the next generation of spacecraft, which may have to be designed for space missions of several years duration, the possible sublimation effects on these materials should be considered.

Another effect of high vacuum is cold welding, which is related to friction and lubrication and to volatility of the oxide coating. Laboratory tests showed that cold welding does not occur above pressure levels of 10^{-7} torr. Cold welding is only possible when the surfaces are free of lubricant and oxide coating. This condition can be caused by evaporation of lubricant and absorbed gases. In moving parts, after evaporation of the lubricant, the oxide coating is abraded from the surface, which causes seizing or cold welding. It seems that the ability to cold weld is to some extent also related to the volatility of the metal or alloy and to the degree of its hardness, since the diffusion rate of the metal through and into the surface layer is controlled by these properties.

The possibility of cold welding can present a problem for the reliable functioning of some mechanical systems and also of some pyrotechnic devices and systems used in spacecraft and missiles. Devices that could be affected by cold welding are, for example, a mortar closure on a mortar tube as used for parachute ejection, shroud joints and release devices, and thrusters and similar devices that are exposed to high vacuum during a certain length of time.

Pyrotechnic devices containing moving components that function within a hermetically sealed enclosure, such as switches and cable and line cutters, and similar sealed devices will not be affected by cold welding.

All known liquid lubricants and fatty acids evaporate, and they are therefore unsuitable for space conditions. Tests showed that liquid lubricants do not even provide adequate lubrication in the lower vacuums of space simulators. Solid lubricants, such as molybdenum disulfide, tungsten disulfide, and the soft metals have given better results. However, the known data about space lubricants are results of simulator measurements made in the pressure range of 10^{-5} to 10^{-6} torr which does not simulate real space conditions, and therefore, these available data cannot be considered completely valid. It can be expected that definite data on lubricant performance in a vacuum will be obtained by conducting tests in a simulator that reaches the low 10^{-10} torr range. At this pressure level, the monolayer formation time is increased to at least several hours which will result in a sufficient time span for observing the metal surfaces.

Organic materials suitable for high vacuum applications are some of the common elastomers, as for instance vinyl-idene fluoride-hexafluoropropane, chlorotrifluoroethylene, butodiene-styrene and isoprene. Plastics suitable for space applications are silicone resins, tetrafluoro-ethylene, polyethylene, polypropylene, and ethylene terephtalate, however, in some cases their application is limited by their resistance to pressure, temperature and nuclear radiation.

Temperatures do not present great problems to pyrotechnic devices used in spacecraft, mainly because most pyrotechnic devices are mounted at such locations in the vehicle where temperatures are within a range of $+50°C$ and $-35°C$, which is within the temperature range that the devices can withstand. Some pyrotechnic devices, such as parachute mortars, which are exposed to the environment during their operation, are even without a protective cover safe within the allowable temperature range, because they only operate in the atmosphere, and while in space, they are protected by an insulating cover.

Radiation environments that can affect spacecraft and their systems are solar radiation and cosmic radiation. It may be necessary also to consider nuclear radiation in the future, because of the potential of nuclear energy for spacecraft propulsion and auxiliary power.

Solar radiation consists mainly of ultraviolet, infrared and X-ray radiation. The ultraviolet radiation is the most destructive portion of the solar spectrum. Infrared radiation, which is actually heat, may be compensated for by suitable surface coatings or reflective surface materials. X-rays of solar origin may have sufficient energy to cause ionization or atomic displacement.

Cosmic radiation, consisting primarily of protons, alpha particles and a small percentage of heavy nuclei, is isotropic, with the exception of radiant energy induced by magnetic fields, such as the Van Allen belts. Cosmic radiation does not present a problem with regard to effects on spacecraft materials.

Metals and alloys are not damaged by radiation, except by high fluxes and energies of nuclear radiation as exist in nuclear reactors. Damage to spacecraft metals by radiation is not to be expected. However, organic materials are susceptible to radiation damage. Organic materials, unlike metals, have a definite molecular structure which can be destroyed by radiation effects, and healing of the damage, similar to annealing in metals, is not possible. Radiation can have the following effects on polymers: reduction of their molecular weight, increase of their vapor pressure, increase of their viscosity and decrease of their mechanical strength. Nylon, polyethylene and Teflon are affected by radiation so that the crystallinity in these polymers is destroyed.

Inorganic materials are, in general, capable of withstanding radiation without damage. However, some properties or inorganic materials may be changed by radiation, as for instance, transparency in glass or in crystalline material may be affected, or it may become opaque, and some electrical insulation materials may become partially conductive. Radiation does not affect ceramic materials seriously, as many practical application examples proved.

Pyrotechnic compositions are usually affected to some extent by nuclear radiation. Some of the effects are a change in the rate of gas production, weight loss, lowered melting point, reduced ignition temperature, altered sensitivity and altered resistance. Yet, many pyrotechnic compositions have remarkably high functional thresholds and experience no damage after an exposure of 10^7 rads and higher. The level where changes in performance become significant, is called "functional threshold". In numerous tests, pyrotechnic devices were exposed to 2×10^7 rads of gamma irradiation and to 2×10^{11} nvt (sulfur), and in other tests to 8×10^{10} rads/second for a total exposure of 9×10^5 rads. All of the pyrotechnic devices passed these tests without firing during irradiation.

Functional thresholds of some pyrotechnic materials are presented in Table 14:2.

As a result of numerous tests and calculations, expected service life

Table 14.2

Functional Thresholds of Pyrotechnic Materials
(10 Percent Loss of Available Energy)

Material	Functional Threshold (rads)
Nitrocellulose	5×10^6
Nitroglycerine	10^7
Lead Azide	5×10^7
Hercules "Hi-Temp"	8×10^7
Hercules "Unique"	10^7
Hercules 5250.95	10^7
42-G Primer	10^7
RDX	8×10^7

(Courtesy of Northrop Corporation, Ventura Division)

Table 14.3

**Shelf Life of Pyrotechnic Compositions
under various Temperature Conditions**

Composition or Material	Continuous 66°C	Continuous 71°C	Ambient (−29°C to 43°C)	66°C to −36°C Cycle
Olin-Mathieson M-42G Primer		1 year		
Hercules "Unique"	250 Days		20 Years	7 Years
Hercules "5250.95"	500 Days		20 Years	14 Years
Hercules "Hi-Temp"	Indefinite		20 Years	

(Courtesy of Northrop Corporation, Ventura Division)

data of pyrotechnic devices were obtained. According to these data, the estimated service life of most devices when exposed to space radiation, temperature, zero G, and meteorites in lunar missions and earth orbits exceeds two years. A shorter service life for these devices has been listed only for exposure to high vacuum in space; the result is 2-3 months at present. With improvements in design, and manufacturing and test methods of pyrotechnic devices, a substantial increase of the service life of these devices can also be expected.

Since exposure to high temperature can affect some properties of pyrotechnic compositions, it is useful for many applications to be familiar with data about these effects. Shelf life data for some pyrotechnic compositions under various temperature conditions are listed in Table 14:3.

Pyrotechnic devices that are to be used in deep sea applications, must be capable of functioning under extreme deep sea conditions and of withstanding the effects of the deep sea environment. The materials used for these devices and all their components that will be exposed to sea water, must be salt water corrosion-resistant. Pyrotechnic underwater devices must also be designed to withstand the high water pressures, depending on the depth level of their application. An important factor to be considered already in the design stage is the required life time under sea water effects. Special care must be given to the hermetic seals that should be provided, for example, in cable cutters to protect the pyrotechnic charges and the cutter blade, piston and O-rings from damage by sea water. A further important requirement is, of course, to use sealed, sea water proof electrical connectors and cables for these pyrotechnic underwater devices.

15   Reliability Theory

With the ever increasing rate of the utilization of pyrotechnics for important functions, a high reliability or pyrotechnic devices and systems became a necessity, because in many cases, the success of a space mission or
of a pilot's ejection from an aircraft in distress, or of the quick emergency
escape of passengers from commercial airplanes by using inflatable escape
slides depends on the reliable operation of pyrotechnic devices and systems.

Reliability can be defined as the capability of a system or device to
perform its required function for a given period of time under specified
environmental conditions. Reliability is inversely related to the expected
rate of failure of a device or system and can be calculated by subtracting
the expected probability of failure from unity. Reliability is expressed by
a decimal figure between 0 and 1, which represents the probability that a
device or system will function as specified. Usually, with the statement of
a reliability value, a "confidence level" is listed. The confidence level is a
confidence percentage which states how sure one is that a certain expected
event will happen. A reliability of 0.9995 with a confidence level of 90
percent has been attained with explosive-actuated devices. For numerous
commonly used devices, such as initiator cartridges, a reliability of 0.999
with a confidence of 95 percent is listed. The confidence level rises with an
increase of the number of successful tests of similar devices or systems.
Reliability requirements for pyrotechnic devices and systems used in
manned vehicles and those used in unmanned vehicles vary approximately
by an order of magnitude. Numbers of tests required to obtain certain
required reliabilities at specified confidence levels are listed in Table 15:1.

Depending on the importance of the mission or application for which

Ta

Number of Tests without Fai

Reliability						Confide
	50	60	70	75	80	85
0.999999	693150	916290	1203970	1386290	1609440	189712
0.99999	69315	91629	120397	138629	160944	18971
0.9999	6932	9163	12040	13863	16094	1897
0.999	693	916	1204	1386	1609	189
0.998	347	458	602	694	805	94
0.997	231	305	401	462	537	63
0.996	173	229	301	346	401	47
0.995	138	183	241	277	321	379
0.994	115	152	201	230	267	31
0.993	99	130	174	198	229	27
0.992	86	114	150	173	200	23
0.991	77	101	134	153	178	21
0.99	69	92	120	138	160	18
0.98	34	45	60	69	80	9
0.97	23	30	40	45	53	6
0.96	17	23	30	34	39	4
0.95	14	18	24	27	31	3
0.94	11	15	20	22	26	3
0.93	10	13	17	19	22	2
0.92	9	11	15	17	19	2
0.91	8	10	13	15	17	2
0.9	7	9	12	13	15	1
0.8	3	4	6	6	7	9
0.7	2	3	4	4	5	6
0.6	2	2	3	3	4	4
0.5	1	1	2	2	3	3

.1

Reliability and Confidence

evel, Percent

90	95	97.5	99	99.5	99.9
2302590	2995730	3688889	4605170	5298320	6907760
230259	299573	368889	460517	529832	690776
23026	29957	36889	46052	52983	69078
2303	2996	3689	4605	5298	6908
1152	1498	1845	2303	2650	3454
768	999	1230	1535	1766	2303
575	747	920	1149	1322	1723
460	598	737	920	1058	1379
383	498	613	765	880	1148
328	427	526	657	755	985
287	373	460	574	660	860
255	332	408	510	586	764
229	298	367	459	527	688
114	149	183	228	263	342
76	99	121	151	174	227
57	74	91	113	130	170
45	58	72	90	103	135
37	49	60	75	86	112
32	42	51	64	74	96
28	36	45	55	64	83
25	32	39	49	57	74
22	29	35	44	51	66
11	14	17	21	24	31
7	9	11	13	15	20
5	6	8	9	11	14
4	5	6	7	8	10

an explosive-actuated device or system is to be used, the required reliability and confidence level should be determined prior to preparing a test plan. In many cases, it would be prohibitive for economic reasons to allocate a very large number of test items in order to attain an unrealistic high reliability for an application of minor importance. It is necessary to know the extreme conditions under which a device will have to operate to provide a design of a reasonable reliability. It should not be assumed that the use of stock items will eliminate the need for qualification tests. Only in some cases where the intended application is almost identical to the application for which the device was originally designed can qualification tests by eliminated. Even a well-proved explosive-actuated device may have a lower reliability value when used in a new application. Different environmental conditions than in the original application can drastically affect the function and the reliability of a device.

Since an explosive-actuated device is a self-contained dynamic system which changes electrical energy into explosive mechanical energy or chemical reaction products, it is inherently more dependent on the proper interaction of its components and precise matching on input and output requirements than a passive device, as for instance, a resistor. To attain the required reliability of each individual device used in the system, and the complete system's reliability. The reliability for a complete system should be determined by using test fixtures and conditions that simulate as exactly as possible the intended end application of the system.

The reliability R_s of a pyrotechnic system consisting of n components is equal to the product of the component reliabilities.

$$R_{c_1}, R_{c_2}, R_{c_3} \quad . \quad . \quad . \quad . \quad R_{c_n}$$

$$R_s = R_{c_1} \cdot R_{c_2} \cdot R_{c_3} \quad . \quad . \quad . \quad . \quad . \quad . \quad \cdot R_{c_n}$$

The overall failure rate is determined by dividing the actual number of failures observed by the total number of tests. The resulting failure rate subtracted from unity gives the reliability, which is a "point estimate", or a reliability value given as a single figure based on the available number of tests. It is relatively simple to compute and utilize any available consistent data. However, the single-figure reliability value gives no information about the degree of confidence that may be placed in this figure as a true measure of the potential performance of the device or system.

The reliability with a chosen confidence coefficient R_g can be calculated from:

$$R_g = 1 - \frac{\text{computation factor } c}{N}$$

where N = the number of tests. For each individual device and system, a graph should be prepared, plotting confidence coefficients relative to the number of failures and to computation factors c, as shown in an example diagram used for determining the reliability and confidence level of a parachute system, Fig. 15:1.

A great difficulty for determining the reliability of explosive-actuated devices by tests is presented by the fact that in many cases the test specimens are one-shot devices which are destroyed in the tests. In addition to

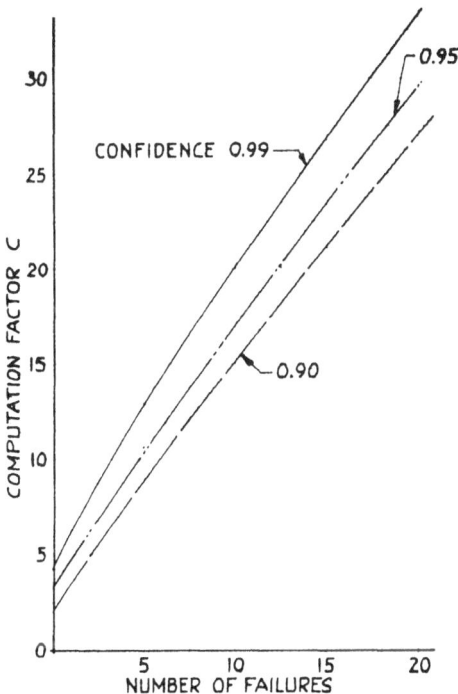

Fig. 15:1 Reliability computation factors vs. number of failures (Ref: "Performance and Design Criteria for Deployable Aerodynamic Decelerators" (ASD-TR-61-579).

tests, a very useful aid for determining the reliability are statistical quality control data and special lot acceptance procedures, which list predicted reliability.

In many cases, great expenses and extensive testing can be avoided by using redundant pyrotechnic systems, which will always result in a high reliability, and also by utilizing qualified devices.

It should be expected that the test specimen size and reliability requirements are closely connected with the problem of reliability calculation for explosive-actuated devices, and that with an increase of the reliability requirements the necessary number of test specimens will increase proportionally. Experience has shown, however, that in some cases where the cost of the individual large device was very high, the size of the test specimen has been decreased at a high reliability level.

The reliability of electro-explosive devices can vary depending on the pulse current used, as shown in a diagram in Fig. 15:2, in which reliability values for a power cartridge are plotted versus various pulse currents.

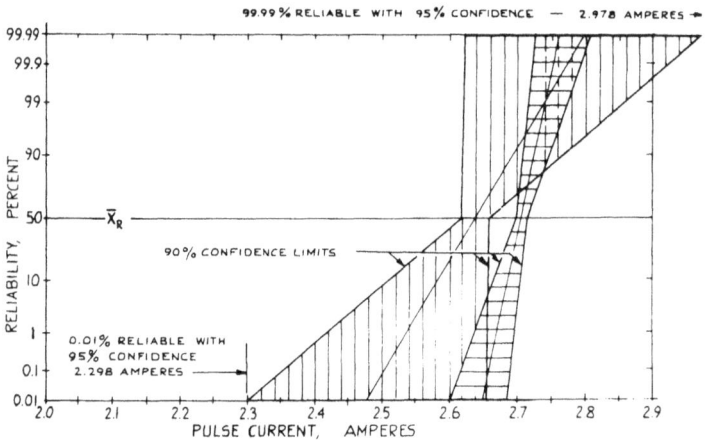

Fig. 15:2 Reliability vs. current for power cartridge (Courtesy of Hi-Shear Corp.).

A reliability comparison of four different initiator systems, as presented in Fig. 15:3, shows various ways for attaining high system reliability values without increasing the number of system components substantially. When comparing System IV with System III, it is interesting to note that because of the high individual component reliability, the cross-

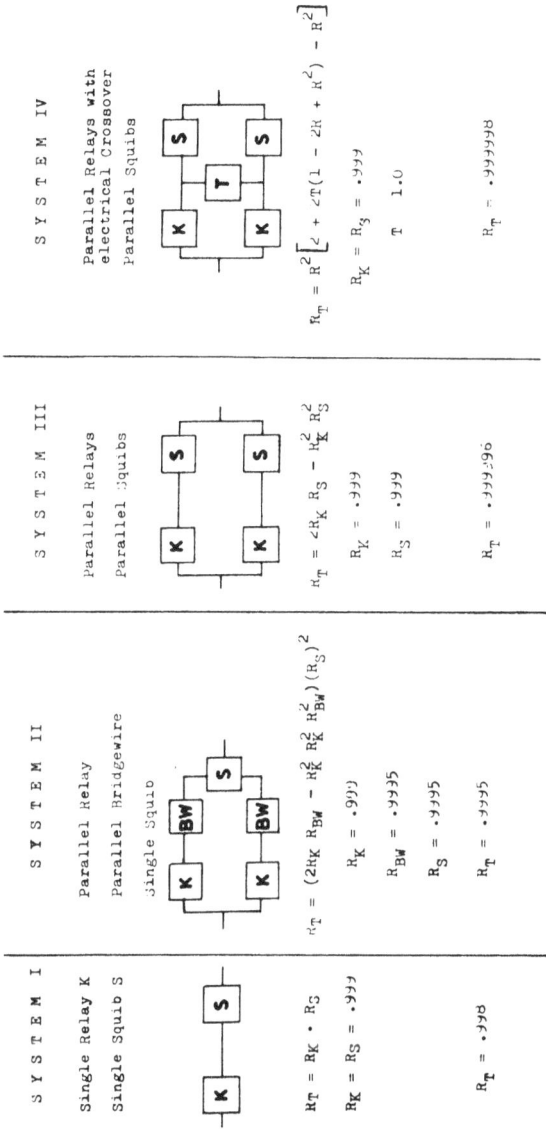

Fig. 15:3 Reliability comparison of initiator systems (Courtesy of Northrop Corp., Ventura Div.).

SYSTEM I

Single Relay K
Single Squib S

$R_T = R_K \cdot R_S$

$R_K = R_S = .999$

$R_T = .998$

SYSTEM II

Parallel Relay
Parallel Bridgewire
Single Squib

$R_T = (2R_K R_{BW} - R_K^2 R_K^2 R_{BW}^2)(R_S)^2$

$R_K = .999$

$R_{BW} = .9995$

$R_S = .9995$

$R_T = .9995$

SYSTEM III

Parallel Relays
Parallel Squibs

$R_T = 2R_K R_S - R_K^2 R_S^2$

$R_K = .999$

$R_S = .999$

$R_T = .999996$

SYSTEM IV

Parallel Relays with
electrical Crossover
Parallel Squibs

$R_T = R^2\left[2 + 2T(1 - 2R + R^2) - R^2\right]$

$R_K = R_S = .999$

$T = 1.0$

$R_T = .999998$

Note: Single Bridgewire used for each squib
Crossover Function T cannot be ignored for Component Values $\leq .999$
No Failure modes considered

over link in System IV results only in a small improvement of the reliability value over System III.

The importance of reliability tests and analyses for optimal application of explosive-actuated devices and systems does not need to be more emphasized. However, the calculated reliability of a device or system cannot be used to predict, on an absolute basis, the performance of a single example of that device or system in a single application. It gives the "odds", but it does not predict the result of any single event. It refers to the rate of successful uses to be expected when a great number of identical devices or systems are used, or when a given device or system is used a great number of times. Thus, concluding, reliability may be defined as the probability of successful operation of a device or system under specified conditions in the long run.

16   Quality Assurance Testing

In the various phases of the development and production of pyrotechnic devices, several types of tests must be conducted to prove that the devices meet the specified requirements. Development tests, which usually consist of visual, dimensional and X-ray inspections, and structural integrity tests, are conducted during the design phase to assist in the design and fabrication of the devices. Acceptance tests are carried out to assure that the devices are fabricated according to the requirements as specified on the drawing. Of major importance are the qualification tests which generally consist of a series of functional and environmental tests. The test objectives in each case are dependent on the type of the device, its application, and its performance requirements under specified environmental conditions.

Since electro-explosive cartridges are commonly used to initiate many different devices, test methods for initiator cartridges are described first.

A typical test program for initiator cartridges includes the following tests:

a. Prefunction inspection and tests,
b. Function tests, and
c. Storage tests

a. *Prefunction Inspection and Tests*

All initiator cartridges comprising a lot must undergo the prefunction and inspection tests. A production lot shall consist of minimum ten and maximum 2,000 cartridges which are manufactured in a continuous production run, and the components, inert and explosive used for the lot shall be from a single batch. A visual inspection must be conducted to prove that the cartridges are free from tool marks, blemishes, sharp edges and burrs, and that the cartridge markings are correct.

Each cartridge must be tested for leakage. The output end of the cartridge is checked for a hermetic seal by using a mass spectrometer leak detector. The leakage rate measured shall not exceed 1×10^{-6} cubic centimeters of helium per second, when tested at 1×10^{-5} mm of mercury.

The dc insulation resistance is checked by applying a potential of 500 volts dc between the contact pins shorted together and the cartridge body for a minimum of one minute. The measured insulation resistance shall be greater than 100 megohms. A Freed megohmmeter Model 1620 or an equivalent instrument can be used for this test.

The ac insulation breakdown is tested by applying a 500 volts ac (RMS), 60 cycle, between the contact pins shorted together and the cartridge for a minimum of one minute. The cartridge shall not show an increasing electrical breakdown, it shall not exhibit a current flow in excess of 250 microamps, and it shall not become inoperative. A Slaughter leakage tester Model No. 122, or an equivalent instrument can be used for these tests.

The resistance of each bridgewire, which shall not exceed 1.1 ± 0.10 ohms, shall be measured by using a 10 milliampere maximum test current. A Cubic Digital Ohmmeter or an equivalent instrument can be used for this test. At the same time, the cartridges shall be inspected for proper orientation of the bridgewire.

When a continuity loop is used in the cartridge, the resistance of the loop, which shall be less than 0.10 ohms, shall be measured by using a 10 milliampere maximum current. A Cubic Digital Ohmmeter or an equivalent instrument can be used for this test. The cartridge shall also be inspected for proper orientation of the continuity loop.

After these prefunction inspections and tests, as described, all cartridges which appear to be acceptable, shall be x-rayed twice with the views perpendicular to one another along the longitudinal axis to detect possible imperfections in the material or method of assembly.

b. *Function Tests*

The function tests, as described in the following, shall be conducted with specimens selected at random from the inspection lot according to the following list:

Lot Size	Number of Specimens
10 – 50	5
51 – 100	6
101 – 200	7
201 – 300	8
301 – 500	10
501 – 800	15
801 – 1300	20
1301 – 2000	25

To perform a no-fire current test, the bridgewires to be tested are to be connected in series, and a direct current of $1.0 \pm {}^{0.1}_{0}$ ampere shall be passed through the bridgewire for $5.0 \pm {}^{0.5}_{0}$ minutes. For this test, the cartridges shall be placed on their sides on a fiberboard or wood surface. During this test, the cartridges shall not fire or be degraded. An E & R Development Constant Current Supply Model No. PSC-1-018 or an equivalent instrument can be used for this test. After the test, the cartridges shall be cooled at room temperature for a minimum of one hour, or for a minimum of 30 minutes under fan-dispensed room air before proceeding to the next test.

To determine the pressure-time curve, the test cartridge is to be assembled into an applicable closed test chamber. A direct current of 5.0 ± 0.1 ampere shall be applied to one bridgewire circuit, and the pressure-time curve shall be recorded. The time span from the start of pressure to peak pressure shall not exceed 5 milliseconds. A Standard Controls Corporation Pressure Transducer Model No. 150-1 and a Tektronix Oscilloscope Model No. 535B or equivalent apparatus can be used for this test.

The ignition time of the cartridge is identical to the time span from application of the firing current to start of pressure rise, which shall not exceed 5 milliseconds.

The next test, the non-shorting test, shall be conducted within 5 minutes after completing the pressure-time test. During the non-shorting test, the current flow is to be measured as follows:

1. Through circuits, exclusive of continuity loop.
2. Between circuits, including continuity loop.
3. Between four pins shorted together and cartridge body.

The non-shorting test is to be conducted prior to removing the cartridge from the test chamber and prior to venting of the chamber. When using a dc potential of 28 volts, the current flow shall not exceed 28 milliamperes at 1000 ohms resistance.

An initiator cartridge which does not pass any of the prefunction tests, as described, is to be rejected and to be removed from the lot.

If in the function tests one cartridge fails to pass the no-fire current test, or the non-shorting test, or fails to fire when current is applied, the whole lot is to be rejected.

c. *Storage Tests*

To determine the functional capability of initiator cartridges after a specified storage time, special function tests are conducted. At present, storage tests with cartridges are performed by keeping a small number of test specimens of a lot, for instance two or three cartridges, for three years in a storage bunker at a temperature range from $0°C$ to $+38°C$ and at a humidity range from 30 to 90%. Tests after this storage time showed that the cartridges did not degrade in any way.

In future storage tests, it is planned to place five cartridges from one lot in a storage bunker. Two cartridges are to be removed from the storage bunker after four years for firing tests, and the remaining three cartridges are to be tested after completion of five years storage time. The obtained performance data will be useful for future applications.

For reliability determination of electro-explosive cartridges, their stimulus level is determined by suitable sensitivity tests. Two frequently used sensitivity test methods are the Bruceton test method and the Run Down method.

The Bruceton test method is most commonly used for sensitivity tests for electro-explosive devices. Prior to beginning a series of tests, the approximate amperage, at which 50% of the test specimens will fire, must be determined by several test shots starting at a current level near the "no fire" point, and continuing at a gradual current level increase until one test specimen fires. The data obtained from these initial shots, which are only required to determine the 50% firing point, should not be used in the final Bruceton test analysis.

The Bruceton tests are conducted at equally spaced test levels. To obtain useful results, it is important to determine the test interval "d" as accurately as possible. The test level for a shot is determined by the result of the previous shot. If a shot fires, the subsequent shot is made at the next lower test level, or if a shot does not fire, the next shot is made at the next higher test level. If the test interval "d" is too large, too few test levels will be used, and if the interval "d" is too small, too many test levels will be utilized. The interval value "d" should be between 0.33 times and 2.5 times the standard deviation. The minimum number of test levels should be eight. At least thirty test specimens should be fired, when using the Bruceton test method, to obtain reliable results. Since testing by applying the Bruceton test method is concentrated about the mean, excellent test results of the mean values are obtained. No other test method will produce equally accurate mean values. However, the Bruceton test method has the disadvantage that insufficient data are gathered in the tails of the distribution, which results in the tendency to underestimate the standard deviation, which means that an electro-explosive device may appear to be more reliable than it really is at the rated firing current, and it may also appear to be safer than it really is to the danger of low-stimulus initiation. Extrapolations can be used to obtain data of the tail distributions.

As an aid for practical use, a flow chart for the Bruceton analysis, various tables containing values as needed for the analysis, the necessary mathematical equations, and a nomenclature for the Bruceton analysis are presented.

Nomenclature

C = Confidence level (in %) specified in analysis
P = Reliability requirement (in %) specified in analysis

I_m	=	Mean current derived from analysis
I_o	=	The lowest current level used in the analysis
d	=	Increment of amperage between test levels
n	=	Either the total number of fires, or of no-fires depending on the outcome of the analysis. It is not the total number of devices fired
A	=	Computational factor
B	=	Computational factor
M	=	Computational factor
S_{I_m}	=	Estimate of the error in the determination of I_m
S	=	Estimate of the standard deviation of the data about I_m
S_s	=	Estimate of the error in the determination of S
Z_p	=	Normal deviate corresponding to P
$S(I_m \pm Z_pS)$	=	Estimate of the error in the determination of $I_m \pm Z_pS$
t	=	Confidence factor determined from C and n
I_F	=	The all-fire current for specified confidence and reliability levels
I_{NF}	=	The no-fire current for specified confidence and reliability levels

More accurate test data on the probability distribution than by the Bruceton method are generated by a different test method known as the Run Down method. The main difference of the Run Down method is that it assumes a very conservative probability distribution, the log logistic distribution. When using the Run Down test method, an initial Bruceton test is conducted with twenty test specimens to roughly determine the characteristics of the probability distribution. Two firing levels are calculated, level 1 ($\bar{x} + 0.4s$), and level 2 ($\bar{x} + 1.3s$). At level 1, fifty specimens are tested. If five or fewer specimens are used, testing is continued until 130 have been tested. At this step, a new firing level is computed ($\bar{x} + 0.2s$), and fifty specimens are tested at this level. When more than five specimens fire at level 1, 130 specimens are tested at level 2. A total of 200 specimens is required for these tests.

The data obtained by the Run Down method are analyzed for a specified reliability and confidence level. The advantage of the Run Down method is that it produces data further out on the probability distribution than the Bruceton method. However, the Run Down

Table 16.1

Values of Confidence Parameter t

$n \backslash C$	75%	80%	90%	95%	97.5%	99%	99.5%
2	1.000	1.376	3.08	6.31	12.71	31.82	63.66
3	.816	1.061	1.89	2.92	4.30	6.96	9.92
4	.765	.978	1.64	2.35	3.18	4.54	5.84
5	.741	.941	1.53	2.31	2.78	3.75	4.60
6	.727	.920	1.48	2.62	2.57	3.36	4.03
7	.718	.906	1.44	1.94	2.45	3.14	3.71
8	.711	.896	1.42	1.90	2.36	3.00	3.50
10	.703	.833	1.38	1.83	2.26	2.82	3.25
12	.697	.876	1.36	1.80	2.20	2.72	3.11
14	.694	.876	1.35	1.77	2.16	2.65	3.01
16	.691	.866	1.34	1.75	2.13	2.62	2.95
18	.689	.863	1.33	1.74	2.11	2.57	2.90
20	.688	.861	1.33	1.73	2.09	2.54	2.86
22	.686	.859	1.32	1.72	2.08	2.52	2.83
24	.685	.858	1.32	1.71	2.07	2.50	2.81
26	.684	.856	1.32	1.71	2.06	2.48	2.79
28	.684	.855	1.31	1.70	2.05	2.47	2.77
30	.683	.854	1.31	1.70	2.04	2.46	2.76
40	.681	.851	1.30	1.68	2.02	2.42	2.70
60	.679	.848	1.30	1.67	2.00	2.39	2.66
120	.677	.845	1.29	1.66	1.98	2.36	2.62
∞	.674	.842	1.28	1.65	1.96	2.33	2.58

Table 16.2

Values of z_p

$P\%$	z_p
80.0	0.84
85.0	1.04
90.0	1.28
95.0	1.64
97.5	1.96
99.0	2.33
99.5	2.58
99.9	3.09
99.95	3.29
99.99	3.72

method has the disadvantageous limitation that it provides valid results only to a reliability of 99.0% while in many cases a reliability of 99.99% at a confidence level of 95% is required for electro-explosive devices. When using the Run Down test method, long extrapolations are necessary to attain the required reliability.

Qualification tests for electro-explosive devices which are initiated by qualified initiators, are planned according to the function of the devices, their intended application, and environmental conditions. Only a small number of firing tests is usually required to qualify these devices, whereas electro-explosive devices which are to be initiated by special built-in initiators must be qualified by a series of firing tests similar to those as described for initiator cartridges, to determine their reliability.

Table 16.3

Values of $\dfrac{1}{\sqrt{n}}$

n	$\dfrac{1}{\sqrt{n}}$	n	$\dfrac{1}{\sqrt{n}}$
1	1.000	23	.208
2	.709	24	.205
3	.578	25	.200
4	.500	26	.196
5	.424	27	.192
6	.408	28	.189
7	.377	29	.186
8	.353	30	.182
9	.333	31	.180
10	.316	32	.177
11	.301	33	.174
12	.289	34	.172
13	.278	35	.169
14	.267	36	.167
15	.258	37	.164
16	.250	38	.162
17	.243	39	.160
18	.236	40	.158
19	.229	41	.156
20	.224	42	.154
21	.218	43	.152
22	.213	44	.151

Table 16.4

Data

1	2	3	4	5	6
I	i	F ( )	NF ( )	in_i	$i^2 n_i$
	8				
	7				
	6				
	5				
	4				
	3				
	2				
	1				
	0				
Sums					

Table 16.5

Results

$$C = \ldots\ldots \qquad\qquad s = \ldots\ldots$$
$$P = \ldots\ldots \qquad\qquad s_{I_m} = \ldots\ldots$$
$$I_o = \ldots\ldots \qquad\qquad s_s = \ldots\ldots$$
$$d = \ldots\ldots \qquad\qquad z_p = \ldots\ldots$$
$$n = \ldots\ldots \qquad\qquad I_m + z_p s = \ldots\ldots$$
$$A = \ldots\ldots \qquad\qquad I_m - z_p s = \ldots\ldots$$
$$B = \ldots\ldots \qquad\qquad s_{(I_m \pm z_p s)} = \ldots\ldots$$
$$M = \ldots\ldots \qquad\qquad t = \ldots\ldots$$
$$I_F = \ldots\ldots \qquad I_{NF} = \ldots\ldots$$

Table 16.6

Calculations for Bruceton Analysis

$$I_m = I_0 + d\left(\frac{A}{n} \pm \frac{1}{2}\right)$$

$$= (\quad) + (\quad) \left[\frac{(\underline{\quad})}{(\underline{\quad})} \pm \frac{1}{2}\right]$$

$$= \dots\dots$$

$$M = \frac{nB - A^2}{n^2}$$

$$= \frac{(\quad)(\quad) - (\quad)^2}{(\quad)^2}$$

$$= \dots\dots$$

$$s = 1.62d(M + 0.0290)$$

$$= 1.62(\quad)[(\quad) + 0.0290]$$

$$= \dots\dots$$

$$s_{I_m} = \left(\frac{1}{n}\right)\left(\frac{6s + d}{7}\right)$$

$$= (\quad)\left(\frac{1}{7}\right)[6(\quad) + (\quad)]$$

$$= \dots\dots$$

$$s_s = \left(\frac{1}{n}\right)\left(\frac{s}{d}\right)(1.1d + 0.3s)$$

$$= (\quad)\left[\frac{(\underline{\quad})}{(\underline{\quad})}\right]\left[1.1(\quad) + (\quad)\right]$$

$$= \dots\dots$$

$$I_m + z_p s = (\quad) + (\quad)(\quad)$$

$$= \dots\dots$$

$$I_m - z_p s = (\quad) - (\quad)(\quad)$$

$$= \dots\dots$$

$$s_{(I_m + z_p s)} = s_{I_m}^2 + z_p^2 s_s^2$$

$$= \dots\dots$$

$$I_F = (I_m + z_p s) + t s_{(I_m + z_p s)}$$

$$= (\quad) + (\quad)$$

$$= \dots\dots$$

$$I_{NF} = (I_m - z_p s) - t s_{(I_m \pm z_p s)}$$

$$= (\quad) - (\quad)$$

$$= \dots\dots$$

TABLE 16.7

Data Recorded from Test Firings

Shot No.	Firing Current (amps)										
1											
2											
3											
4											
5											
6											
7											
8											
9											
10											
11											
12											
13											
14											
15											
16											
17											
18											
19											
20											
21											
22											
23											
24											
25											
26											
27											
28											
29											

TABLE 16.7 (cont.)

Shot No.	Firing Current (amps)										
30											
31											
32											
33											
34											
35											
36											
37											
38											

"X" denotes fired; "O" denotes did not fire

Quality assurance test programs for electro-explosive devices, such as cutters, thrusters, pin pullers, bolt releases, valves, gas generators, and similar devices usually consist of three major types of tests:
- Non-destructive acceptance tests
- Environmental tests, and
- Functional tests.

Non-destructive acceptance tests of electro-explosive devices usually include a visual examination, a radiographic inspection, a bridgewire circuit resistance test, a dielectric strength test, and an insulation resistance test, similar to those as used for testing of initiator cartridges. As a part of a non-destructive test program, hermetic seal integrity testing and helium leak rate testing are also conducted on devices that have inherent leakage problems.

Environmental tests are of great importance in the qualification procedure for electro-explosive devices, which are utilized extensively in spacecraft, missiles, aircraft, and underwater vehicles and systems under extreme environmental conditions. In environmental tests, the expected environmental conditions under which the electro-explosive devices and systems must function reliably, should be duplicated as precisely as possible to obtain realistic test results. However, a difficulty is presented by the fact that all environmental conditions that are encountered in space flight or under water cannot simultaneously be simulated in presently available test facilities. At present, only vacuum and extreme temperature conditions can be duplicated simultaneously in test chambers. The capability of electro-explosive devices to withstand the effects of shock, vibration and acceleration must be tested separately.

To qualify electro-explosive devices for a typical space mission application, they are tested to meet the following requirements:

High-temperature storage, non-operating:	$+60°C$
Low-temperature storage, non-operating:	$-45°C$
High temperature, operating:	$+93°C$
Temperature-vacuum, operating:	$93°C, 1.33 \times 10^{-9}$ PSIA for 30 minutes
Pressure, arcing and corona:	1.5×10^{-2} PSIA to 7.0×10^{-2} PSIA for 30 minutes
Shock:	1750 G for 0.25 ms
Acceleration:	15 G
Vibration:	20-130 cps at 0.1 g^2/cps 200-1000 cps at 0.5 g^2/cps
Salt water immersion	36 hours

To qualify electro-explosive devices for applications under extreme environmental conditions, as listed, requires extensive testing under a well-planned test plan, which frequently includes sand and dust tests, humidity tests, salt spray tests, and drop tests, and also firing tests under low and high temperature conditions.

A major portion of a test program for electro-explosive devices consists of development and evaluation tests as needed to achieve the desired reliability. The most important tests, however, are the formal qualification tests, which are conducted to demonstrate to the customer that the devices and systems, as designed and manufactured, are capable of meeting all contract requirements. In many cases, a single qualification test performance proves the design adequacy and the manufacturing capability, and thus qualifies the electro-explosive devices and systems for their intended application.

Table 16.8

Flow Chart for Bruceton Analysis

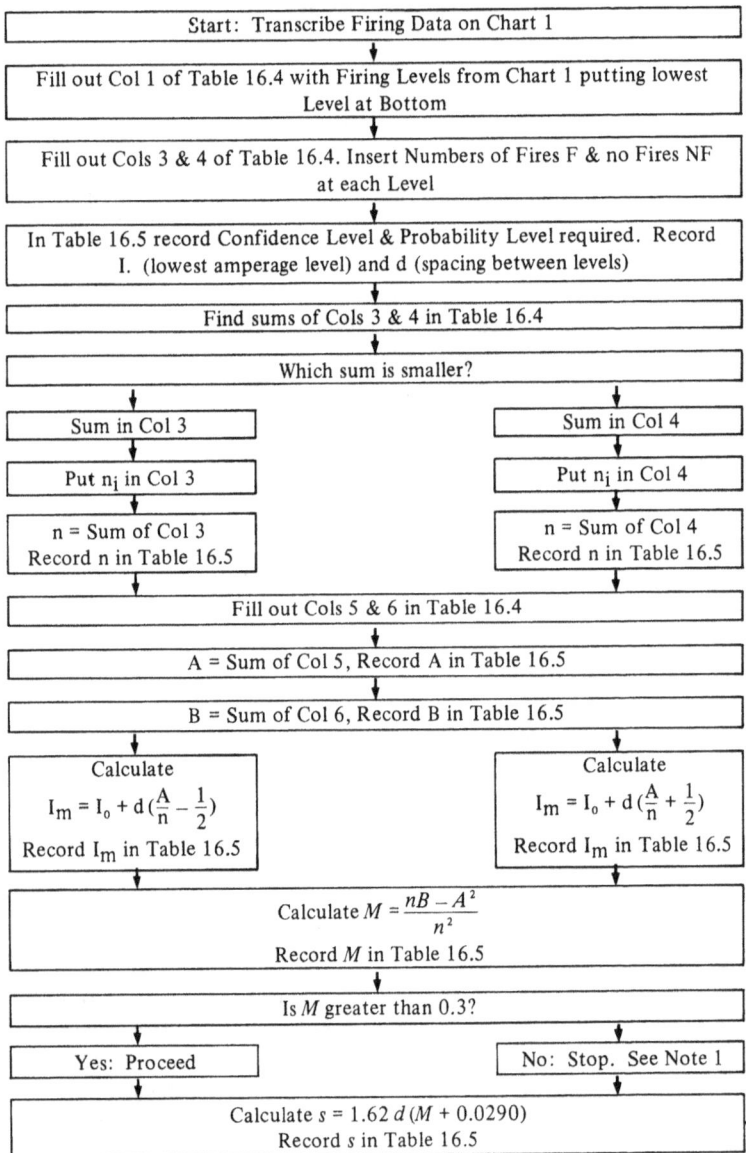

Start: Transcribe Firing Data on Chart 1

↓

Fill out Col 1 of Table 16.4 with Firing Levels from Chart 1 putting lowest Level at Bottom

↓

Fill out Cols 3 & 4 of Table 16.4. Insert Numbers of Fires F & no Fires NF at each Level

↓

In Table 16.5 record Confidence Level & Probability Level required. Record I. (lowest amperage level) and d (spacing between levels)

↓

Find sums of Cols 3 & 4 in Table 16.4

↓

Which sum is smaller?

Sum in Col 3	Sum in Col 4

↓ ↓

Put n_i in Col 3	Put n_i in Col 4

↓ ↓

n = Sum of Col 3 Record n in Table 16.5	n = Sum of Col 4 Record n in Table 16.5

↓

Fill out Cols 5 & 6 in Table 16.4

↓

A = Sum of Col 5, Record A in Table 16.5

↓

B = Sum of Col 6, Record B in Table 16.5

Calculate $I_m = I_o + d\left(\dfrac{A}{n} - \dfrac{1}{2}\right)$ Record I_m in Table 16.5	Calculate $I_m = I_o + d\left(\dfrac{A}{n} + \dfrac{1}{2}\right)$ Record I_m in Table 16.5

↓

Calculate $M = \dfrac{nB - A^2}{n^2}$ Record M in Table 16.5

↓

Is M greater than 0.3?

Yes: Proceed	No: Stop. See Note 1

↓

Calculate $s = 1.62\, d\,(M + 0.0290)$ Record s in Table 16.5

Is d less than $3s$?	
Yes: Proceed	No: Stop. See Note 2

Calculate $s_{I_m} = (\frac{1}{\sqrt{n}}) (\frac{6s + d}{7})$

Record s_{I_m} in Table 16.5

Is d less than $2s$?	
Yes: Proceed	No: Stop. See Note 2

Calculate $s_s = (\frac{1}{\sqrt{n}}) (\frac{s}{d}) (1.1d + 0.3s)$

Record s_s in Table 16.5

Which Type Firing is specified?	
No fire	All fire

Using P from Table 16.5
look up z_p in Table 16.2
Record z_p in Table 16.5

Calculate $I_m - z_p s$ Record in Table 16.5	Calculate $I_m + z_p s$ Record in Table 16.5

Are Confidence Levels required?	
No	Yes

Calculate $s_{(I_m \pm z_p s)} = \sqrt{s_{I_m}^2 + z_p^2 s_s^2}$

Record in Table 16.5

Using C & n from Table 16.5
look up t in Table 16.1
Record t in Table 16.5

Calculate I_{NF} Record in Table 16.5	Calculate I_F Record in Table 16.5

End of Calculations

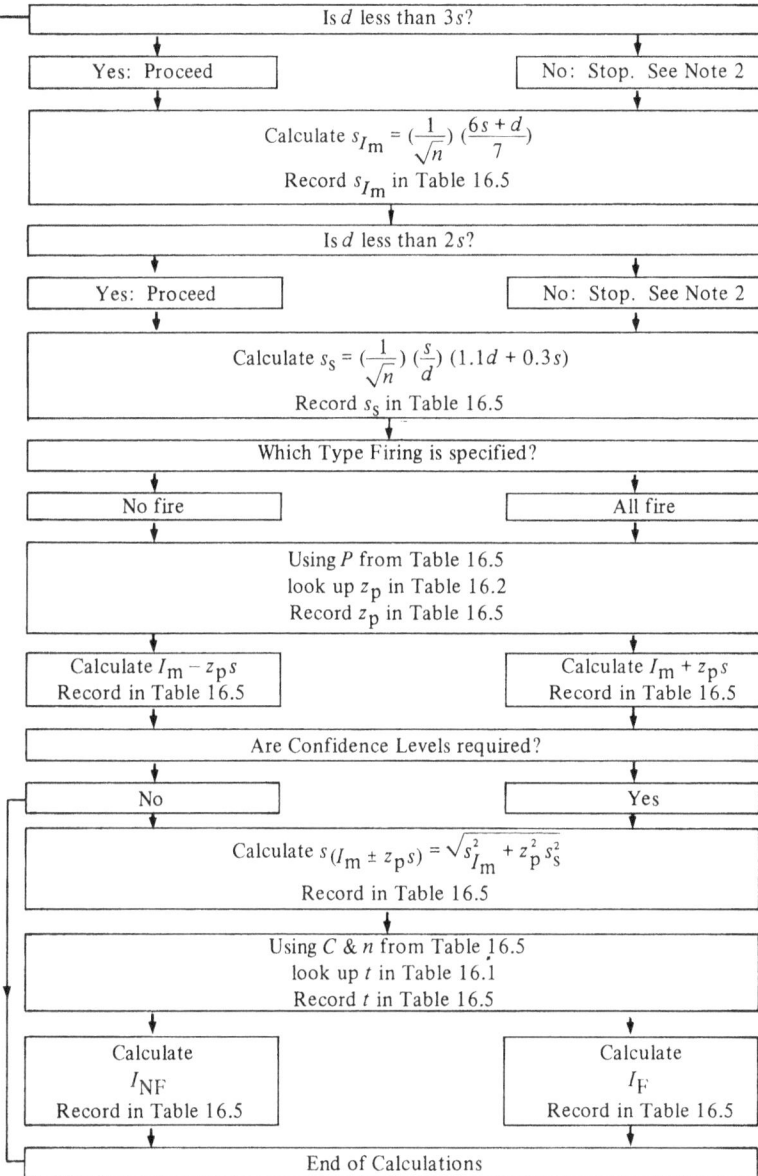

Notes: 1. Use larger value of n. 2. Use smaller value of d.

(Ref.: "The Bruceton Test Method", Holex Incorporated)

17 Present and Future Requirements

Considerable accomplishments have been made in recent years in the development and application of pyrotechnic devices and systems in aircraft, spacecraft, missiles, underwater and also on the ground, as shown in previous chapters of this book.

Great difficulties have been overcome in producing highly reliable pyrotechnic devices and systems, which will operate when needed, after having been exposed to unprecedented environmental conditions. Yet, standing still would be equal to going backward. There is a wide open field before us for further developmental work and for more useful applications of pyrotechnic devices and systems. Some of the most promising applications are pointed out in the following:

A. *Aircraft Applications*

1. Conventional Aircraft

With the increase of the size of commercial aircraft and with the increase in the number of passengers per aircraft, the possibility of injuries or even of casualties in case of landing gear failure or fire in the passenger compartment also increases. At present, an emergency evacuation from an airliner on the ground depends primarily on the proper functioning of the exit doors and of the inflatable escape slides which are provided at the exit doors. However, the escape slides can only serve their purpose, when the exit doors can be opened without difficulty and without delay. In case of a one-sided landing or of a failure of the landing gear on one side, it could be expected that the fuselage gets slightly twisted, which could easily result in jamming of the exit doors, thus making a quick emergency escape of the passengers impossible. According to FAA reports, in the case of fire or dense smoke onboard, a safe escape should be accomplished within less than five minutes by using all emergency exits.

To provide sufficient emergency exits, it is suggested to seriously consider a continuation of development and application tests of fuselage blow-out panels, utilizing linear-shaped charge or similar pyrotechnics to be initiated from the cockpit or from a similar safe central location in the aircraft. Each of these emergency exits should be equipped with an inflatable escape slide.

2. Helicopters

To provide a means for a safe descent of crews and passengers of a trouble-stricken helicopter, it is recommended to explore possible applications of pyrotechnic devices and systems. Since the available time for any action and for any emergency system to function is rather short in case of a helicopter malfunction, probably no conventional mechanical system could be considered for this task. Therefore, this appears to be a worthwhile application for pyrotechnic devices and systems which operate instantaneously on command. It is recommended to conduct feasibility studies of such applications and to develop some promising emergency system concepts. Only some brief theoretical studies of possible problem solutions have been made, but nothing has been accomplished yet in a practical way to decrease the danger rate of helicopter flying.

If a propeller assembly must be blown off in an attempt to let the craft with crew and passengers descend and land safely, as has been recommended in one proposal, there is still a considerable danger presented by the separated propellers flying uncontrolled in the air. Yet, this danger could be reduced or eliminated by providing an automatically-opening gas-jet thrust device at the side of the propeller hub, or at a similar suitable location, to move the dangerous obstacle far enough away from the manned craft. At a sufficient altitude and at a suitable attitude of the craft, bailout of the crew members using manual parachutes, and of the passengers using automatic parachutes could be considered, if no efficient emergency decelerator can be provided to descend and land the manned vehicle safely.

Another more complex emergency descent system for helicopters, which has been suggested, consists of a pyrotechnic separation system for the propeller assemblies, and of a cluster of large parachutes, which should be deployed after the troubled propeller assemblies and the tail assembly have been blown off. However, this concept does not seem to be of sufficient practical value at present, because in most emergency cases, it would require a considerable time between separating the malfunctioning assemblies and deploying the large parachutes. Yet, as the problems appear to be unsurmountable at present, there might come up an ideal solution in the near future. It can be expected that in most emergency systems that will be developed, pyrotechnic devices and systems will be utilized ideally because of their advantageous instant functioning, high reliability, and small size.

At present, the advantageous features of pyrotechnics are utilized primarily in military aircraft systems, such as ejection seats, ejection capsules, life raft inflation systems, and signalling devices, such as colored flares. Since the reliable functioning of pyrotechnic devices and systems in military aircraft has been proved many thousand times and since many lives have been saved by using these systems, it is recommended to consider the use of pyrotechnic devices and systems for more applications also for civilian aircraft. Small, simple gas generators could be used to inflate large life rafts for airline passengers. Civilian aircraft could possibly also be equipped with additional emergency signalling devices in form of colored flares, and with location aid devices, similar to SOFAR bombs, for locating a distressed aircraft after ditching on the ocean.

B.   *Spacecraft Applications*

1.   Space Rescue Systems

An urgent need exists for the development of reliable space rescue systems which are capable or returning astronauts from a malfunctioning spacecraft to earth or of transferring them to a different spacecraft. Various concepts of rescue systems for astronauts have been proposed in recent years. However, an ideal rescue system has not been developed yet, nor has a suitable concept been defined.

One proposal suggests a tandem arrangement of two Gemini spacecraft as a space rescue vehicle capable of returning five astronauts, including the two-man rescue crew, to earth, after docking with the disabled spacecraft. Another concept consists of a small one-man capsule which is to be ejected from the spacecraft in emergency situations, similar to an ejection seat from a disabled aircraft. A third rescue system proposal suggests an unfoldable one-man reentry vehicle, consisting of a heat-shield-equipped Mylar plastic bag which is rigidized by plastic foam generated by two cylinders during the initial phase of leaving the spacecraft. This vehicle would be equipped with a small rocket motor for achieving the correct deorbit trajectory and vehicle attitude, and a parachute would be provided for the final descent to earth.

The first proposed concept has the disadvantage that the vehicle must be launched from earth after a rescue request signal has been received. The necessary launch preparations and the flight to the disabled spacecraft would probably in most cases require too long a time to be of real help in case of a serious emergency.

Both other rescue systems concepts, as described, provide for the rescue of only one man, while in present spacecraft, the crew consists of three men. In case of a serious emergency, either one three-man rescue vehicle or three one-man vehicles would be required. If three rescue vehicles would have to be used, they would have to leave the disabled spacecraft simultaneously or in a very close sequence, presenting the problem of the possibility of a collision, and thus, of a major failure of the intended recue operation.

Considerable development work will have to be done in the near future to provide instantaneously deployable and reliable space rescue vehicles and systems. It can be expected that pyrotechnic devices and systems will be utilized extensively in these urgently needed rescue vehicles, for example, for ignition, control, structure rigidizing, parachute deployment, riser separation, and as location aid devices.

2. Sterilization of Explosives

To assure that space explorers will find possible extraterrestrial life unchanged when they visit other planets, inadvertent seeding of earth's micro-organisms must be avoided. Therefore, spacecraft that are to land on other planets must undergo a biological sterilization. All devices, instruments and materials, including the explosive materials used in the spacecraft must be sterilized. In general, radiation, ultrasonics, chemical methods, or dry heat can be used to sterilize materials. However, it was found that explosive materials can be sterilized only by dry heat without damaging effects to the material, and that the best suitable sterilization method is to place the explosive materials in a heat chamber at a temperature of 150°C for 36 hours. Numerous explosive materials have been investigated for their sterilization suitability and their properties. It was found that the majority of the explosive materials is not capable of withstanding the required sterilization cycle. Explosive materials which can be sterilized by the described method are listed in Part I, Chapter 1, Section K of this book under the heading "Explosive Materials for Future Space Applications".

Since the presently available data on suitability of explosive materials for sterilization are the results of some exploratory tests of a limited scope, it is recommended to continue such suitability tests with the objective to broaden the range of explosive materials which are suitable for sterilization and to determine the best suitable materials for applications in spacecraft that will land on other planets.

3. Booster Recovery

At present, it would not be economical to recover expended boosters of large vehicles such as Saturn, after completion of their mission, as detailed studies showed. However, the recovery of considerably larger boosters, as may be used for the next generation of spacecraft, could be economical. Besides parachute recovery systems, lifting surfaces, such as folded-out wings of various shapes and rotors have been considered for booster recovery. Rotors appear to be promising recovery devices for boosters, because very good results have been obtained with rotors as recovery devices for small rocket probes. If a rotor recovery system is to be applied to a booster, the problem has to be solved how to provide sufficient storage space for the folded rotors and for the deployment mechanism in the booster shell. Deployment of the recovery system, either rotors or wings, could be accomplished by various pyrotechnic devices, such as shroud release systems, pin pullers, thrusters, switches, valves, and similar devices. To prevent the booster from rotating with the rotors during the descent, counteracting gas jet nozzles could be provided at the booster shell. More detailed studies will have to be made to determine, at what size level a booster recovery will be economical, and whether a sea or land recovery should be considered.

C. *Missile Applications*

With the trend to constantly improve pyrotechnic devices and their performance capabilities, it can be expected that the results of these improvements and of new development work will have advantageous effects on missile systems mainly by improving the missile's performance accuracy and by simplifying the rather complicated missile systems.

Another expected result of these developments is a reduction of the required test lot sizes of pyrotechnic devices, which will result in a reduction of the final cost of the devices used in missiles and other vehicles.

D. *Underwater Applications*

At present, only a limited number of pyrotechnic devices and systems is used in underwater applications. Since parallel to extensive oceanography studies, an ever increasing number of practical underwater research projects and exploratory missions is being conducted, it is recommended to consider the utilization of pyrotechnic devices and systems for underwater applications on an increased scale.

The main differences of the environmental conditions that exist underwater, as compared to a high-altitude environment, for example, are the external hydrostatic pressure which varies with the depth, the salt water, and the temperature. Because of these different environmental conditions, available shelf items of pyrotechnic devices cannot be utilized for underwater applications. The casings must be made from seawater-resistant material, all seals and O-rings must be capable of withstanding the underwater environment, and special water-tight electrical connectors and cables are required for these applications, when the pyrotechnic devices are exposed to the underwater environment.

In some types of devices, such as cutters, valves, explosive bolts, pin pullers, switches, gas generators, and similar devices, the design of the inner components including the selected initiator charges and time delay trains will not have to be changed, but only the housing and other external hardware components will require a redesign to be capable of withstanding the underwater environmental conditions.

Part V

Explosive Production
Methods

Large and complex-shaped parts from new high-strength and heat-resistant
metals that were primarily developed for space, missile and advanced
aircraft applications, are ideally formed and fabricated by using explosive
metal-working methods. Because of their limited capability, conventional
machinery and equipment can in most cases not be considered for the
forming and fabricating of parts from these materials. Besides forming, the
controlled high-velocity energy released from an explosion is used for
welding, forging, cutting, piercing, work-hardening, extrusion, riveting, and
powder compaction.

Extensive research and development work has been done to provide
safe and economical explosive metal-working production methods, which
are used advantageously and economically for low-volume production, for
production of large and complex shapes, and for fabrication of parts from
hard-to-work materials, such as refractory metals. Closer tolerances of the
workpiece's dimensions can be obtained with explosive metal-working
methods than with conventional methods, because of a reduced springback
resulting from the explosive method. Another advantage of explosive
metal-working methods is that they require rather simple tooling which
makes especially the explosive forming method more economical than
conventional methods.

Some disadvantages of explosive metal-working methods are hazard and noise problems, restrictions to use and storage of explosives in urban areas and, in some cases, high cost for empirical testing required to develop operational information data.

Explosive metal-working operations can be classified in two categories: operations which use direct-contact charges, and those operations in which stand-off charges are utilized. Contact charges are usually applied for controlled work-hardening, welding, extruding, cutting, forging, and metal powder compacting, whereas stand-off charges are used for forming, sizing, and flanging of sheet and plate.

As the names indicate, in a contact operation, the explosives are placed in direct contact with the metal, while in a stand-off operation, the explosive charge is placed at a certain distance from the workpiece. In the stand-off operation, the energy from the charge is transmitted to the workpiece through a medium, usually water. Depending on the application, other materials, such as sand and oil may be used as a medium.

The explosives used for metal-working operations are carefully selected and sized for the individual application. High explosives, such as TNT, PETN, and dynamite. are primarily used in open systems, while low explosives, such as smokeless powder and black powder, are mainly utilized in closed systems. A typical open system is, for example, used for explosive forming of a semispherical dome by utilizing a female die and a stand-off explosive charge in water as a medium. A closed system, on the other hand. ususally consists of two closed die-halves, which enclose the workpiece. The closed system is ideally used for bulge-forming of tubular parts, and for the forming and piercing of cones and similar complex-shaped parts.

Tools used for explosive metal-working are generally less expensive than those used for conventional production methods. A major cost advantage results from the fact that only a die, but no punch is required for explosive forming or piercing. Suitable die and tool materials for explosive metal-working operations are Kirksite, ductile iron, and epoxy-faced concrete. Kirksite is best used for forming parts from light-gage material under low explosive pressure. Since Kirksite has the tendency to increase in size with use, the number of parts formed with one die from this material should be limited to fifty. Ductile iron is an ideal die material of good strength and is capable of withstanding explosive shock. Its only disadvantage is size limitation. Concrete dies with a dense and smooth epoxy facing are primarily used for forming large parts the size of which exceeds the size of ductile iron plate material.

To determine the usefulness and economical aspects of explosive metal-working methods for industrial applications, especially in production line operations, a careful and detailed analysis of the advantages and disadvantages of these unconventional production methods is required. It can be expected that explosive metal-working methods will find increased application in industrial production, but they will probably not be used as routinely as, for example, hydraulic presses or similar conventional equipment in the near future. An ideal utilization in many cases will probably be a combination of both the conventional metal-working methods and explosive methods, especially for fabrication of complex-shaped parts and of parts from high-strength materials, such as refractory metals.

18   Explosive Forming

The explosive forming method is applied advantageously to metal parts of large size or unusual shapes, and to parts from hard-to-work materials, which cannot be readily fabricated by conventional methods.

Depending on the shape and size of the part to be formed, either a confined or an unconfined system is used for explosive forming. The confined or closed system is preferably utilized for the forming of light-gage material to close tolerances, because under the sustained pressure, the material tends to set to the die. A confined explosive forming system consists of a separable die which completely enclosed the energy source. For bulge-forming operations, low explosives, such as black powder or propellant-type materials, are used, while high explosives are utilized for forming and piercing operations.

A cross-section through a confined explosive forming system is shown in Fig. 18:1.

The confined system has the disadvantages of size limitation, gas erosion of the die, and possible die shrapnel hazards. Another disadvantage is that required wall thicknesses of the die and the overall die dimension reach uneconomical proportions, when the diameter of the part to be formed exceeds 2 inches. For parts of an irregular, other than tubular shape, the uneconomical die wall thickness and other die dimensions will be reached at an even smaller diameter of the part.

Fig. 18:2 shows a typical part made by utilizing the confined explosive forming method. The elbow-shaped part shown in the center is explosively bulged from the seamless tube of 6061 aluminum alloy having an outside diameter of 8.625 inches and a wall thickness of 0.322 inches, as shown on the left. The outside diameter after the explosive forming is

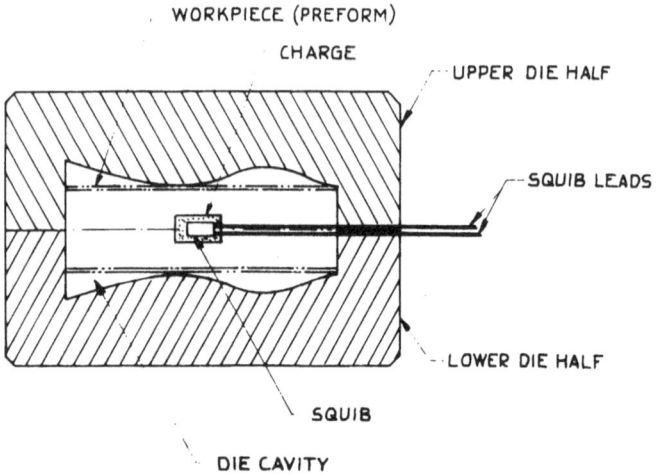

Fig. 18:1 Confined explosive forming system.

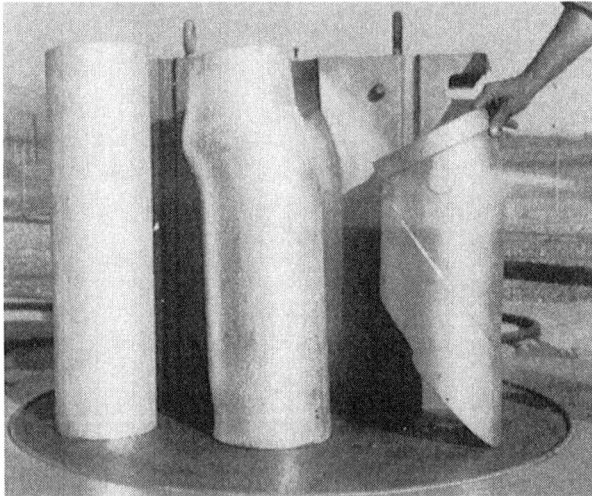

Fig. 18:2 Explosively-bulged fuel outlet elbow parts (Courtesy of North American Rockwell Corp.).

$10.000 \pm {}^{0.000}_{0.014}$ inches. After explosive forming, the part is trimmed and chem-milled on the outside, as shown on the right. A considerable cost and weight reduction and an increase in strength and reliability as compared with the previously used welded part were achieved by explosive forming.

Fig. 18:3 Explosively-formed gun barrel bore evacuators (Courtesy of North American Rockwell Corp.).

Complex-shaped gun barrel evacuators bulged from medium-gage material, as shown in Fig. 18:3, are other examples of the capability of the confined explosive forming method.

The ideal suitability of the confined forming method for producing intricate-shaped parts with high accuracy is demonstrated by the explosively-bulged spiral tube, as shown in Fig. 18:4.

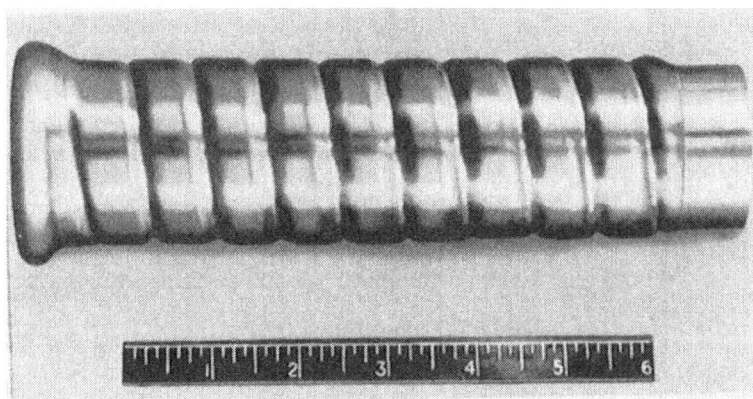

Fig. 18:4 Explosively-bulged spiral tube (Courtesy of North American Rockwell Corp.).

Fig. 18:5 Explosively-punched cones (Courtesy of North American Rockwell Corp.).

A good example for the capability of the explosive forming method to perform punching operations with great exactness are the punched cones as shown in Fig. 18:5.

The unconfined explosive forming system offers the advantages over conventional metal-working methods and over the confined system that parts of very large size and complex shapes can be produced and that close tolerances in the order of 0.002 inch can be obtained by using rather simple tooling. With regard to large and complex-shaped parts, the capability of unconfined explosive forming systems exceeds that of conventional forming methods. Springback is minimized in explosively-formed parts. Tough high-strength materials and refractory metals can be formed more readily by explosive techniques than by conventional forming methods. Low production runs can be produced more economically by applying explosive forming methods than by other processes, because the need for expensive dies and punches is eliminated.

Depending on the application of the charge, explosive metal-forming methods can be classified in two categories:

a. Pressure-forming methods using contact charges of deflagrating explosives, and

b. Shock-forming methods utilizing stand-off charges of detonating explosives.

In pressure-forming methods using contact charges, most of the energy is transmitted directly into the work-piece and appears in the form of a high-intensity stress pulse, whereas in shock-forming methods using stand-off charges, the released energy from the detonating explosive is transmitted through the medium to the workpiece. Both the pressure-forming method and the shock-forming method are mainly used in un-confined explosive forming systems.

An unconfined explosive forming system, as shown in Fig. 18:6, usually consists of a single female die and an explosive charge. The plate or sheet to be formed is clamped on the top surface of the die by a clamp ring or another suitable clamping device. The explosive charge with the detonator is mounted above the workpiece. Between the workpiece and the die, a sealing ring is provided for sealing the die cavity, which is connected to a vacuum line. The complete assembly is immersed in a water tank. To attain the best efficiency of the system, the explosive charge should be detonated as far below the water surface as possible. A recom-mended distance between the charge and the water surface is at least twice the stand-off distance. When the charge is fired under water, a shock wave is generated and moves out from the explosive, and a rapidly expanding gas bubble follows. If the explosive charge is located too close to the water surface, the gas bubble could vent to the surface without sufficiently impinging on the workpiece, and valuable energy could be lost.

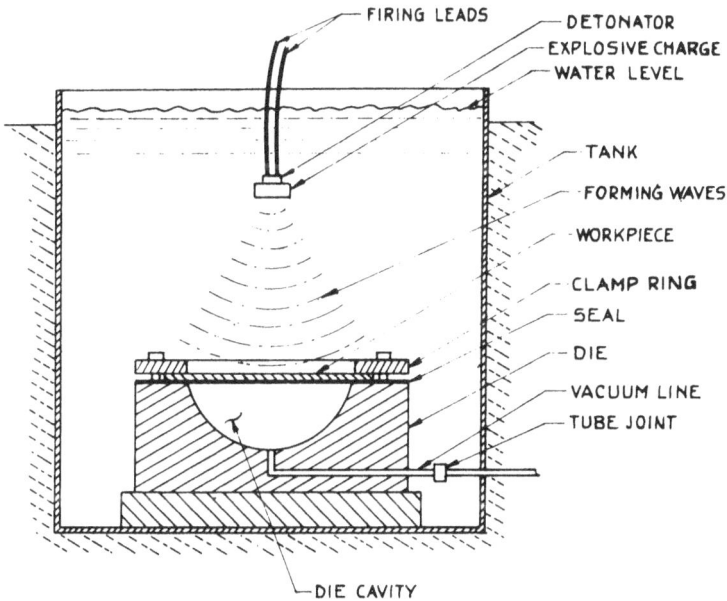

Fig. 18:6 Unconfined explosive forming system.

High explosives to be used in stand-off forming operations should be capable of releasing consistent energy, and they also should be safe, inexpensive, and readily available. Recommended charge materials for these applications are TNT, pentolite, and Composition A-3. The latter material has the advantage that it can be machined after pressing it to a specified density.

Deep-draw parts, such as domes, can best be formed by utilizing a charge which is centrally located at a stand-off distance of one-sixty of the diameter of the die opening. The required charge weight for such deep-draw operations can be calculated by using the following equation:

$$We(1 - \cos\phi) = \frac{D^2 h_o K}{n + 1} \left[2 Ln \left\{ 1.5 - \frac{1}{2(1 + 4\frac{w^2}{D^2})} \right\} \right]^{n - 1}$$

where W = weight of the explosive, lbs
         e = specific energy of the explosive, in lbs./lb.
         ϕ = arc tan $\frac{D}{2L}$
         D = diameter of die opening
         L = stand-off distance, in.
         h_o = initial plate thickness of workpiece, in.
         w = draw depth at center, in.
         K = strain hardening coefficient of plate material, PSI
         n = strain hardening exponent of plate material

Table 18.1

Specific Energies of Common High Explosives

Explosive	Specific Energy x 10^{-6} inch lbs/lb
TNT	15.5
RDX	21.7
PETN	23.2
Composition A-3	19.7
Composition B	20.8
Nitroguanidine	12.1
Pentolite	20.4
Detasheet	16.8
40% Straight Dynamite	9.7
60% Extra Dynamite	11.5

(Ref.68)

Table 18.2

Plastic Strain Hardening Relationships

$$\sigma = K\epsilon^n$$

Material	Strain Hardening Relationship		Source of Data
	K	n	
a. **Aluminum Alloys**			
6061-0	35,000	0.24	Dow Chemical
6061-T-6	60,000	0.075	,, ,,
2014-0	46,500	0.244	Martin Co. Tests
2024-0	58,000	0.25	ASTME, Battelle
1100-0	26,000	0.32	Battelle
Reynolds R-301 annealed	48,450	0.211	Marin
Alcoa 24-S annealed	55,900	0.211	,,
b. **Mild Steels**			
1020	76,000	0.12	Dow Chemical
,,	100,000	0.29	Battelle
0.05% Carbon rimmed-annealed	77,100	0.261	Marin
0.05% Carbon killed-annealed and temper rolled	73,100	0.234	,,
Same as above, completely decarburized-annealed	75,000	0.284	,,
0.05/0.07% Phosporous low carbon steel annealed	93,330	0.156	,,
c. **Stainless Steels**			
PH 15-7 Mo	660,000	0.829	ASTME
17-7 PH	205,000	0.32	Battelle
304 (austenite)	160,000	0.29	,,
,,	225,000	0.51	Dow Chemical
430 (17% Cr, Ferritic)	143,000	0.224	Marin
d. **High-Strength Steels**			
Maraging steel (18% Ni)	270,000	0.158	Martin Co. Tests
AM 350	770,000	0.87	Battelle
4130	120,000	0.35	,,
A-286	195,000	0.390	,,
USS 12 Mo V	180,000	0.22	ASTME
Vascojet 1000	155,000	0.15	,,
L 105	310,000	0.386	,,
Ladish D6-AC	387,000	0.073	Felgar, Space Tech. Lab.

(Table 18.2, continued)

Material	Strain Hardening Relationship		Source of Data
	K	n	
e. **Refractory Metals**			
René-41	315,000	0.39	Battelle
Columbian (10 Mo -10 Ti)	152,000	0.10	ASTME
Inconel-X	220,000	0.39	,,
Hastelloy-X	250,000	0.426	,,
f. **Titanium Alloys**			
Ti 6Al-4V	170,000	0.08	Battelle
Titanium (beta)	162,000	0.05	ASTME
75-A	128,000	1.10	Dow Chemical
g. **Copper**			
Annealed copper	75,000	0.38	Battelle
h. **Brass**			
Brass, soft	106,000	0.48	Johnson

(Ref. 68)

Sources:

ASTME,	"High Velocity Forming of Metals"
Battelle,	"Metal Deformation Processing", D.M.I.C. Report 226
Marin, J.	"Behavior of Engineering Materials", Prentice-Hall
Johnson, W.	"Placity for Mechanical Engineers", Van Nostrand
Felgar, R. P.,	"Plastic Analysis of Pressure Vessel Instability", EM 10-24, Engineering Mech. Dept., Space Technology Laboratories.

Specific energies of common high explosives are listed in Table 18:2, and plastic strain hardening relationships are presented in Table 18:2.

Some experimental data on the explosive forming of domes of three different materials and of various sizes and material thicknesses are listed in Table 18:3.

The presented data apply to one-shot explosive forming of domes

Table 18.3

Experimental Data on the Explosive Forming of Domes

Dome Diameter in.	Blank Material	Initial Blank Thickness in.	Type of Explosive Charge	Specific Energy of Charge in. lbs/lb	Polar Deflection in.	Stand-off Distance Ratio L/D	Weight of Charge lbs	Source of Data
6	18% Ni Maraging Steel	0.034	Pressed TNT	15.5×10^6	1.93	1 : 6	0.0174	Martin-Marietta Co.
6	"	0.065	Pressed TNT	15.5×10^6	2.4	1 : 6	0.042	"
12	2014-0 Al	0.05	Comp A-3	19.7×10^6	3.6	1 : 6	0.01	"
51	2014-0 Al	0.20	Comp A-3	19.7×10^6	20.43	1 : 5.1	2.78	"
51	2014-0 Al	0.25	Comp A-3	19.7×10^6	20.43	1 : 6.2	0.894	"
79	SAE 1020 Steel	0.50	60% Xtra Dynamite	11.5×10^6	43.5	1 : 7.9	30	North American Rockwell Corporation
121.5	2014-0 Al	0.75	Comp A-3	19.7×10^6	47	1 : 5.7	26.8	Martin-Marietta Co.

(Ref.:   )

359

made from suitable ductile materials. A forming depth limitation is given by the maximum allowable strain of the material. In cases where a dome must have a greater depth, a multi-shot explosive forming operation with intermediate anneals should be considered. In an ideal deep-draw process, no local thinning of the workpiece will take place, which can be achieved by keeping the maximum strain for any one shot below the strain at onset of necking.

Fig. 18:7 Explosively-formed dome from high strength steel (Courtesy of North American Rockwell Corp.).

A large explosively-formed dome from high-strength steel for a deep-diving vessel application is shown in Fig. 18:7.

In some unconfined explosive forming systems used for production of slightly molded large parts, a woven net of explosive cord is utilized as a charge. Typical parts formed by using this method are the integrally-stiffened waffle panels, as shown in Fig. 18:8, gores for tank ends, as shown in Fig. 18:9, and sun dials, as presented in Fig. 18:10.

The waffle panels, having a size of 76 by 96 inches, are machined in flat condition from a 1.5-inch thick aluminum plate. To support the vertical ribs during forming, the spaces between the ribs are filled with Cerrobend, a low melting point alloy, after machining. By using steam, the Cerrobend is easily melted out after forming. Tolerances of ±0.060 inch on the contour have been obtained.

Fig. 18:8 Integrally-stiffened panels before and after forming (Courtesy of North American Rockwell Corp.).

Fig. 18:9 Removing of explosively-formed gore from the die (Courtesy of North American Rockwell Corp.).

The gore sections, as shown in Fig. 18:9, are explosively formed from flat sheets of 2017 aluminum up to 1 inch thick. Twelve of such gores, each having a width of approximately 10 feet and a length of about 18 feet, are welded together to form tank ends of 33 feet diameter.

361

Fig. 18:10 Explosively-formed sun dials (Courtesy of North American Rockwell Corp.).

Fig. 18:11 Six-inch thick die liner explosively formed in one shot (Courtesy of North American Rockwell Corp.).

An impressive example of the capability of explosive forming methods is the six-inch thick die liner from mild steel, which was formed in one shot, shown in Fig. 18:11.

The high quality obtained by using low-cost tooling is demonstrated

Fig. 18:12 Aluminum mirror formed on low-cost plaster die (Courtesy of North American Rockwell Corp.).

Fig. 18:13 Explosively-formed pressure bags (Courtesy of North American Rockwell Corp.).

by the aluminum mirror, as shown in Fig. 18:12. This mirror, which is to be used in solar simulators, was formed on a low-cost plaster die.

Pressure bags, as shown in Fig. 18:13, are explosively formed and embossed on low-cost fiberglass tooling from preformed cones made from 0.010-inch thick 321 stainless steel. These bags, which have a diameter of 7 feet and a height of 5 feet, are used as tooling for furnace brazing of rocket engines.

Fig. 18:14 Explosively-formed architectural panels (Courtesy of North American Rockwell Corp.).

The architectural panels, as presented in Fig. 18:14, demonstrate that explosive forming methods are suitable for economical production of commercial products, especially when only a low production run is required.

Concluding, it should be stated that explosive forming can provide a useful method of metal-working with a minimum expenditure for facilities, as the examples have shown: Explosive forming does not offer a solution to all difficult forming problems, but it can be utilized ideally in many cases, where conventional methods and equipment are inadequate, or where the number of parts to be produced will not warrant the expenditure for more elaborate tooling.

19   Explosive Welding

The development of explosive welding methods has reached a state that they can be used as economical production methods for special applications. Explosive welding is advantageously used as a suitable method for welding dissimilar metals that cannot be welded together by conventional methods. For example, steel can be welded to aluminum, titanium to steel, brass to steel, steel to lead, and lead to lead by the explosive welding method. Difficulties in welding of metals, such as lead, having a low melting point, are eliminated by explosive welding. Good results were obtained in numerous explosive welding tests with a great variety of metals.

Dissimilar metals, which are metallurgically incompatible, can ideally be joined by explosive welding. In some cases, if a conventional welding method would be used, the applied heat would cause the two metals to melt and to form a brittle alloy, which would make the weld useless. Other important advantages of explosive welding are that explosively-welded joints usually have higher strength properties than fusion-welded joints and that heat-treated materials can be explosively welded without causing a decrease in properties.

Generally, metallurgical bonding is achieved by bringing the atoms of two pieces of metal into such a close contact that their normal forces of interatomic attraction become operative and thus produce a bond. However, the surfaces of metals are covered with a natural film of oxides, nitrides and absorbed gases, which would not permit bonding even under very high pressure, if it would not be removed from the surfaces. Bonding would similarly be precluded by any protective surface coating. The natural film is removed in the process of explosive welding by "surface jetting", which results when the high pressure generated by a chemical

365

explosion forces two metal surfaces to collide with one another at a high velocity and at a certain angle, and develops high pressure at the collision interface. By this action, a portion of the surfaces becomes fluid and is expelled, and thus, the surface films are removed, providing for intimate contact of the metal surfaces. An explosive weld is obtained at a relative velocity of 500 to 1000 feet per second. If relative to the properties of the metals, the impacting velocities are too high, the metal at the interface will not have sufficient time to move before it is trapped by the other metal, and as a result, bonding will not take place. On the other hand, if the impact velocities are too low, no interaction at the interface will be achieved.

The impact of the metals to be welded begins at the initiating end of the explosive and continues across the joint in a wave motion as it forms the weld. As a result, a wavy interface is obtained, which offers the advantages over flat interfaces of conventional bonds that a larger bond area is provided, the possibility of crack propagation is minimized, and the resistance of structural joints to thermal cycle failure is increased, because the stresses in joints between metals, which have substantial differences in thermal expansion, are distributed over a larger volume. In a typical weld interface, a wave has a height of approximately 0.002 inches and a length of about 0.006 inches.

The quality of the weld zone can be affected by small discontinuous pockets, which, caused by some dissimilar metal combinations, may be brittle. The formation of such undesirable brittle pockets can be eliminated or minimized by selecting suitable parameters for the explosive welding process.

An advantageous feature of explosive welding is that it is essentially a "cold" welding process, since heat generated during the explosive welding is insignificant and dissipates quickly. This feature makes the explosive method suitable for welding strain-hardened and heat-treated materials without affecting their properties. Another advantage of the explosive welding method is that it is a relatively inexpensive process, since very simple equipment and inexpensive explosives are used. However, the method must be properly applied to be economical.

Some high-strength and high-hardness materials, as for example, hardened tool steels, stellite, and beryllium are not suitable for explosive welding because of their low impact strength.

The principle of explosive welding is explained schematically by the examples of a plate joint and of a tube joint, shown in Fig. 19:1.

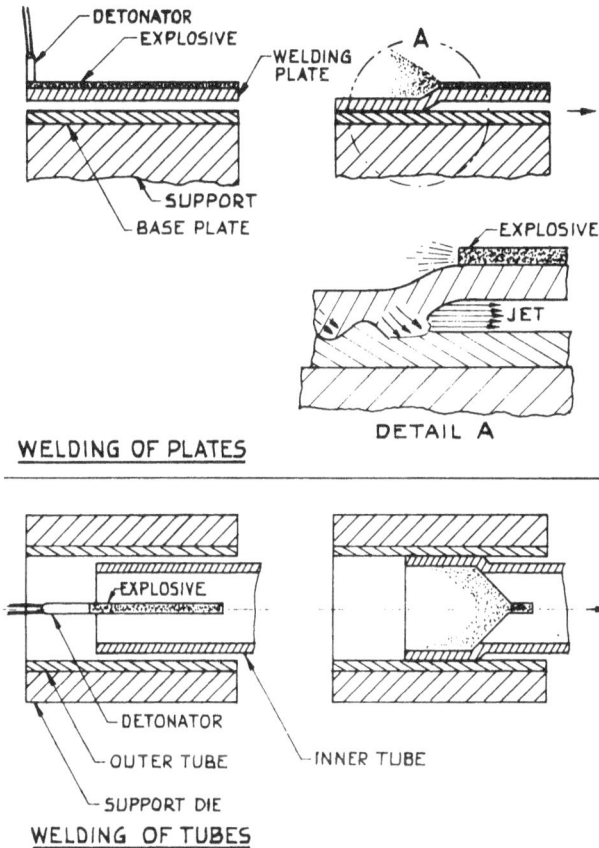

WELDING OF PLATES

DETAIL A

WELDING OF TUBES

Fig. 19:1 Explosive welding of plates and tubes.

Detonating explosives are generally used for explosive welding. Tests and practical applications showed that for explosive welding operations, granular explosives of low detonation velocity are best used to keep the peak pressure down and to prevent shock damage and excessive deformation of the parts. Commonly used explosives for welding are nitroguanidine, amatol, and dynamite. The low detonation propagation rate in each of these explosives results from both the composition and the low packing density. The detonation velocity of these granular explosives can be varied from 8000 to 20,000 feet per second by sizing the packing

367

density accordingly. A dynamite which consists mainly of a mixture of nitrostarch and ammonia-nitrate, but which does not contain glycerin, proved suitable as an explosive for welding. Its detonation velocity can be varied from 8000 to 12,000 feet per second, and it develops a peak pressure of up to 700 psi. This explosive, which has the form of dry powder, does not require unusual precautions for handling, and it is rather insensitive to impact. Detonation of these explosives is achieved by electric initiators.

Explosive welding provides an excellent method for producing transition joints from tubes to tube sheets as used, for example, in heat exchangers. A greater variety of materials and also thinner tubes can be used in such applications, when explosive welding methods are employed. Other important applications of explosive welding are tube joints between compatible and incompatible materials. Lap joints and tapered joints are used. Tubes in sizes from 1/4 inch to 8 inches diameter are joined by explosive welding.

Plate material is ideally welded by explosive methods. However, seam welding did not find a wide application yet, in spite of the fact that special machinery has been developed for explosive welding of pipes and rings.

Explosive welding or bonding is ideally suited for cladding. Plate and sheet in sizes up to 10 x 30 feet have been clad by this method, and tubes up to 35 feet long were also clad to provide the required corrosion resistance. Special cladding techniques have been developed for the production of aluminum-clad steel mainly for navy operations, and for corrosion-resistant titanium-clad steel pressure vessels.

Explosion-bonded clad metals in a variety of materials and sizes are commercially available from E. I. du Pont de Nemours & Co.

It can be expected that explosive welding and bonding will find wider application in the near future as an economic production method because of the obvious advantages over current conventional production methods.

20 Explosive Riveting

In many fastening applications where insufficient back-up space and/or limited access precludes the use of conventional rivets, explosive blind rivets can be used advantageously for structural and non-structural purposes. The application of explosive energy to blind rivet fasteners was pioneered by Du Pont engineers two decades ago. Explosive rivet designs have constantly been improved, and highly efficient, economical methods of explosive riveting and suitable tools and equipment have been developed.

A typical explosive rivet, which has a similar external shape and dimensional proportions as a conventional rivet, contains a carefully sized small explosive charge in its sealed shank. An electrode located in the rivet head provides a circuit path to the explosive charge. By introducing a small electric charge at the electrode, the explosive charge is fired instantly, causing the shank of the rivet to expand both in the workpiece and below, thus locking the rivet securely in place. The controlled explosion is completely contained in the sealed shank, eliminating any blast and noise. Only a "click" is heard, indicating that the rivet is set. A secure, neat fastening, which requires a minimum of back clearance, is achieved by this simple, one-step operation.

Tools and equipment for manual insertion of the explosive rivets and for automatic insertion are available. A self-contained rivet gun, powered by a standard 90-volt dry-cell battery, which is capable of firing over 100,000 rivets on one charge, can be used for low-volume operations. For actuation of explosive rivets in constricted locations, which are inaccessible for the rivet gun, a special thimble, which can be placed on a probe, is utilized.

For high-volume operations, an automatic rivet application system, consisting of a rivet gun and a rivet supply cart, is used. A continuous supply of explosive rivets, which are automatically positioned at the nose of the rivet gun, is provided from the supply cart through a plastic tube by air pressure. The rivet-feed system is operated by compressed air from the plant, and the rivet gun is powered by a standard 110-volt current.

The rivet gun is operated by simply placing it over a pre-drilled hole in the workpiece, locating the exposed rivet stem in the hole and pushing the gun down. By the push-down force exerted on the rivet gun, the workpieces are pressed together, and at a predetermined push-down load, the electrical circuit through the rivel electrode and through the explosive charge is closed, thus initiating the explosion in the rivet shank and setting the rivet, which is accomplished within 10 millionths of a second.

In the automatic riveting system, firing of one rivet triggers the instantaneous delivery of the next rivet from the supply cart to the rivet gun, eliminating manual placement of the rivets. Sixty rivets can be installed per minute by using the automatic riveting system.

Tests showed that even with the simple manual riveting system, explosive rivets are set at a higher speed than is obtained in conventional riveting.

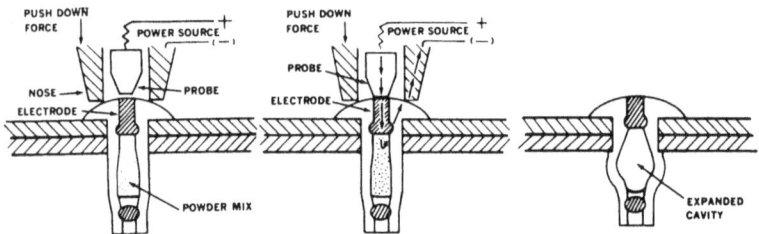

Fig. 20:1 Sequence of explosive riveting Du Pont instant rivet system.

A sequence of setting an explosive rivet is presented in Fig. 20:1.

Explosive rivets are widely used in the aircraft industry in cases where the use of conventional rivets is precluded. Some typical applications are shown in Fig. 20:2. Explosive rivets for aircraft applications are primarily made from aluminum alloys, such as 2017-T, 2117, and 5056 aluminum alloy. For high-temperature applications, explosive rivets are made from special stainless steel alloys, low carbon nickel alloys and columbium.

There are numerous applications for explosive rivets in other

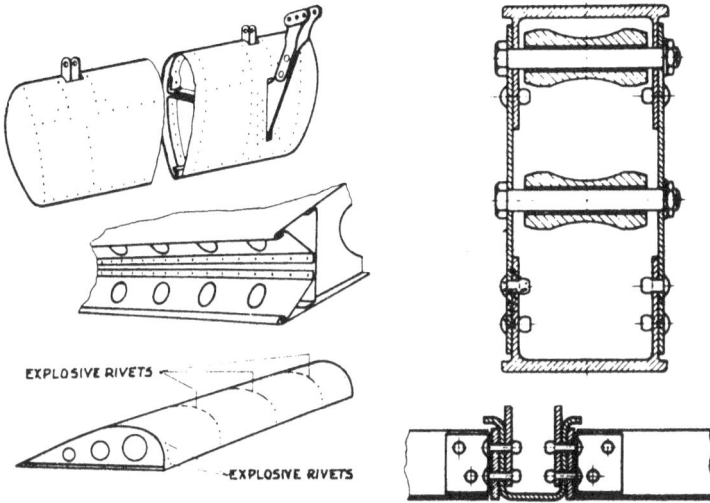

Fig. 20:2 Typical applications of explosive riveting.

industries, for example for connecting brake lining to automobile brake shoes, or to join machinery parts.

A different and very simple type of explosive rivets, than described above, which was not equipped with an electrode, was used in the German aircraft industry in the forties. In these explosive rivets made from aluminum alloy, the explosive charge, which extended from the free end a short distance into the grip area of the shank, was sealed off by a pressed-in end closure disc. An electrically-heated hand tool, similar to an electric soldering iron, was used to fire the charge, and thus close the rivet joint by explosively expanding the shank end.

Based on the experience with those explosive rivets, the development of explosive rivet bolts emerged in Germany. These bolts, made from steel, were designed on the same principle as the simple explosive bolts. They were used in lieu of conventional steel bolts in special applications. Standard sizes of these explosive rivet bolts were from 1/4 to 3/4 inch diameter, and grip lengths from 1/8 to 5/16 inch.

Explosive rivets are used in diameter ranges from 0.098 to 0.202 inch, and in grip lengths from 0.025 to 0.565 inch.

21   Explosive Cutting

Various explosive methods for cutting of metal parts have been developed and are primarily used in applications where conventional cutting methods, such as shearing or flame cutting, cannot be utilized. The area where the cut is to be made may be inaccessible for conventional cutting tools and equipment, the shape of the cut to be made may be too complex, or the properties and characteristics of the material to be cut may preclude conventional cutting.

A simple method of performing explosive cutting operations consists simply of fastening a flat strip of explosive material on the metal to be cut, and of detonating the explosive. Parting of the material results from the interaction of stress waves induced by the detonation. The thickness of the explosive material to be used in such operation is to be sized depending on the properties of the metal. To obtain satisfactory results with this cutting method, the width of the flat explosive strip should be approximately twice the thickness of the metal to be cut. This simple explosive method can be employed where no high cutting accuracy is required.

Better results in cutting accuracy and depth than with the simple method, as described above, can be obtained by using flexible linear-shaped charge (FLSC). A description of FLSC is given under the heading "Explosive Cord" in Part I, Chapter 1, Section D of this book.

Characteristic features of the flexible linear-shaped charge are that it consists of a seamless metal sheath with a cross-sectional shape of either a Vee or an arc in which a long column of explosive material is contained, and that it can be bent into any shape without affecting its functioning. The sheath is usually made from aluminum, lead, brass, or copper. In both the Vee- and arc-shaped types of FLSC, the explosive charge extends to the midpoints of the sides of the sheath.

To perform a cutting operation, the FLSC is bent exactly according to the line marked on the workpiece for the intended cut, and it is placed on the metal so that the apex of the Vee or arc is exactly above the marked line. The size, charge weight, sheath material, cross-sectional shape, and stand-off distance must be carefully selected for an individual application to obtain the best results. Standard initiators or blasting caps can be used to initiate the charge. When detonated, the forces resulting from the detonation are directed along a plane which bisects the angle or arc formed by the explosive. The resultant force plane combined with the hot metal particles from the metal sheath provide excellent cutting ability.

For explosive cutting operations in high-temperature environments, Du Pont TACOT flexible linear-shaped charge can be utilized. TACOT, which has approximately 96 percent of the explosive power of TNT, can withstand temperatures of $+315°C$ for one to nine hours without impairment of its explosive function.

Special devices for explosively cutting of circular holes have been developed by Du Pont. Such a hole cutter consists of a molded plastic disc containing a ring of Flexible Linear-Shaped Charge (FLSC), a primer and a safe-arm mechanism. Du Pont FLSC hole cutters are available in sizes to cut a hole of 2 inches and of 6 inches in diameter. To cut a circular hole, the cutter is attached to the workpiece. After careful checking for safe operating conditions, the cutter is armed, and the FLSC is initiated. The explosive charge used in the FLSC has to be sized according to the material and thickness to be cut.

A different method for explosively cutting holes into metal is represented by one example, the punched cones shown in Fig. 18:5 in Part V, Chapter 18 of this book under the heading "Explosive Forming". For the hole punching operation, the confined explosive forming method was used, utilizing a closed die in which holes in the sizes, as later required in the part, are provided at the exact locations. This cutting method is preferably utilized when a great number of holes must be cut in complex-shaped metal parts, as these cones are.

This confined explosive cutting method has the advantage that it requires only one short to cut a great number of holes, which may be of any size and shape. A disadvantage of this method as compared to other explosive cutting methods is, however, that an expensive die is required.

In a very simple explosive cutting method that does not require dies or similar equipment, the flexible explosive Du Pont "Detasheet", which is described under the heading "Explosive Sheet" in Part I, Chapter 1, Section E of this book, is utilized. "Detasheet", which is composed of an integral mixture of PETN (pentaerythritol tetranitrate) and elastomer binder, is available as pliable sheets, cords and extruded shapes in a variety of thicknesses and diameters. "Detasheet" offers the advantages that it is flexible over a wide range of temperatures, is not modified by repeated flexing, is easy to cut and handle, and safe to use, and retains the explosive properties of PETN.

The "Detasheet" was developed to meet the military specification MIL-E-46676MU. The unique properties of "Detasheet" flexible explosive include toughness and durability, uniformity of detonation velocity, and a very high degree of safety. It is extremely insensitive to shock and will not detonate upon impact of 0.30-caliber rifle bullets from a distance of 40 feet. Additional advantages of "Detasheet" are that it is completely waterproof, highly resistant to water erosion and relatively unchanged by extreme hydrostatic pressures.

"Detasheet" is applied to any complex pattern quickly and accurately by simply glueing with adhesive or by means of adhesive tape. A variety of thicknesses of this flexible explosive provides a wide range of explosive weight per square inch of surface area, as required for different applications that call for varying amounts of explosive force per unit area. Several thicknesses of "Detasheet" may be laminated to tailor the charge according to specific requirements.

Pin-point metal-cutting is made possible by using "Detasheet" with a minimum weight of explosive. Cuts of unusual and complex patterns that are impossible to be made by other methods, can be made by utilizing "Detasheet". "Detasheet" is also used for metal hardening. A detonator of No. 8 strength, containing 6.9 grains PETN, or greater, is to be used for reliable initiation of "Detasheet" C. When installing the detonators, it is important to provide an intimate bond between the detonator and the explosive, which may be achieved by using masking tape or similar adhesive material.

Concluding, it may be stated that, at present, there is no explosive method available that will meet all cutting requirements, but the best suitable method for an individual application must be determined by a careful analysis, under consideration of complexity, safety and ease of operation, degree of cutting accuracy, tooling and equipment requirements, set-up and production time, and cost.

Part VI

Appendix

REFERENCES

1. H.H. Koelle, ed., *Handbook of Astronautical Engineering*, McGraw-Hill Book Company, 1961.
2. H. Ellern, *Military and Civilian Pyrotechnics*, Chemical Publishing Company.
3. H. Ellern, *Modern Pyrotechnics*, Chemical Publishing Company, 1961.
4. G.B. Sutton, *Rocket Propulsion Elements*, John Wiley & Sons, 1964.
5. F.B. Pollard and J.H. Arnold Jr., *Aerospace Ordnance Handbook*, Prentice-Hall.
6. J. Bebie, *Manual of Explosives, Military Pyrotechnics and Chemical Warfare Agents*, The MacMillan Company.
7. *Apollo Spacecraft News Reference, Command and Service Module*, Space Division, North American Rockwell Corporation, and NASA Manned Spacecraft Center.
8. *Saturn V News Reference*, Boeing Launch Systems Branch, McDonnell Douglas Astronautics Company, IBM and Rocketdyne Division, North American Rockwell Corporation.
9. *Apollo Spacecraft News Reference*, Grumman Aircraft Engineering Corp. and NASA.
10. *Apollo X Final Report, Extended Apollo Systems Utilization Study, Volume 19, Earth Landing System*, Northrop Corporation, Ventura Division, and North American Aviation Inc., Space & Information Systems Division, Report No. SID 64-1860-19.
11. *Performance and Design Criteria for Deployable Aerodynamic Decelerators*, American Power Jet Company, Technical Report No. ASD-TR-61-579, December 1963, Air Force Flight Dynamics Laboratory, Research and Technology Division, Air Force Flight Systems Command, Wright Patterson Air Force Base, Ohio.
12. Robert F. Reinking, *Designing with Explosive Devices*, Machine Design, July 4, 1968.
13. H.J. Fisher, *Design Considerations for Electro-Explosive Devices*, Paper presented to National Aero-Nautical Meeting, Washington, D.C., April 1963, published by the Society of Automotive Engineers (SAE).
14. J.A. Grow, *Explosive Actuators*, Machine Design, February 4, 1965.
15. O. Romaine, *Why Explosive Devices?*, Space/Aeronautics, March, 1963.
16. George G. Herzl, *Designing for Space*, Machine Design, May 28, 1970.
17. Karl O. Brauer, *Present and Future Applications of Pyrotechnic Devices and Pyrotechnic Systems for Spacecraft*, IAF Paper No. SD 81, presented to the 19th Congress of the International Astronautical Federation, New York, October 1968.
18. E.E. Kilmer, *Heat-Resistant Explosives for Space Applications*, Journal of Spacecraft, Vol. 5, No. 10, October, 1968.

19. L.V. Hebenstreit, *High-Performance Inflation Systems*, Paper presented to the Fifth Annual Helicopter Rescue Symposium, Philadelphia, September 1964.
20. L.V. Hebenstreit and T.T. Hadeler, *Pressurization Means for Inflatable Structures*, Paper presented to National Aerospace Engineering and Manufacturing Meeting, Los Angeles, October 1962, published by Society of Automotive Engineers (SAE).
21. Russell A. Pohl, *A Midair-Deployed Buoyancy Suspension System for the Briteye Battlefield Illumination Flare*, Journal of Aircraft, Vol. 6, No. 4, July-August 1969.
22. R.J. Richards, *Solid-Propellant Cool-Gas Generating Systems*, Paper presented to 77th Flight Safety, Survival and Personnel Equipment Symposium, Las Vegas, October 1969.
23. John J. Ridgeway, *Explosive Anchors for Sea Mooring*, UnderSea Technology, December 1970.
24. *Explosive Ground Anchor*, Patent Brief, Design News, March 16, 1966.
25. Sir James Martin, *Ejection Seats*, Special Anniversery Issue 100 Years Royal Aeronautical Society, Journal of the Royal Aeronautical Society, Great Britain.
26. *Aircraft Escape System (Yankee)*, News Trend, Machine Design, March 19, 1970.
27. Hammond R. Moy, *Advanced Stabilized Ejection Seat Development Program*, Report No. SEG-TR-67-51, Douglas Aircraft Company.
28. Jim Hong and E.A. Newquist, *Ejectable Nose Crew Escape System*, Report No. LR 16551, February 1963, Lockheed Aircraft Corporation.
29. *Pilot Rescue System completes Flight Test*, Machine Design, January 7, 1971.
30. *Pilot Flies Away from Crashing Airplane*, Machine Design, July 24, 1969.
31. Constantin Sabin Ioan, *Parachuting at Supersonic Speeds*, Rumanian Periodical *Stiinta Tehnica*, translated from Russian, Tech. translation FTD-TT 62-1307, February 20, 1963.
32. *Explosive Emergency Exit System Tested*, Aviation Week & Space Technology, November 2, 1970.
33. *Explosive Blow-Out Plane Doors to save Airline Passengers*, Product Engineering, November 23, 1970.
34. James Wargo, *Safety Experts propose Systems to eject Airline Passengers*, McGraw-Hill *World News*, Detroit, Reprint in Product Engineering, February 2, 1970.
35. Richard G. Snyder and Col. John P. Stapp, *Emergency In-Flight Evacuation from Future Air Transport Aircraft*, Paper presented to 7th Annual Meeting of the Survival and Flight Equipment Association, Las Vegas, October 1969.
36. Stuart M. Levin, *Air Safety, Surviving the Crash*, Space/Aeronautics, May 1968.
37. *Air-Inflated Stairway*, (Inflatair), Machine Design, August 2, 1969.
38. William H. Simmons, *Apollo Spacecraft Pyrotechnics*, NASA Report, NASA Manned Spacecraft Center, Houston.
39. J.F. McCarthy, J. Ian Dodds, and R.S. Crowder, *Development of the Apollo Launch Escape System*, Journal of Spacecraft, Vol. 5, No. 8, August 1968.
40. Kenneth L. Christensen and Russell M. Narahara, *Spacecraft Separation*, Space/Aeronautics, July 1966.
41. J.W. Kiker, J.B. Lee, and J.K. Hinson, *Earth Landing Systems for Manned*

Spacecraft, NASA Paper, National Aeronautics and Space Administration, Washington, D.C., Paper presented to the Flight Mechanics Panel of the Advisory Group for Aeronautical Research and Development, Turin, Italy, April 1963.

42. William B. Pepper and Randall C. Maydew, *Aerodynamic Decelerators – an Engineering Review*, Journal of Aircraft, Vol. 8, No. 1, January 1971.

43. T.W. Knacke, *The Apollo Parachute Landing System*, Northrop Corporation, Ventura Division, Paper No. TP-131, presented to the AIAA Second Aerodynamic Decelerator Systems Conference, El Centro, California, September 1968.

44. T.W. Knacke, *Systems Considerations*, Northrop Corporation, Ventura Division, Technical Publication No. 59.

45. *Safing, Arming and Fuzing Concepts*, Missile Design & Development, March 1959.

46. R.M. Knox, S.J. Minton, and E.B. Zwick, *Space Ignition*, AIAA Paper No. 66-609, presented to AIAA Second Propulsion Joint Specialist Conference, Colorado Springs, June 1966.

47. L.I. Knudsen, *Electrical Ignition of Rocket Engines and Motors*, SAE Paper No. 682D, Society of Automotive Engineers, presented to National Aeronautical Meeting, Washington, D.C., April 1963.

48. Paul N. Laufman, *Exothermic Ignition Systems for Solid Rocket Motors reduce Shock Loadings*, Space/Aeronautics, April 1970.

49. R.S. Brown and Ralph Anderson, *Aft End Ignition for Solid Propellant Motors*, Space/Aeronautics, January 1966.

50. Kurt R. Stehling, *Prime Missile Headache: Clean Stage Separation*, Space/Aeronautics, August 1959.

51. John R. Mock, *Guide to Environmental Tests*, Materials Engineering, June 1970.

52. L.J. Bonis, *Effects of the Space Vacuum on Metals*, Space/Aeronautics, June 1965.

53. John J. Tierney, *Subsystem Tests add to Data on Radiation Resistance*, Space/Aeronautics, April 1961.

54. Wayne M. Gauntt and J. Derbyshire, *Space Vehicle Corrosion*, Ordnance, May-June 1964.

55. Herbert D. Peckham, *Problems in Sensitivity Testing of One-Shot Electro-Explosive Devices*, Paper presented to IEEE Aerospace Conference, Houston, June 1965.

56. Robert A. Yereance, *Reliability Facts and Factors*, Series of articles, Systems Design, December 1964 through June 1965.

57. Ernest J. Stecker, *Safety of Electro-Explosive Devices*, Paper presented to 50th Air Force-Industry Conference, Riverside, California, June 1962.

58. Robert T. Williams, *Reliability Predictions in Design*, Machine Design, April 27, 1961.

59. Melville Leonard and Lorena O'Connor, *Brining 'em back alive from Space*, Space Digest, May 1966.

60. Norman J. Bowman, and E.F. Knippenberg, *The Sterilization of Pyrotechnic Devices*, AIAA Paper No. A66-25288, American Institute of Aeronautics and Astronautics.

REFERENCES

61. *Other World Bacteria Worry Space Scientists*, Machine Design, April 1965.
62. Kurt R. Stehling, *Economics is Key Factor in Booster Recovery*, Space/ Aeronautics, April 1961.
63. Donald E. Krantz, *Explosive Forming on a Production Basis*, Metals Engineering Quarterly, American Society for Metals, November 1968.
64. George de Groat, *HERF, Metalworking's New Frontier*, Part I, *Explosive Forming*, American Machinist/Metalworking, Manufacturing, Special Report No. 526, September 1962.
65. Walter A. Beck, *Explosive Forming: Handling the Parameter*, The Tool and Manufacturing Engineer, September 1969.
66. W.W. Rasmussen, *Production Factors in Designing for Explosive Forming*, Machine Design, April, 1966.
67. D.E. Strohbecker, R.J. Carlson, S.W. Porembka, Jr., and F.W. Boulger, *Explosive Forming of Metals*, DMIC Report No. 203, Defense Metals Information Center, Battelle Memorial Institute.
68. A. Ezra, *Principles and Procedures for the Explosive Forming of Large Domes from Flat Blanks*, American Society of Tool and Manufacturing Engineers, Technical Paper No. MF 69-186.
69. R. Gorcey, J. Glyman, and E. Green, *Progress Report on Developments in Explosive Forming*, Machine Design, April 13, 1961.
70. Robert W. Carson, *High-Energy-Rate Forming*, Special Report, Product Engineering, about 1969.
71. T.Z. Blazynski, *Air Cushion Effect in the Explosive Forming of Metal Sheet*, Magazine *The Engineer*, London, January 10, 1969.
72. *Metal Forming, High-Energy-Rate Forming*, Report No. NASA SP-5015, Conference on New Technology, Lewis Research Center, Cleveland, Ohio, June 1964.
73. H.G. Otto, *Versuche über Vielfach-Lochungen dünner Bleche mittels Sprengstoffs*, Technische Mitteilung T 43/67, Deutsch-Französisches Forschungsinstitut Saint-Louis, France.
74. Francis J. Lavoie, *Explosive Welding*, Machine Design, July 10, 1969.
75. B. Crossland, J.D. Williams, and V. Shribman, *Developments in Explosive Welding*, Aircraft Engineering, Great Britain, December 1968.
76. Thomas J. Enright, William F. Sharp, and Oswald R. Bergmann, *Explosive Bonding Dissimilar Metals*, Metal Progress, July 1970.
77 Robert H. Wittman, *Explosive Bonding Ready to come into Wider Use*, Space/ Aeronautics, November 1967.
78. Ronald J. Carlson, *Explosive Welding*, Design News, July 21, 1965.
79. Charles C. Simmons, *Explosive Metalworking*, DMIC Memorandum 71, Defense Metals Information Center, Battelle Memorial Institute, Columbus, Ohio.
80. *Explosive Forming of Refractory Metals*, Chromalloy Corporation, Manufacturer's report.
81. *Konstruktionsrichtlinien fur Sprengnietung*, He N 11559 Sprengniete aus Leichtmetall, und He N 11560 Sprengnietbolzen, Ernst Heinkel Flugzeugwerke G.m.b.H., Seestadt Rostock, March 1942.
82. R.C. Allen, *Non-Electric Stimulus Transfer Systems and Through-Bulkhead Ignition*, Technical Bulletin AE 62-3-A, McCormick-Selph.

381

83. R.C. Allen, *NESTS Components*, Technical Bulletin RE-264-1, McCormick-Selph.
84. *Small Column Insulated Delays for Precision Pyrotechnic Delays and Ordnance Distribution Systems*, Technical Bulletin 67 M001, McCormick-Selph.
85. George Stevens, *Advanced Electro-Mechanical Safe/Arm Initiator incorporating One-Ampere, One-Watt, No-Fire Capability*, Technical Bulletin, McCormick-Selph.
86. *History and Development of Martin-Baker Ejection Systems*, Booklet by Martin-Baker Aircraft Co. Ltd.
87. *Use of High-Shear Ordnance Products in the N.A.A.* X-15A-2 External Tank Recovery System, Report 2-179923, Hi-Shear Corporation, Ordnance Division.
88. *Maintenance and Overhaul Manual, Cool Gas Generator*, No. F-41062, Walter Kidde & Co., Inc., 1963.
89. *USAF B-58 Hustler*, General Dynamics, Fort Worth.
90. *Apollo Qualified Man-Rated Modular Pyrotechnic Devices*, Space Ordnance Systems, Inc., 1967.
91. *Firing Characteristics*, Data Sheet 901, Atlas Chemical Industries, Inc.
92. *Explosives: An Introduction to the Explosives used in Explosive Components*, Data Sheet 903, Atlas Chemical Industries, Inc.
93. *Reliability*, Data Sheet 904, Atlas Chemical Industries, Inc.
94. *Gas Generators, compact and convenient energy Sources*, Hercules Incorporated.
95. *Facts about All-Metal Separable Connectors for Leakproof Sealing, Marman Conoseal Joints*, Aeroquip Corporation, 1966.

NOMENCLATURE

amp	ampere
Btu	British thermal unit
C	centigrade, Celsius
cc	cubic centimeter
cm^2	square centimeter
cps	cycles per second, (vibration)
cu.in	cubic inch
erg	unit of work
ft	foot
g	gram
G	force equal to earth gravity
Hz	hertz, cycles per second
in	inch
lb	pounds
m	meter
mc	microfarad (=mfd)

mil	1/1000
mm	millimeter
ms	millisecond
mv	millivolt
psi	pounds per square inch
psia	pounds per square inch absolute
psig	pounds per square inch gage
rad	radiation dose absorbed
sec	second
torr	mm Hg at $0^{\circ}C$

ABBREVIATIONS

ac	Alternating current
APS	Ascent Propulsion System, Lunar Module
APU	Auxiliary power unit
ASI	Apollo Standard Initiator
CDI	Confined detonating fuse
CM	Command Module, Apollo
CSM	Command and Service Module, Apollo
dc	Direct current
DIPAM	Diamino-hexanitrobiphenol
EBW	Exploding bridgewire
EDNA	Ethylene-dinitramine
FAA	Federal Aviation Agency
FLSC	Flexible linear shaped charge
G	Gravity
He	Helium
HF	High frequency
Hg	Mercury
HNS	Hexanitrostilbene
KDNBI	Potassium dinitrobenzoferoxan
LES	Launch Escape System, Apollo
LM	Lunar Module, Apollo
LMNR	Lead mononitroresorcinate
LSC	Linear shaped charge
MDI	Mild detonating fuse
NASA	National Aeronautics and Space Administration
NESTS	Non-electric stimulus transfer line (trade name)
OAO	Orbiting Astronomical Observatory, satellite

383

O.D.	Outside diameter
PETN	Pentaerythritol tetranitrate
PNC	Plastisol nitrocellulose
RCS	Reaction Control Subsystem, Apollo
RF	Radio frequency
RDX	Cyclotrimethylene trinitramine
SBASI	Single Bridgewire Apollo Standard Initiator
SCID	Small column insulated delay, (trade name)
SLA	Spacecraft-LM Adapter, Apollo
SM	Service Module, Apollo
SOFAR	Sound-fixing and ranging
TBI	Through-bulkhead initiator
TEGDN	Triethylene glycol dinitrate
TMETN	Trinitrotoluene
UHF	Ultrahigh frequency
v	Velocity
VHF	Very high frequency

GLOSSARY

Abort — Premature termination of the launch or mission of a spacecraft or missile because of equipment failure or other problems.

Actuator — A device that transforms chemical energy into a mechanical motion.

All-fire current — The minimum electric current that will initiate an explosive-actuated device.

Ambient — The normal surrounding environmental conditions, such as pressure and temperature.

Ascent engine — A thrust engine used for ascent of a spacecraft, as for example, the Apollo Lunar Module's upper stage, called "ascent stage", in which the ascent engine is also used for flight adjustments and for prelanding abort.

Ascent stage — The stage of a spacecraft designed to ascend for the return flight after a completed lunar or planetary surface mission, as for example, the upper stage of the Lunar Module, which contains the crew, controls, and ascent engine, and is used to return the crew to the Apollo Command Module in lunar orbit.

Attitude — The position or orientation of a vehicle determined by the inclination of its axes to a certain reference.

Auto-ignition — Self-ignition (of pyrotechnic compositions, for example).

Blasting cap — A device consisting of a small, short cylindrical housing containing an explosive charge, which is detonated by an electric bridgewire or by a safety fuse, and which is used for initiating a high explosive.

GLOSSARY

Braid – Woven bare wire.

Bridgewire – A fine wire which, depending on the design type, either heats up or explodes when electric current is applied; used to initiate explosive-actuated devices.

Bulkhead – a dome-shaped segment which encloses the end of a propellant tank; also a structural wall of a pressurized aircraft compartment.

Capacitance – the ratio of the electric charge applied to a device to the resultant charge of potential.

Cladding – A method of applying a layer of metal onto another metal by a continuous welded joint of both metals.

Command – A signal or pulse used to initiate a function or a sequence; used in missile and spacecraft operations.

Command Module – The control center of Apollo, used as living quarters for most of the lunar voyage.

Command and Service Module – Combined Command Module and Service Module which during the lunar surface mission of the Lunar Module remain in lunar orbit and are not separated from each other until shortly before re-entry into earth's atmosphere.

Composition – A pyrotechnic or an explosive mixture of several materials.

Crimped joint – A joint of thin metal parts obtained by folding the edges of one part over and around the edges of another part.

Descent stage – The lower portion of the Lunar Module, containing the descent engine, propellant tanks, landing gear and storage sections. For the lift-off from the moon, it serves as launching platform for the ascent stage, and it remains on the lunar surface.

Detonator – An explosive or device which is initiated by a primer and is used to initiate another explosive which can be less sensitive and larger.

Dielectric strength – The maximum potential gradient that an insulating material can withstand without breakdown, expressed as voltage gradient.

Docking – The closing and mating together two spacecraft, following rendezvous.

Docking probe – An extendable device attached to the Docking Ring on the Apollo Command Module for engaging a drogue on the Lunar Module.

Docking tunnel – Cylindrical tunnel through which Apollo crew transfers between Lunar Module and Command Module.

Ejection seat – Aircraft pilot's seat which is ejected by actuating the ejection system in case of emergency.

Exothermic material – Material evolving heat.

Exploding bridgewire – A wire which heats to a high temperature and explodes when subjected to a high-voltage high-energy pulse.

Explosive train – An in-line arrangement of explosive and pyrotechnic elements of different sensitivities and other properties.

Fragmentation – Breaking up into several pieces.

Function time – Time span between the application of initiating energy to a device and completion of its operation, or pressure rise in the device.

Gas generator – A device for producing gas of a predetermined temperature by burning solid propellant.

385

Gimbal – A mounting arrangement with two or three mutually perpendicular and intersecting axes of rotation on which a rocket engine or other device can be mounted, and which allows it to swivel or move in two or three directions.

Grain – The configuration of solid propellant preformed to a particular geometric shape and size, for use in rocket motors and gas generators.

Guidance system – A system which measures and evaluates flight information, correlates this with target data, converts the results into operations required to achieve the desired flight path, and transmits this data in the form of commands to the flight control system.

Gyroscope – A device using angular momentum of a spinning rotor to sense angular motion of its base about one or two axes at right angles to the spin axis.

Header – A device's insulated end through which electrical connections pass, as typically used in cartridges.

Heat sink – A contrivance for the absorption or transfer of heat away from a critical part or parts.

Helium – Gas used to pressurize propellant tanks and to force propellant into feed lines. Helium is also used in leak tests of explosive-actuated devices.

Hypergolic propellants – Propellants which ignite spontaneously upon contact with an oxidizer. This eliminates the need for an ignition system in liquid-propellant rocket engines.

Igniter – A device consisting of a contained powder charge, with bridgewire and electrical leads, that produces a flame and slight pressure, used to ignite other materials.

Incendiary material – Spontaneously igniting combustible material, used in incendiary destructors and in similar applications.

Initiator – A device consisting of a small explosive charge which is detonated by electric current or impact and, in turn, detonates a larger less sensitive charge.

Insulation resistance – The electrical resistance between the terminals of a device and the exterior body or exposed metal parts.

Ionization – A process making the total electronic charge of a substance unequal to its positive charge.

Jetevator – Ring-shaped rotatable rocket nozzle deflector mounted at the nozzle periphery.

Lanyard – A line used for mechanical initiation of some explosive-actuated devices. A typical application is a lanyard-actuated reefing line cutter, where the firing pin is released by pull of the lanyard.

Lunar Module – The vehicle consisting of the ascent and descent stages, used to transport two astronauts from the Apollo Command Module in lunar orbit, to land on the lunar surface, to provide a base of operation on the moon, and to return the astronauts to the Command Module.

Manifold – A component or device providing multiple connections.

Match – A non-contained powder charge, ususlly formed around two electrical leads, which, when energized, produces a flame for ignition.

Melting point – The temperature level at which a material melts.

Mortar – A device consisting of a short cylindrical tube containing a piston-type sabot and a breech assembly equipped with a pressure cartridge at the aft end, used for deployment of drogue chutes and parachutes.

No-Fire level - The maximum level of electric energy input that will in no case, within a specified time, initiate an explosive-actuated device.

O-Ring   A sealing component, usually made from compressible rubber-like material.

Oxidizer   Substance that supplies the oxygen necessary for the burning of propellants and other materials.

Parallel redundancy — An arrangement of two components, methods, or systems working at the same time to accomplish the same task, although either one could perform the operation alone.

Pellet – Propellant in shaped, bound or compressed form.

Percussion primer – A device in form of a small capsule, containing a prime explosive charge which is ignited by the impact of a firing pin.

Pitch – Movement of a vehicle from its lateral axis.

Port – An internal or external terminus of a passage in a device. A typical example are thrust reversers of rocket engines.

Prime explosive – The ignition material in which the bridgewire is embedded in an initiator.

Purging – Removing residual fluid or gas from a fuel tank or line.

Pyrotechnic Devices – Explosive-actuated devices, specifically, devices that burn rather than produce a shattering effect.

Redundant   A second means for accomplishing a given task.

Re-entry – The return of a spacecraft into the earth's atmosphere.

Re-entry vehicle - A spacecraft designed to withstand the heating associated with re-entry.

Rendezvous – Meeting of spacecraft in orbit at a planned time and place.

Retrorocket   A rocket which produces thrust opposed to the vehicle's forward motion.

Roll   The rotation of a vehicle about its length axis, which is usually designated with "Z-axis".

Sabot – A piston-type component in a mortar provided for parachute or drogue chute deployment.

Service Module   The compartment of the Apollo spacecraft that contains propellant, navigation equipment, propulsion system and other equipment required for space flight.

Sheath   A protective cover as used for flexible linear shaped charges, made from suitable metal.

Solar radiation   Energy radiated from the sun.

Space environment   The conditions existing in space, i.e. a vacuum, temperature from -273°C to $+540^\circ$C, zero gravity state, and radiation, etc.

Spin motor - A small rocket motor arranged tangentially on a spacecraft or stage for providing spin motion about the vehicle's length axis for flight stabilization.

Squib   A device consisting of a contained powder charge and a bridgewire for initiation. When energized, a squib produces heat and slight pressure.

Stability   The ability of an explosive material to withstand long periods of storage under adverse conditions.

Staging   Stage separation, i.e. separation of a vehicle stage, such as an expended booster stage, from a spacecraft or missile.

Sublimation – The transformation of a material from a solid state to a gaseous state without passing through a liquid state.

Thrust – The pushing force developed by a rocket engine, measured by multiplying the propellant mass flow rate by the exhaust velocity relative to the vehicle, expressed in pounds.

Thrust chamber - The combustion chamber of a rocket engine, where by burning propellant in the presence of an oxidizer high-velocity gases are produced which will exit through the nozzle, and thus produce thrust.

Thrust vectoring – An attitude control for rockets wherein one or more rocket engines are gimbal-mounted so that the direction of the thrust force may be changed in relation to the center of gravity of the vehicle to produce a turning movement.

Trajectory – The flight path traced by a vehicle under power or as a result of power.

Ullage – The volume above the surface of the liquid in a tank, partially a function of temperature.

Ullage maneuver – A quick thrust of the vehicle made prior to firing the engine of a spacecraft, resulting in shifting the propellant to the bottom of the tanks, thus providing the necessary conditions for proper feeding.

Umbilical – The connecting service lines for electrical power, liquids and gases between the launch tower and the spacecraft, or between two stages of the vehicle.

Vernier rocket – A small swivel-mounted auxiliary rocket motor used for flight path correction maneuvers.

X-Axis – Spacecraft axis associated with yaw maneuvers in which the spacecraft turns or twists about its Y-axis.

Yaw – Movement of a vehicle from its longitudinal axis.

Y-Axis – Lateral axis running through the spacecraft; associated with pitch maneuvers, in which the spacecraft turns or twists about its Y-axis.

Z-Axis – Fore-aft axis running through the spacecraft; associated with roll maneuvers, in which the spacecraft turns or twists about its Z-axis.

Index